Philosophy and Medicine

VOLUME 117

Founding Co-Editor
Stuart F. Spicker

Senior Editor

H. Tristram Engelhardt, Jr., *Department of Philosophy, Rice University, and Baylor College of Medicine, Houston, Texas*

Series Editor

Lisa M. Rasmussen, *Department of Philosophy, University of North Carolina at Charlotte, North Carolina*

Assistant Editor

Jeffrey P. Bishop, *Gnaegi Center for Health Care Ethics, Saint Louis University, St. Louis, MO, USA*

CATHOLIC STUDIES IN BIOETHICS

Series Founding Co-Editors

John Collins Harvey, *Georgetown University, Washington, D.C., U.S.A.*
Francesc Abel, *Institut Borja de Bioetica, Center Borja, Barcelona, Spain*

Series Editor

Christopher Tollefsen, *University of South Carolina, Columbia, SC, U.S.A.*

Editorial Advisory Board

Joseph Boyle, *St. Michael's College, Toronto, Canada*
Sarah-Vaughan Brakman, *Villanova University, Villanova, PA, U.S.A.*
Thomas Cavanaugh, *University of San Francisco, San Francisco, CA, U.S.A.*
Mark Cherry, *St. Edward's University, Austin, TX, U.S.A.*
Ana Smith Iltis, *Wake Forest University, Winston-Salem, NC, U.S.A.*

More information about this series at http://www.springer.com/series/6414

Frank Sobiech

Ethos, Bioethics, and Sexual Ethics in Work and Reception of the Anatomist Niels Stensen (1638-1686)

Circulation of Love

 Springer

Frank Sobiech
Faculty of Catholic Theology
Julius Maximilians University
Würzburg, Germany

ISSN 0376-7418 ISSN 2215-0080 (electronic)
Philosophy and Medicine
ISBN 978-3-319-32911-6 ISBN 978-3-319-32912-3 (eBook)
DOI 10.1007/978-3-319-32912-3

Library of Congress Control Number: 2016940515

Printed on acid-free paper

This Springer imprint is published by Springer Nature
The registered company is Springer International Publishing AG Switzerland

Circulation of Love
In memory of Dr. med. Paul Takashi Nagai
(1908–1951)

Foreword

Precisely 25 years after the beatification of Niels Stensen (1638–1686) by the late Pope John Paul II, this first ever study on the Danish anatomist, bishop and vicar apostolic combining the perspectives of medical ethics and theology is being published in answer to the challenges of medical ethics. I am grateful to Frank Sobiech for tackling this task. His historically substantiated work not only provides points of departure for further research, but indeed also lays a foundation for the discussion of medical ethics, including but not limited to aspects of ecumenical Christianity.

Stensen, who broke new ground as an anatomical researcher prior to his ordination, for example with the first anatomical confirmation of the female ovary in humans, is compelling by virtue of the Christian ideas on humanity which he developed during the course of his life, first as a Lutheran, then as a seeker and finally, after his conversion, as a Catholic.

Just as Stensen, who set great store by merciful pastoral care, likened the pastor's role to that of a hospital's head physician characterized by individual concern for every patient, Pope Francis sees the Church as a "field hospital after a battle" (Santa Marta, August 19, 2013) whose duty is to heal each individual's wounds.

The nucleus of the study consists of seven chapters examining questions of sexual ethics, the beginning of human life and development in the womb from Stensen's perspective as an anatomist and bishop as well as in historical comparison. Stensen views the human body as a tool of the soul, attributing to it an essential role as "interpreter of the love of God" which he contrasts with, as phrased by Benedict XVI, the "debasement of the human body" (encyclical "Deus caritas est", 5).

Sobiech also deals with Van Rensselaer Potter – who coined the term "bioethics" in his eponymous book –, and concludes the study by comparing Potter's freethinking, secularist concept of life to Stensen's.

The ethos of Niels Stensen may be regarded by every scientist and practicing physician, and likewise by every bishop, as a paradigm for their own behavior.

Rome, on the feast of St. Vincent de Paul 2013

Gerhard Ludwig Müller

Preface

The fact that commemoration of the figure and work of Niels Stensen (1638–1686) in medical literature has never entirely ceased since the end of the seventeenth century – in particular following the complete edition of Stensen's anatomical and geological writings by Vilhelm Maar more than 100 years ago – as well as the publication of various smaller contributions in medical periodicals throughout the past two decades clearly document that the Danish[1] anatomist and geologist Stensen is considered a part of the cultural memory of medicine as a whole.

Dr. med. Niels Stensen, the first modern natural scientist ever to be beatified – in a ceremony conducted by Pope John Paul II ([*1920] 1978–2005) in Rome on October 23, 1988 –, contributed significantly to the Scientific Revolution during the early modern period which marked the beginning of a new era in the field of medicine. Based on analyses of Stensen's own writings as well as literature published in the fields of medicine and history of medicine since the early modern period, the historical-theological study presented herein describes for the first time Stensen's ethos, its implications for medical ethics and theology and its reception by physicians and natural scientists from the seventeenth century until today. The study's goal is to contribute to the interdisciplinary perception of Stensen. In light of its interconnections with theology and the Magisterium discussed on the following pages, it becomes obvious that medical science in the early modern period still largely awaits methodical examination.

In his address (January 27, 2012) to the plenary meeting of the Congregation for the Doctrine of the Faith, Pope Benedict XVI (2005–2013) spoke of the challenges posed by "the great moral questions regarding human life, the family, sexuality, bioethics, freedom, justice and peace" on the path of ecumenism, stating that it would be important to speak "with one voice" on this topic based on the foundations in Scripture and living Tradition of the Church "to decipher the language of the

---

[1] Cf. the appreciation of Stensen in Ejvind Slottved and Ditlev Tamm, *Copenhagen*, esp. pp. 61–65.

Creator in his creation."[2] For these questions, particularly those regarding the essence of professional ethics in medical science and practice as well as the principles of bioethics, the scientist Niels Stensen as a person, and likewise his work, provide astoundingly up-to-date answers despite lying more than 325 years in the past.

I wish to cordially thank the Prefect of the Congregation for the Doctrine of the Faith, His Eminence Gerhard Ludwig Cardinal Müller, for kindly writing the foreword.

This study is the revised English edition based on the 2nd German edition *Radius in manu Dei: Ethos und Bioethik in Werk und Rezeption des Anatomen Niels Stensen (1638–1686)* (Münster: Aschendorff, 2014) as vol. 17 of the series *Westfalia sacra* (editor: Prof. Dr. Reimund Haas, Cologne/Münster).[3] The 1st edition (2013) was presented in the presence of His Excellency Msgr. Czeslaw Kozon, bishop of Copenhagen, at the "Anniversary Seminar" in Copenhagen on November 23, 2013, celebrating the 25th anniversary of Stensen's beatification and organized by the *Academicum Catholicum*, the *Sankt Andreas Bibliotek* (est. 1648) and the Cathedral parish of St. Ansgar in Copenhagen.[4] On December 10, 1988, to commemorate Stensen's beatification, a memorial plaque was affixed in the gallery before the Great Hall (Aula Magna) of the seat of the *Pontifical Academy of Sciences*, the *Casina Pio IV*, in the Vatican Gardens (Fig. 1).[5] I thank the President of the *Pontifical Academy of Sciences*, Prof. em. Dr. Werner Arber (Basel), for his kind permission to publish a close-up of this memorial plaque.

My heartfelt thanks go out to Dr. Sebastian Olden-Jørgensen and his wife Charlotte for their hospitality and our pleasant conversation during my research visit to Copenhagen in June 2011. Dr. Helge Clausen (Århus) guided me through the *Sankt Andreas Bibliotek*, Dr. Troels Kardel informed me about his and Dr. Paul Maquet's translation project, and with Fr. Dr. August Ziggelaar SJ in the *Sankt Knuds Stiftelse* on June 17, 2011, I once again experienced international ecclesial fellowship. I also wish to thank Rector Dr. Hans-Peter Fischer for the cordial acceptance into the community of the *Pontificio Collegio Teutonico di Santa Maria in Campo Santo* during my visits in Rome and the Vatican City. The Florentine journalist Dr. Maria Francesca Gallifante (Basilica of San Lorenzo) and Msgr. Angiolo Livi (1914–2014), the then Prior of the Basilica of San Lorenzo in Florence, bestowed unforgettable experiences in Stensen's spiritual hometown on me.

Dr. Finn Aaserud, director of the Niels Bohr Archive in Copenhagen, digitized the correspondence between Niels Bohr and Gustav Scherz. My thanks go out to Prof. Dr. Robert Kretz and Prof. em. Dr. Günter Rager (University of Fribourg,

---

[2] "Le grandi questioni morali circa la vita umana, la famiglia, la sessualità, la bioetica, la libertà, la giustizia e la pace […] con una sola voce […] a decifrare il linguaggio del Creatore nella sua creazione." Acta Apostolicae Sedis 104 (2012), 108–111, here 110 (hereafter cited as AAS).

[3] Sobiech, *Radius in manu Dei.*

[4] Cf. http://nielssteensen.dk/anniversary-seminar; Rütz, "Seminar".

[5] Pontificia Academia Scientiarum, *Memorial Plaque*, pp. 27–29 (text of the memorial plaque with two images); Marini-Bettòlo, "Background Notes".

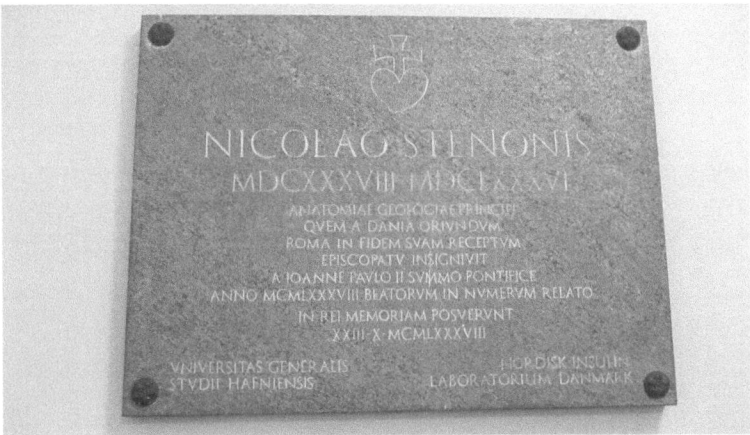

Fig. 1 The memorial plaque in the gallery before the Great Hall (Aula Magna) of the seat of the *Pontifical Academy of Sciences*, Casina Pio IV, Vatican City
The engraved inscription reads "Nicolao Stenonis/ MDCXXXVIII – MDCLXXXVI/anatomiae geologiae principi/quem a Dania oriundum/Roma in fidem suam receptum/episcopatu insignivit/a Ioanne Paulo II summo pontifice/anno MCMLXXXVIII beatorum in numerum relato/in rei memoriam posuerunt/XXIII – X – MCMLXXXVIII/Universitas generalis studii Hafniensis/ Nordisk Insulin Laboratorium Danmark" (To Niels Stensen, 1638–1686, excelling in anatomy and geology, whom, descending from Denmark, Rome took charge of and honored with the bishopric, added to the numbers of the blessed by Pope John Paul II in the year 1988, the University of Copenhagen and Nordisk Insulin Laboratorium Denmark mounted [this memorial plaque] in the memory of that event, [which took place on] October 23, 1988)

Switzerland) for information about Prof. Dr. med. Adolf Faller (1913–1989), and to Prof. em. Dr. Carl Schirren (Hamburg) for insights into his former activities. Helge Clausen, Troels Kardel and the *Edizioni Studio Domenicano* (Bologna) gifted me with books; and the secretaries of the *Académie des Sciences, Arts et Belles-Lettres de Caen* (Caen), Nicolas Rajaomilison and Claude Roche, procured the difficult-to-obtain essay by Léon Tolmer for me at no charge.

I thank the translator Stephan Stockinger (Vienna) for his dedication to our joint translation project, and the St. Ansgarius-Werk of the archdiocese of Cologne, the Ansgar-Werk of the diocese of Münster and the Ansgar-Werk of the dioceses of Osnabrück and Hamburg for their financial assistance. Angelika E. Paul (Spokane, WA) was the first to read the manuscript.

Unless otherwise specified, all dates mentioned in this study refer to the Gregorian calendar with each year beginning on January 1.

Würzburg, Germany   Frank Sobiech
March 2016

Contents

About the Author

Frank Sobiech, ThD, born 1972, is a postdoctoral researcher at the Faculty of Catholic Theology of the Julius Maximilians University of Würzburg, Germany. He earned his doctorate in Catholic theology from the Faculty of Catholic Theology of the Westfalian Wilhelms University of Münster, Germany, and has degrees in Catholic theology, history, Latin philology and law.

About the Book

The Danish anatomist Dr. med. Niels Stensen (1638–1686), the first modern natural scientist ever to be beatified – by Pope John Paul II in Rome in 1988 –, made essential contributions to the Scientific Revolution of the early modern era. It was Stensen who first proved the function of the female ovaries, which at the time were still commonly referred to as testes. In Copenhagen, where Stensen was instrumental in the new academic midwives' exam as so-called Royal Anatomist, he crowned and ended his scientific career in 1673 with the public dissection of a female corpse. A sculpture unveiled in 1963 in front of the Health Science Faculty library of the University of Copenhagen commemorates this event. In 1675, Stensen was ordained a priest in Florence, where he had converted to the Roman Catholic faith in 1667; in 1677, he was ordained a bishop in Rome and appointed Vicar Apostolic of the Northern Missions. This medico-historical study delineates for the first time ever Stensen's ethos and its medico-ethical and theological implications and reception by physicians and natural scientists from the seventeenth century until today. Its focal points are questions of bioethics, especially with regard to human reproduction, sexual ethics, the beginning of life and the ensoulment of the embryo, as well as the discussion of a case of plagiarism and the boundaries of pastoral care. Despite dating back more than 300 years, Stensen's character and his work offer up surprisingly topical answers to current questions on the nature of professional ethics in medical science and practice.

Chapter 1
Basic Premises

With regard to its objective of investigating "Ethos and Bioethics in Work and Reception" of the Danish anatomist and bishop Dr. med. Niels Stensen (1638–1686),[1] this study represents the elaboration of the assumption introduced in the author's dissertation that detailed investigation of Stensen's spirituality, ie the specific nature of Stensen's life of faith, would allow his profiles as physician, geologist, controversial and pastoral theologian to be better assessed and appreciated.[2] To this end Stensen's own writings as well as literature published in the fields of medicine and history of medicine since the early modern period are evaluated in order to trace the intellectual profile of the anatomist and bishop and adequately assess the question of his function as a role model for today. Possibilities, boundaries and dangers of work in medical research and as a medical practitioner during the second half of the seventeenth century are thus elucidated.

---

[1] An overview is provided by the following encyclopedia articles (not listed are publications on the medical eponyms). See section ABBREVIATIONS, B in the appendix for full bibliographical details on these sources: (1) Gustav Scherz, "Niels Stensen (1638 bis 1688 [sic!])"; in: DbÄ, pp. 113–116; (2) BiogrHM, s.v. "Niels Stensen (1638–1686)", pp. 138–139; (3) Gustav Scherz, "Stensen, Niels"; in: DSB 13 (1976), pp. 30–35; (4) Egill Snorrason, "Stensen, Niels"; in: DBL[3] 14 (1983), pp. 85–93; (5) DDPhS, s.v. "Stensen (Steno, Stenosis [sic!]), Niels (Nicolaus)", cols. 1876–1878; (6) John Henry, "Stenone, Nicolò"; in: DBSM 4 (1989), pp. 137–138; (7) Gerhard Bettendorf, "Stensen, Niels"; in: EndokrG, pp. 564–565; (8) David Oldroyd, "Steno, Nicolaus (Niels Stensen)"; in: ESR, pp. 618–619; (9) August Ziggelaar, "Stensen (Steensen, Steno), Niels"; in: DHCJ 4 (2001), pp. 3636–3637; (10) Francesco Abbona, "Steensen, Niels (1638–1686)"; in: DISF 2 (2002), pp. 2099–2110; (11) Fye, "Stensen"; (12) BEdtM 2 (2002), s.v. "Steno, Nicolaus, eigentl. Niels Stensen", p. 602; (13) HGK, s.v. "Steensen, Niels", pp. 634–635; (14) Régine Pouzet, "Sténon, Jean-Nicolas"; in: DPR, p. 950; (15) Ralf Bröer, "Stensen, Niels"; in: ÄLex.[3], p. 311; (16) Carl Henrik Koch, "Stensen, Niels (aka Nicolai Stenonis, Steno)"; in: DMB 5 (2007), pp. 1188–1189; (17) Barbara I. Tshisuaka, "Stenon (Ste[e]nsen, Stenonis, Stenson, Steno), Nicolas (Niels)"; in: EMG 3 (2007), pp. 1358–1359; (18) Reimund B. Sdzuj, "Stensen, Steensen, Steno, auch: Stenonis, Niels, Nicolaus"; in: Killy[2] 11 (2011), pp. 238–241; (19) Frank Sobiech, "Stensen, Niels (Steen[sen], Nicolaus Steno[nis], Niccolò Stenone)"; in: NDB 25 (2013), pp. 251–253.

[2] Sobiech, *Herz, Gott, Kreuz*, pp. 23–24.

© Springer International Publishing Switzerland 2016
F. Sobiech, *Ethos, Bioethics, and Sexual Ethics in Work and Reception of the Anatomist Niels Stensen (1638-1686)*, Philosophy and Medicine 117, DOI 10.1007/978-3-319-32912-3_1

1.1   Structure of the Study

Chapter 1 consists of a brief summary of Niels Stensen's life and comments on the applied methodology, sources and the current state of research, followed by an outline of the medical ethos, which is comprised of the physician's ethos as well as the ethos of medical research, and its relationship with theology from the early modern 'republic of physicians' until today. The main body of the study is spread over Chaps. 2 and 3, and conclusions are discussed in Chap. 4.

The structure of the study correlates the individual chapters with one another: while Sect. 2.1 explains the foundations of Stensen's ethos based on events which shaped him during his academic studies in anatomy, Sect. 2.2 deals with the relationships between anatomist, physician and patient in a 'horizontal' perspective, ie in their respective correlations with each other, as well as in a 'vertical' perspective in terms of their correlation with God. As such, Sect. 2.2 is intended as a hinge of sorts preparing the reader for Sect. 2.3, which focuses on questions of bioethics – in particular those of human reproduction, sexual ethics, the beginning of life and the ensoulment of the embryo as well as pastoral border cases. As Stensen's research on human reproduction forms an essential part of his contribution to the Scientific Revolution of the early modern period and is likewise highly relevant from a theological point of view, it is particularly suitable for defining Stensen's ethos as an anatomist as well as for drawing connections to his time as a priest and bishop.

This focal point is set within an outer frame of Sects. 2.2.1 and 2.3.3 and an inner frame of Sects. 2.2.2 and 2.3.2: while the outer frame deals with the profession of physician first from a medical, then from a theological perspective, the inner frame examines the nature and duties of anatomists and anatomy, their relationship to the Creator and the questions about health, illness and death Stensen saw himself confronted with as a result. Chapter 3 develops a character sketch of the anatomist and beatified bishop Stensen based on outside perceptions of him and sets the stage for the question of the significance of his ethos for today discussed in Chap. 4.

1.2   The Life of Niels Stensen the Anatomist[3]

The eventful life of the anatomist Niels Stensen is like a mirror of the times in the seventeenth century, a period characterized by scientific highs in the field of medicine as well as by theological controversies – ie extraordinary piety on the one hand, which occasionally served as a uniting factor across confessions, and various

---

[3] This chapter is an extended version of the biographical section of Sobiech, "Stensen, Niels (Steen[sen], Nicolaus Steno[nis], Niccolò Stenone)"; in: NDB 25 (2013), pp. 251–253. Cf. the chapter summarizing the state of academic discourse on Stensen's life and expanding it in particular with regard to his religious development and conversion in Sobiech, *Herz, Gott, Kreuz*, pp. 29–90, here 39–68. For a more comprehensive chronology of Stensen's life cf. the information in Scherz, *Biographie*, vol. 1 (English trans. with new introduction and color plates: Kardel/ Maquet, *Biography and Original Papers*, pp. xxvii–xxxviii, 1–2, 5–344); Scherz, *Biographie*, vol. 2 (English summary: Kardel/Maquet, *Biography and Original Papers*, pp. 3–4).

seemingly irreconcilable confessional differences on the other. Born on January 1, 1638 st.v. (January 11 st.n.) in Copenhagen, Stensen grew up in uniformly Lutheran Denmark. Following his classical schooling at the Latin school *Vor Frue Skole* (School of Our Lady) in Copenhagen from c. 1648 to 1656, he studied medicine in Copenhagen from 1656 to 1659, then went to Amsterdam in 1660 and completed his studies in Leiden in 1663/1664 with his graduation as Dr. med. While in Amsterdam, Stensen discovered the *Ductus parotideus* (Ductus Stenonianus), the excretory duct of the parotid gland, and clashed with his professor Gerard Leonard Blasius (1624/1625–1692), who attempted to take credit for the discovery. In 1662/1663, Stensen established the muscular structure of the human heart, thereby disproving the traditional theory of the heart as the seat of the soul. This discovery as well as his anatomical studies on the brain made Stensen known throughout Europe. Passed over during the appointment of new chairs at the University of Copenhagen after his return from Leiden in 1664, he traveled to France and eventually (1666) to the Grand Duchy of Tuscany, where he continued his medical studies and provided, among other results, the first precise description of the human ovaries.

As Stensen had gradually estranged himself from his Lutheran confession since 1660 in favor of a more rationalistic, deistic stance, he eventually converted to Catholicism on All Souls' Day 1667 in Florence, where he enjoyed generous sponsorship by the Grand Dukes Ferdinando II (1610–1670) and Cosimo III (1642–1723) up to and beyond his death. The design of his heart-cross coat of arms (Fig. 1.1) dates from this time. From 1672 to 1674, Stensen served as a so-called "Royal Anatomist" in Copenhagen. Consecrated as secular priest in Florence in 1675 and as Titular Bishop of Titiopolis in Rome in 1677 as well as being appointed Vicar Apostolic of the Northern Missions, he then worked at the court of the converted Duke Johann Friedrich of Braunschweig-Lüneburg (1625–1679) in Hanover from 1677 to 1680, during which time he maintained contact with Gottfried Wilhelm Leibniz (1646–1716) and others. As suffragan bishop of the Prince-Bishopric of Münster (1680–1683/1684), against the resistance of the aristocratic cathedral chapter but supported by Prince-Bishop Ferdinand of Fürstenberg (1626–1683), Stensen engaged in extensive reforming activity in keeping with the Council of Trent (1545–1563), which he expounded in his pastoral theological treatise, the booklet *Parochorum hoc age [...]* (Florence 1684 [corr. 1683]; partial Italian translation Florence 1684, 2nd edition 1685). In order to protest the imminent simoniacal election of the Elector of Cologne and Archbishop Maximilian Heinrich of Bavaria (1621–1688) as bishop of Münster – an election never acknowledged by Rome as a result of Stensen's reports –, he left the city on the morning of the election day. From 1683 to 1685 he worked in the free imperial city of Hamburg, and from December 1685 onwards for the converted Christian Louis I, Duke of Mecklenburg-Schwerin (1623–1692), at the palace chapel in Schwerin. There he engaged in groundwork for the creation of a Catholic community and eventually died on November 25, 1686 st.v. (December 5 st.n.) (cf. Table 1.1).

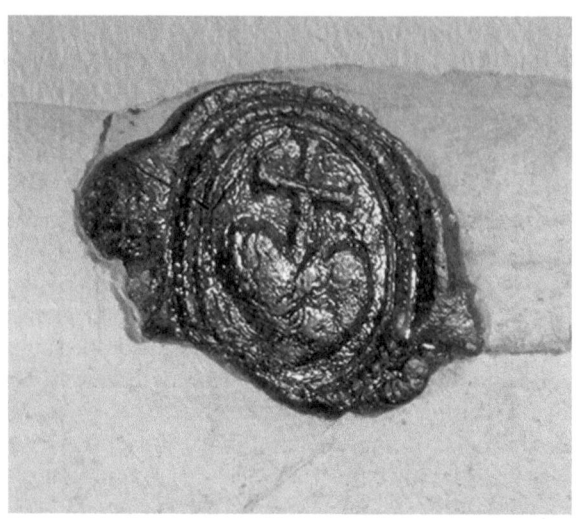

Fig. 1.1 Wax seal with heart-and-cross coat of arms on a letter by Niels Stensen (Florence, September 2, 1670) to Heinrich Meibom Jr. in Helmstedt
This is Stensen's seal emblem from the time after his conversion in 1667, cf. Bruun, *Niels Stensen-breve*, p. 144n77
LBH, Ms. XLII 1902, fol. 16^v

Table 1.1 Selected biographical data of Niels Stensen

1638–1659	Childhood and youth in Copenhagen
01/01/1638 st.v.	Birth in Copenhagen
1648(?)–1656	Latin student at the Copenhagen *Vor Frue Skole* (School of Our Lady)
1656–1659	Student at University of Copenhagen
1660	Brief sojourn in Rostock and student in Amsterdam, where:
04/07/1660	Discovery of the *Ductus parotideus*
1660–1663	Student in Leiden
07/27/1660	Enrollment
12/04/1664	Graduation as Dr. med.
1664–1665	Research travels through France
c. Feb. 1665	*Discours sur l'anatomie du cerveau*, Paris
09/16/1665	Departure from Paris for Italy
1666–1672	Scientist at the Medici court, Florence
May/June 1666	Residence in Rome
June 1667	Residence in Rome
11/07/1667	Conversion in Florence
1668–1670	Extensive geological field trip from Italy to Lower Hungary
Nov. 1668	Residence in Rome
1672–1674	"Anatomicus regius" (Royal Anatomist) in Copenhagen
01/29/1673 st.v.	*Prooemium demonstrationum anatomicarum in Theatro Hafniensi anni 1673*
1675–1677	Tutor of heir to the throne Ferdinando III at the Medici court in Florence

(continued)

Table 1.1 (continued)

04/13/1675	Ordination as priest
May–Sept. 1677	Residence in Rome
08/02/1677	Appointment as Titular Bishop of Titiopolis and Vicar Apostolic of the Northern Missions
08/06 & 09/13/1677	Audiences with Pope Innocent XI (1676–1689)
09/19/1677	Consecration as bishop
1677–1680	At the ducal court in Hanover
1680–1683/1684	Suffragan bishop of the Prince-Bishopric of Münster
1680/1681	Dean and parish priest of the collegiate church St. Ludgeri in Münster
09/01/1683	Departure from Münster
1683–1685	Residence in Hamburg
1685–1686	Residence in Schwerin
11/25/1686 st.v.	Death

1.3 Applied Methodology and Sources

With its concept of analyzing a physician of the early modern period and his writings from the perspective of "ethos" and structuring systematically the insights thus gained, the presented study treads relatively sparsely cultivated ground. While the first edition (1959) of the biography of Stensen by Max Bierbaum does mention the "scientific ethos"[4] of the anatomist and theologian, it fails to pursue the concept systematically. Applying the relatively young technical term of "bioethics" to the concrete manifestation of an early modern physician's ethos in one of his fields of research and the ethical questions he dealt with, however, represents a true first in terms of historical treatises to the best of the author's knowledge.

The structure of the study is determined by the inner criteria of this ethos, which appears in interdependency with Stensen's research as well as his philosophic orientation and scientific methodology. The latter three topics are dealt with insofar as they are of significance for Stensen's ethos, and all sources which provide information on this ethos directly or indirectly are consulted. The systematization in the presentation of the subject matter is based on the structure of the individual constituents of this ethos, with questions of bioethics being given particular consideration. The constituents are ascertained using the instrument of the historical-philological method. As the professional ethics of medical science and practice in the early modern period are largely determined by the behavior of individual physicians as members of the 'republic of physicians', and thus by social structures, excursions into the field of sociology[5] to illuminate Stensen's life and thought processes are occasionally interspersed.

---

[4] Bierbaum, *Stensen*, p. 117 = Bierbaum/Faller, *Stensen*, p. 113 = Bierbaum/Faller/Traeger, *Stensen*, p. 114.

[5] The sources for these excursions are the sociologists Robert K. Merton, whose knowledge of medicine is documented by his co-editorship of the study Merton/Reader/Kendall, *The Student-Physician* (Cambridge, MA 1957, 2nd ed. 1969), and, with a medical degree, Georges Canguilhem.

Stensen did not develop an explicit scientific set of ethics; the main features of his ethos can however be outlined by assembling and analyzing pertinent statements in his body of work under consideration of its development during Stensen's life and additional historic contextualization. The task thus posed is similar to that undertaken in the investigation of Stensen's spirituality from a theological perspective: his ethos as anatomist and natural scientist can be viewed as feature-complete by the time of his ordination as priest in 1675. The gradual development of Stensen's spirituality into its final shape in 1686, however, intensified only after 1667,[6] the year of his conversion, and thus occurred only later. Stensen's ethos and spirituality continually shaped each other since his student years. For the most part, statements by Stensen which are to be considered expressions of his creation spirituality are not presented in this study as they have already been dealt with by the author in his Münster Th. D. thesis *Herz, Gott, Kreuz. Die Spiritualität des Anatomen, Geologen und Bischofs Dr. med. Niels Stensen (1638–86) [...]* (Heart, God, Cross: The Spirituality of the Anatomist, Geologist and Bishop Dr. med. Niels Stensen [1638–86]).[7]

Due to the fact that Stensen's ethos was shaped by his education and a multitude of events in the course of his life, and equally had to prove itself under varied circumstances, biographical details will be discussed in the respective depth required. In addition, Stensen's work – like most other writing – contains direct and indirect clues illuminating the author's character and lifestyle, which in turn have significant influence on the practice of his ethos and are therefore dealt with as appropriate within the study.

The presented study therefore represents an analysis of Stensen's entire body of work, ie his research and teaching papers, his anatomical discoveries, his manuscripts and correspondence, examining his underlying ethos as physician and natural scientist against the background of his vita and the time he lived in, particularly the literature of the so-called 'republic of physicians' in the seventeenth century. Because Stensen only became a physician as a "physician of the soul",[8] ie a priest-physician with a focus on the salvation of the souls of humans, and worked exclusively as a scientist before his ordination as priest, statements on the professional ethics of physicians are found less often in his writings on anatomical research than in his later correspondence starting in the mid-1670s, his spiritual records and sermons. Stensen most often comments indirectly on medical ethics by developing the ethos of the priest and parish priest on the basis of the physician's, so that the latter can be deduced from the former.

Here the only pastoral theological work by Stensen to be published in print plays an important role, for it displays his background of experience with the physician's ethos – well-founded though not acquired in personal practice, as he had frequently spent time in hospitals for research ever since his years of study in Leiden. One may assume that Stensen closely observed the physicians in their actions and behavior

---

[6] For details see Sobiech, *Herz, Gott, Kreuz*, p. 26.

[7] See note 64 below.

[8] As a term Sobiech, *Herz, Gott, Kreuz*, p. 81.

towards their patients, just as he was accustomed to doing in all other situations and environments. The work in question is the booklet *Parochorum hoc age [...]* (Parish priests' duties [...]; Florence 1684 [corr. 1683]), based on his occupation as suffragan bishop of the Prince-Bishopric of Münster. Only a scant few copies of two editions of the partial Italian translation of this Latin original appear to have been preserved in Northern and Central Italy.[9] Like the Latin edition, published anonymously at Stensen's explicit request, they do not specify Stensen as their author.

The first edition of the partial translation bears the title *L'obbligo de['] parochi dimostrato con evidenza. Da zelante, e dotto Prelato. Opera utilissima. Tradotta dal Latino d'ordine. Per istruzione, e profitto de['] Parochi della sua Diocesi* (The duties of parish priests presented clearly. By a diligent and learned prelate. A very useful work. Properly translated from Latin. For the education and benefit of the parish priests of his diocese; Florence 1684). This edition was published by the bishop of Fiesole Dr. iur. utr. Filippo Neri Altoviti ([* 1634] 1675–1702), who had translated it after having received the Latin original from Grand Duke Cosimo III de' Medici ([* 1642] 1670–1723).[10] Apart from the existence of one copy each in Castiglion Fiorentino and Merate, it had apparently been unknown since the middle of the eighteenth century, evidenced by the fact that it is not mentioned in the biography of Stensen[11] by the Florentine printer and librarian Domenico Maria Manni (1690–1788), Lector for Tuscan Language at the seminary of the Florentine archbishopric since 1736.

The second edition, bearing the shorter title *L'obbligo de' parochi dimostrato con evidenza. Da zelante, e dotto Prelato. Opera utilissima* (Florence 1685) and mentioned by Manni,[12] is extant in three copies in Arezzo and Castiglion Fiorentino. It identifies the bishop of Arezzo, Dr. iur. utr. Giuseppe Ottavio Attavanti ([* 1643] 1683–1691), who metaphorically describes the treatise as a "nautical map" for parish priests in his foreword, as its publisher.[13] In literature after Manni, this edition was either given a different name or was not recognized as Stensen's work.

The final two sections of the Latin original with rules for parish visitations by the parish priests are replaced with a reference to synod decrees and a relevant text by

---

[9] These copies are located in the *Biblioteca Città di Arezzo,* the *Biblioteca Comunale,* Castiglion Fiorentino, and the *Biblioteca Comunale A. Manzoni,* Merate.

[10] Altoviti, *Obbligo,* pp. 3–5 (undated foreword by Altoviti "A Proposti, Pievani, Priori, e altri Parochi della sua Diocesi, salute, e Pastoral Benedizione"), here 4. The last two censorial notes are from 09/22/1684 (ibid., p. 6). On Altoviti: HC 5 (1952), p. 201. The booklet is not mentioned in Bruni, *Editori,* pp. 397–398; DOAP 2 (1852), s.v. "Parochorum Hoc Age", p. 315 r. col. lists only ParHAg.

[11] Manni, *Stenone* (Libro IV, Cap. X), pp. 314–317. On Manni: Giuseppe Crimi: "Manni, Domenico Maria"; in DBI 69 (2007), pp. 94–100.

[12] Manni, *Stenone* (Libro IV, Cap. X), pp. 314–317, here 316–317.

[13] Attavanti, *Obbligo,* foreword by Attavanti from 11/01/1685 (unpaginated) "A' Sacerdoti che anno cura d'Anime nella Diocesi Aretina", here the final page before p. 1: "Carta nautica", created by secretary Curzio Tanucci. On Attavanti: HC 5 (1952), p. 98; Tafi, *Vescovi,* no. 88 s.v. "Giuseppe Ottavio Attavanti", pp. 147, 149–150, on the booklet esp. 147 and 150, without naming Stensen as the author. It is not mentioned in Bruni, *Editori,* pp. 415–418.

the (likewise not explicitly named) missionary and preacher Paolo Segneri SJ (1624–1694).[14] Nothing is known about the number of copies published of either Italian edition. Like the original, they were examples of functional pastoral writing and were probably scarcely preserved for precisely this reason, though one may assume they enjoyed a broad reception in practice towards the end of the seventeenth century. Cardinal Gregorio Barbarigo (1625–1697),[15] who had ordained Stensen a bishop in Rome on September 19, 1677 and was attempting to foster increased diligence among the parish clergy by promoting the creation of appropriate professional literature, wrote a letter to Cosimo III from Venice on April 28, 1684, thanking the Grand Duke for sending him a (Latin) copy and praising its contents.[16] Stensen's pastoral theological tenets, shaped not least by his experience with the behavior of medical practitioners caring for their patients, thus came to have an impact on the professional ethos of parish priests in several Italian dioceses.

1.4   State of Research[17]: Niels Stensen in the History of Medicine and Medical Ethics

Owing to the overwhelming emphasis on history of science and scientific methodology in examinations of Niels Stensen's anatomical work – which in the following are outlined starting at the beginning of the twentieth century with a focus on their relevance to the topic of this study –, explicit statements with regard to Stensen's ethos are few and far between. That Stensen has been appreciated regularly in medical journals and books from the 1870s until today – even in the former Eastern Bloc during Cold War times, eg in censored medical journals in the GDR in the 1980s, a fact which might provide for interesting comparisons with the respective 'western' journal essays –, shows that his life and oeuvre as an anatomist are part of the cultural memory of medicine as a whole and are deserving of deeper source-based study with regard to Stensen's ethos.

The basis for in-depth engagement with Stensen as an anatomist is the complete edition *Nicolai Stenonis opera philosophica* (350 numbered copies, 2 vols., Copenhagen 1910) by the Copenhagen medical historian Edvard Vilhelm E. Maar

---

[14] ParHAg., p. 47, line 11–p. 52, line 22 and instead Altoviti, *Obbligo*, p. 58, lines 9–12 (preliminary remarks) & p. 58, line 13–p. 65, line 33 (Rammemorazione. Del debito, ch'hanno i Parochi di pascer l'Anime con la Parola di Dio. D'un divoto Religioso)=Attavanti, *Obbligo*, p. 49, lines 1–4 (preliminary remarks) & p. 49, line 5–p. 56, line 15 (Rammemorazione […]); additionally p. 56, lines 16–20 (Gregory the Great, cf. Étaix/Morel/Judic, *Homélies*, pp. 361–399, here 396 [no. 18]). Cf. Segneri, *Opere*, pp. 455–460; on Segneri: Herman H. Schwedt, "Segneri, Paolo d. Ä."; in: LThK³ 9 (2000), col. 399.

[15] On Barbarigo: Gino Benzoni, "Gregorio (Gregorio Giovanni Gaspare Barbarigo) Barbarigo, santo"; in: DBI 59 (2002), pp. 247–252.

[16] Gios, *Lettere*, pp. LVI–LVII, 70–71 (no. 72); Sobiech, *Herz, Gott, Kreuz*, pp. 80, 110.

[17] Complementary to the respective sections in Sobiech, *Herz, Gott, Kreuz*, pp. 4–23.

(1877–1940). This edition includes all anatomical and geological writings by Stensen known at the time in transcriptions preserving the original early modern notation, preceded by an introduction and supplemented by three appendices, the most extensive of which are the lecture notes of Holger Jakobsen, the *Exercitia academica*. Each of the treatises and appendices is furnished with preliminary remarks and explanations pertaining to specific questions compiled in an accompanying commentary.[18] The term "philosophical", chosen freely as an umbrella term and not corresponding to the title of any actual writing by Stensen, is to be understood as a synonym for "natural scientific" as per its usage in the early modern period.[19] Maar's work remains the definitive new edition of Stensen's anatomical and geological oeuvre known at the time of its publication.

In addition, the editions *Nicolai Stenonis opera theologica* and the *Nicolai Stenonis epistolae et epistolae ad eum datae* should be considered as complementary sources.[20]

Another significant edition is *Steno on Muscles*,[21] published by Troels Kardel in 1994. This work summarizes Kardel's studies on Stensen's "new method"[22] of research on muscles – unrecognized for more than 300 years – and includes a commentarial introduction followed by reprints (with English translations) of the early modern prints of:

1. Stensen's letter sent to Thomas Bartholin (1616–1680) from Leiden on April 30, 1663, published in the latter's *Acta medica & philosophica Hafniensia* – referred to in the following as "Acta Hafniensa" – and entitled *Nova musculorum & cordis fabrica* (New Structure of the Muscles and Heart),[23] and
2. Stensen's *Elementorum myologiae specimen, seu musculi descriptio geometrica. Cui accedunt canis carchariae dissectum caput, et dissectus piscis ex canum genere* (Specimen of Elements of a Myology [= science of muscles] or Geometric Description of the Muscle. Following It Are the Dissected Head of a Great White Shark and a Dissected Dogfish; Florence 1667, Amsterdam 1669) – referred to in the following as "Myology"[24] – under omission of both appendices.[25]

---

[18] Maar, *Opera philosophica*, vols. 1 and 2 (nos. I–XXXIII) with the appendices XXXIV–XXXVI in vol. 2, pp. 283–310 as well as an overview of all editions of natural scientific writings by Stensen published until then ibid., pp. 351–359. On Maar: Egill Snorrason (Axel Hansen), "Maar, Edvard Vilhelm Emil"; in: DBL³ 10 (1982), p. 303.

[19] King, *Medical Enlightenment*, p. 96. The term "philosophia" for medicine is rare in the early modern era according to Richard Toellner, "Medizin III. Frühe Neuzeit"; in: HWPh. 5 (1980), cols. 984–992, here 984–985.

[20] Larsen/Scherz, *Opera theologica*, vol. 1; Larsen/Scherz, *Opera theologica*, vol. 2; Scherz, *Epistolae*, vol. 1; Scherz, *Epistolae*, vol. 2.

[21] Kardel, *Steno on Muscles*.

[22] Myol., p. 64, line 36 = Kardel, *Steno on Muscles*, p. 88, line 3: "nova methodo".

[23] In Maar's edition: NovMusc. = E 13. Cf. NovMusc. (Notes) preliminary remarks, p. 254, r. col.

[24] In Maar's edition: Myol. Cf. ibid. (Notes) preliminary remarks, p. 319.

[25] CanCap. and PiscCan.

An edition[26] of Stensen's Parisian *Discours* commented by Raphaële Andrault followed in 2009, and a chronologically ordered edition[27] of Stensen's works translated into French by Birger Munk Olsen was published in 2010.

The English edition[28] completed in 2013 by Troels Kardel and Paul Maquet includes, alongside other additions, translations supplemented with updated comments on myology and corrections of:

1. The first of two volumes of the biography of Stensen[29] by the Austrian-born Copenhagen researcher on Stensen Fr. Dr. phil. Gustav Scherz CSsR (1859–1971)[30] and
2. the works included in Maar's edition (without the appendices XXXIV–XXXVI), in part using older translations by the authors and others,[31] as well as Stensen's Amsterdam disputation *Disputatio physica de thermis*, dated July 8, 1660 and discovered by Gustav Scherz in Philadelphia as late as 1959[32]; for better orientation the page numbers of Maar's edition are added on the margins respectively.

With regard to secondary literature, there is a noticeable spike in publications illuminating various aspects of Stensen's work during the 1930s overall, not least due to the Stensen anniversaries in 1936 and 1938:

On September 24, 1930, during the 8th "Congresso internazionale di storia della medicina" (Rome, September 22–27, 1930), Danish surgeon and medical historian Valdemar Meisen (1878–1934)[33] gave an account of Stensen's life and research in which he emphasized the anatomist's inaugural lecture in Copenhagen held on January 29, 1673 st.v.: "Of particular interest in this regard [meaning the fact that Stensen led his life consistently until his death] I find his opening lecture as Anatomicus regius at the University of Copenhagen; of interest for its allegorical and poetic language, for its admiration of nature and its wonders, but also for its ethical character and its humble praise of the Lord, Creator of All Things."[34]

---

[26] Andrault, *Stensen*. In Maar's edition: Discours. – Aucante, *Descartes*, lays out René Descartes' medical philosophy and on p. 242 briefly mentions Stensen's *Discours*, which despite the errors by Descartes that it likewise exposed, eg in regard to the relationship between body and soul, had nevertheless defended "a worthy Cartesian statement of faith" (pourtant [...] une digne profession de foi cartésienne).

[27] Olsen, *Sténon*.

[28] Kardel/Maquet, *Biography and Original Papers*; a newly typeset version of Maar's Latin texts under http://extras.springer.com (via ISBN 978-3-642-25078-1 as key; PDF file, VII and 361 pages).

[29] Scherz, *Biographie*, vol. 1.

[30] On Scherz: Gottfried Roth, "P. Dr. Gustav Scherz †"; in: Clio med. 6,2 (1971), pp. 159–160.

[31] Listed in detail in Kardel/Maquet, *Biography and Original Papers*, p. ix (Preface).

[32] On the history of the discovery see Scherz, "Niels Stensen's First Dissertation", pp. 247–248, 253–255, 258–259; in Kardel/Maquet, *Biography and Original Papers*, p. 346 n. a, "1960" should be corrected to "1959".

[33] On Meisen: Otto Carl Aagaard, "Meisen, Valdemar"; in: DBL³ 9 (1981), pp. 494–495.

[34] Meisen, "Stenone", p. 188: "Interessante, dal punto di vista psicologico, mi pare in questo riguardo la sua lezione inaugurale come Anatomicus Regius all'Università di Copenaghen, interes-

On the occasion of the 450th anniversary of the University Library of Copenhagen, Meisen described Stensen's life in shortened form in a publication edited by him in 1932, *Prominent Danish scientists through the ages*, pointing out Stensen's "restless and yearning spirit".[35] A brochure created in preparation of an exhibition donated to the *Museum of Science and Industry* in Chicago by the Danish government in 1933 featured introductions of Danish scientists from the past and present including Niels Stensen and nuclear physicist Niels Bohr.[36] In 1938, the 300th anniversary of Stensen's birth was commemorated with various scientific lectures on Stensen at the University of Copenhagen.[37] In an essay published in *Annals of medical history* (New York) in the same year, American medical historian Dr. med. Anne Tjomsland (1880–1968)[38] stated her opinion that Stensen appeared in an exceptional manner to have been free of all superstitious notions prevailing at the time.[39]

At the 14th "Congresso internazionale di storia della medicina" (Rome and Salerno, September 13–20, 1954), the Copenhagen Professor of History of Medicine Dr. med. Edvard Gotfredsen (1899–1936)[40] offered a brief overview of Stensen's life and ended his address by expressing his hope that a worthy monument to Stensen, which according to Gotfredsen was in a planning stage at the time, would be erected within a few years in Copenhagen, the anatomist's city of birth (Fig. 1.2).[41]

The 16th congress of the *Société Internationale de Chirurgie* (est. 1902) took place in Copenhagen from 24 to 30 July, 1955. Its schedule lists Stensen as the event's patron, and the participants' nametags were adorned with his portrait (Fig. 1.3). In keeping with this theme, Edvard Gotfredsen ends his foreword to the schedule booklet with the programmatic words from Stensen's *De solido intra solidum naturaliter contento dissertationis prodromus* (Precursor to a treatise on solids naturally contained within solids; Florence 1669, Leiden 1679, Pistoia 1763; English translation: London 1671, French partial translation: Dijon/Paris 1757), which establishes geology as a history of Earth: "Science as well as the Church revere Stensen for his principle which he expresses with this motto in his geological treatise: To overthrow what is false, to lend steadfastness to what is true, to cast a bright light on what is unclear."[42]

---

sante per il linguaggio allegorico e poetico, per l'ammirazione della natura e delle sue meraviglie, ma anche per il suo carattere etico e per il suo cantico di lode umile al Maestro, creatore di ogni cosa."

[35] Meisen, "Steno", p. 42: "restless and yearning spirit".

[36] Committee for Selecting and Procuring the Collective Scientific Exhibit, Donated by the Danish Government to the Museum of Science and Industry at Chicago, Danish Scientists, s.v. "Nicolaus Steno (Niels Steensen)", pp. 26–27; ibid., s.v. "Niels Bohr", p. 38. The brochure is relatively unknown.

[37] On the lectures, see Scherz, *Fama sanctitatis*, pp. 47–48.

[38] On Tjomsland: Saul Jarcho "Anne Tjomsland, 1880–1968"; in: JHMAS 24 (1969), p. 482.

[39] Tjomsland, "Stensen", p. 495.

[40] On Gotfredsen: Egill Snorrason, "Gotfredsen, Edvard"; in: DBL³ 5 (1980), pp. 247–248.

[41] Gotfredsen, "Stenonis", p. 153.

[42] Société Internationale de Chirurgie, *Guide*, front flap text: "Tant la science que l'Eglise vénèrent Sténon pour le principe qu'il exprima dans son oeuvre géologique par cette devise: falsa evertenda,

Fig. 1.2 Sculpture "Steno" (1962) by Gottfred Eickhoff (1902–1982) in front of the *Natur- og Sundhedsvidenskabelige Fakultetsbibliotek*, Nørre Allé 49, Copenhagen
Unveiled on October 25, 1963 (Scherz, Brain Research, p. 2), the sculpture shows Niels Stensen during his inaugural address on January 29, 1673 st.v. from 2 p.m. in the *Domus Anatomica*, Frue Plads, Copenhagen

Fig. 1.3 Niels Stensen as a bishop: cover of the program booklet for the 16th congress of the *Société Internationale de Chirurgie* on July 24–30, 1955 in Copenhagen
The portrait drawing, also featured on the name badge attached to the booklet for identification of congress participants, is based on the oil portrait (pictured in b/w e.g. in Scherz, *Epistolae*, vol. 2, p. IV), by Christian August Lorentzen (1749–1828) located in the *Medicinsk-Anatomisk Institut* of the University of Copenhagen (*Panum-komplekset, Det Sundhedsvidenskabelige Fakultet*)
KB, Ny kgl. Samling 4963 4°, VIII

Gustav Scherz, whose correspondence in multiple languages and spanning several decades with a large international network of physicians is archived in the "Royal Library" (Det Kongelige Bibliotek) in Copenhagen,[43] hosted the "International Historical Symposium on Nicolaus Steno and the Brain Research in the Seventeenth Century", attended by 57 scientists from 14 countries, together with medical historian Prof. Dr. Charles D. O'Malley (1907–1970)[44] of the *University of California* (Los Angeles) on August 18–20, 1965.[45]

---

vera stabilienda, obscura eludidanda [*corr.*: elucidanda]." Gotfredsen limited himself to these three elements, presumably due to the tricolon thus created; in Prodromus, p. 183, lines 20–22, here 21–22 = Scherz, *Geological Papers*, p. 136, lines 15–16, here 16, Stensen includes the further element "incognita producenda" (to advance the unexamined).

[43] KB, Tilg. 621.

[44] On O'Malley: Frederick N. L. Poynter, "Obituary Charles Donald O'Malley 1 April 1907–7 April 1970"; in: MedH 14,3 (1970), pp. 320–321.

[45] Program in Dansk medicinsk-historisk Selskab, *Symposium* (unpaginated).

Fig. 1.4 Rear of the sculpture "Steno" (see Fig. 1.2); in the background the southern part of Nørre Campus with the *Sundhedsvidenskabelige Fakultet*
The inscription reads "anatomicus/geologiae fundator/n[atus] 1638/servus Dei/m[ortuus] 1686" (Anatomist/Founder of Geology/Born 1638/Servant of God/Died 1886)

In his closing speech for this symposium entitled "Nicolaus Steno the Humanist", on August 20, Fr. Scherz – with reference to the threat posed to mankind by the increasing specialization necessary for research, resulting in a loss of focus on the human subject – enumerated several essential aspects of Stensen's humanism. This humanism, Scherz stated, manifested itself (1) in the comprehensive scientific perspective which Stensen had always amended his individual studies to reflect, (2) in his philosophical stance favoring succinct brevity and leaving no room for self-promotion, along with his striving for mutual dependency between certainty and truth, and (3) in the "man full of God" (Vir Deo plenus) (cf. Fig. 1.4), whose inaugural lecture at Copenhagen was a hymn to the world's beauty and the connection between any science and all of creation as well as the Creator himself.[46]

Appropriate reference will be made in due course to the anatomist and medical historian Adolf Faller (Fribourg) as well as to other physicians and medical historians who dealt with Stensen's anatomical oeuvre.

---

[46] Scherz, *Humanist*, p. 295. The term "Vir Deo plenus" is taken from the 1687 epitaph on Stensen in the Florentine Basilica of San Lorenzo, cf. Sobiech, "Capella Stenoniana", pp. 75–76.

In his speech "Il rapporto tra filosofia e medicina nella storia del pensiero" (The Relationship Between Philosophy and Medicine in the History of Ideas),[47] delivered at the conference "Etica dell'atto medico" (Ethics of Medical Activity) organized by the *Società Internazionale Tommaso d'Aquino* (est. 1976) on November 23–24, 1990 at the *Pontificia Università S. Tommaso d'Aquino* of the Dominican Order in Rome, Andrea Porcarelli noted that Stensen had not only proven anatomical mistakes made by René Descartes, but also cast doubts on a way of thinking which formed the basis for those anatomical mistakes by claiming to always have a solution to every problem.[48]

Three international medical symposia on salivary gland research dedicated to Stensen were held in the wake of his beatification.[49]

Paolo Perrini, Giacomo Tiezzi, Nicola Montemurro, Alessandro Weiss, Ludovico Lutzemberger and Giuliano Francesco Parenti (Neurosurgical Clinic, Università degli Studi di Pisa) presented their joint poster entitled "Niccolò Stenone 1638–1686: scienziato, neuroanatomico e santo" on October 16, 2009 at the "58° Congresso Nazionale della Società Italiana di Neurochirurgia (SINch)" (Lecce, October 14–17, 2009).[50] During the past two decades, Stensen's life and work have been summarized and appreciated in various internationally published medical periodicals, usually in joint essays by several researchers.[51]

On November 25–26, 2011, a formal ceremony, an exhibition and a festive academy entitled "Seliger Niels Stensen. Naturwissenschaftler und Glaubensbote in Zeiten des Aufbruchs" "Blessed Niels Stensen. Natural Scientist and Messenger of the Faith in Times of Change" were organized in Schwerin, Stensen's last place of

---

[47] Lobato, "Presentazione", p. 5.

[48] Porcarelli, "Rapporto", pp. 80–82.

[49] These were the "Symposium on Salivary Glands dedicated to Niels Stensen (Niccolò Stenone)" in the course of the "Tenth European Anatomical Congress" from 09/17 to 09/21/1995 in Florence with a historical overview of Stensen's relevant research (Riva/Testa Riva, "Stensen"), the "Second Symposium on Salivary Glands, dedicated to Niels Steensen (Niccolò Stenone)" from 05/23 to 05/25/1997 in Cagliari (Riva/Tandler/Murakami/Steward, *Second Symposium*), and hitherto lastly the "3rd International Symposium on Salivary Glands in honor of Niels Stensen. The 36th SEIRIKEN Conference" from 10/20 to 10/24/2006 in Okazaki (Japan) chaired by Masataka Murakami, Hiroshi Sugiya and Alessandro Riva; the program of this symposium can be found on the website of the Japanese *National Institute for Physiological Sciences* under http://www.nips. ac.jp/stensen.

[50] http://www.sinch.it/nuovo/congressi-nazionali-sinch-2009: 58° Congresso Nazionale SINch. Lecce 14/17 ottobre 2009. Programma, edited by Antonio Montinaro (Presidente del Congresso) (PDF file, 83 pages, of which 80 pages paginated), p. 55 (Sessione Poster VI – 4a Parte Miscellanea, P263).

[51] For the period 2002–2015: Holomanova/Ivanova/Brucknerova, "Stensen"; Van Besien/Van Besien, "Stenon"; Perrini/Lanzino/Parenti, "Stensen"; Tubbs/Mortazavi/Shoja/Loukas/Cohen-Gadol, "Stensen"; Tubbs/Gianaris/Shoja/Loukas/Cohen[–]Gadol, "Fallot"; Benter/Benter/Benter, "Stensen"; Parent, "Stensen"; Strkalj, "Stensen"; De Micheli Serra/Izaguirre Ávila, "Saint"; Sobiech, "Capella Stenoniana"; Sobiech, "Science, Ethos, and Transcendence".

activity, commemorating the 325th year of his death.[52] The anthology[53] of previous contributions presented there includes an article on Stensen's Leiden disputation of 1661, which Gerhard Schlegel was able to publish in *Das deutsche Gesundheitswesen. Zeitschrift für klinische Medizin. Organ der Gesellschaft für Klinische Medizin der DDR* in 1982 in cooperation with two medical doctors at the Wilhelm-Pieck-Universität in Rostock (renamed back to its original name Universität Rostock in 1990) and in which Stensen's definition of the goal of anatomy (cf. Fig. 1.5) and his motto are cited.[54]

On February 12–13, 2015, the international conference "Steno and the Philosophers" organized by Raphaële Andrault and Mogens Lærke took place at the *Institut des études avancées de Paris.*[55]

Various dissertations dealing with the history of medicine have addressed Stensen's life and work since the 1960s: (1) Heike Kohlsaat discusses Stensen's findings on the skin and its adnexa.[56] (2) Josephus Gerardus Vugs (1920–1996) focused his study on the classification of Stensen's Parisian *Discours* in terms of the history of science and his own Dutch translation of the text.[57] (3) Marco Marzollo (1915–2002) wrote a medical-historical study similar to a dissertation about Stensen's Myology including an Italian translation.[58] (4) Hans-Michael Bonse covers Stensen's initial description of the Tetralogy of Fallot, also including a translation.[59] (5) Ulrike Heida introduces Stensen in the circle of his contemporary colleagues, stating that his personality "[must] be understood as a synthesis of particular human and scientific qualities which remains exemplary even in our time."[60] (6) Eva-Maria Wicklein emphasizes the "basic moral and ethical attitude" which Stensen was able to "easily conjoin with scientific enthusiasm" in her summary of his scientific activity following his conversion.[61] (7) In his work dealing with Stensen's studies in the Netherlands, Max-Joseph Kraus explicitly mentions

---

[52] Program: Thomas-Morus-Bildungswerk   Schwerin,   *Pilgerreise*,   pp.   175–179   (Die Jubiläumstagung 2011 im Erzbistum Hamburg), here pp. 175–177.

[53] Thomas-Morus-Bildungswerk Schwerin, *Diener der Wahrheit*.

[54] Wischhusen/Schlegel/Ehler, "Stensen", p. 2066 (21). On publication in the GDR, see Thomas-Morus-Bildungswerk Schwerin, *Diener der Wahrheit*, p. 14. On Stensen's motto, see note 241, in Chap. 2 below.

[55] Institut des études avancées de Paris, 17 quai d'Anjou, F-75004 Paris, http://paris-iea.fr.

[56] Kohlsaat, "Haut".

[57] Vugs, "Stensen", pp. 22–29, 97–101, 137–250, 281–282 (Résumé). The final chapter "Stensens persoonlijkheid. Zijn geestelijke evolutie" (pp. 251–271) contains no explicit conclusions on Stensen's ethos as an anatomist. – A Dutch translation of Stensen's motto from the inaugural address is carved into Vugs' gravestone at the *Begraafplaats De Lichtenberg*, Deken Dr Dirckxweg 25 in Bavel (Netherlands): "'Mooi is datgene wat we zien, mooier is wat wij kennen[,] het allermooiste is wat wij niet weten' ("Zalige") Nicolaus Steno"; cf. the color photograph at http://www.online-begraafplaatsen.nl/graf/189397/294900/Josephus-Gerardus-VUGS-1920-1996.

[58] Marzollo, "Musculo". On Marzollo: Ateneo di Salò, *Marzollo*, pp. 37–41.

[59] Bonse, "Tetralogie", here pp. 18–33, German translation of Stensen's treatise *Embryo*.

[60] Heida, *Fachkollegen*, p. 116.

[61] Wicklein, *Konversion*, p. 91; cf. ibid., pp. 5, 90.

Fig. 1.5 Cross (c. 1938) by Marcel Feuillat (1896–1962) and marble plaque with engraving (1988) in the autopsy room of the Anatomical Institute of the University of Fribourg
The quotation is from Prooemium, p. 254, lines 35–37

his own desire "to experience enrichment in my own moral emptiness [sic!] by retracing the life of a scientist acting in an ethically responsible fashion, and especially to support development of a synthesis between my own Christian intellectual world with the freedom to act which I wish to acquire".[62]

The following have contributed thus far to the in-depth development of this study with regard to methodology and content: (1) Troels Kardel, who describes Stensen as a realist and philosopher of science "sui generis" from the perspective of philosophy of science in *Steno. Life – Science – Philosophy*[63]; (2) the theological-historical dissertation *Herz, Gott, Kreuz* by the author himself, which chooses Stensen's discovery that the heart is a muscle as its point of departure,[64] describes and analyzes his religious development and conversion as well as – within the framework of its historical-systematical presentation – his spirituality, shaped as it was to a degree by natural science, and its sources, being the first to evaluate Stensen's entire body of work; (3) Stefano Miniati, who in his dissertation combines elucidations on Stensen's oeuvre from the perspectives of history of science and history of religion, arguing that the core of Stensen's "public ethic" is already manifest in the excerpt from the book *Ioseph Aegypti prorex descriptus et morali doctrina illustratus* (Joseph Viceroy of Egypt, Described and Explained on the Basis of Moral Doctrine; Antwerp 1641) by the Bavarian court preacher Jeremias Drexel SJ (1581–1638) found in Stensen's Copenhagen student's diary.[65]

1.5   Medical Ethos in the Seventeenth Century

The period from the second half of the seventeenth century up to and including the first half of the eighteenth century is described as the crisis from which modern experimental thinking developed.[66] Medicine established itself as an empirical science supported by the new rational world view and empiricism as the method of insight.[67] In this vein, Stensen emphasizes that during his century anatomists made a plethora of discoveries some of which appeared monstrous when compared to previous knowledge,[68] and that it would be up to future accounts to determine how much medical practice owed to these anatomical experiments.[69] This bespeaks his

---

[62] Kraus, "Niels Stensen in Leiden", p. 9.

[63] Kardel, *Steno*, p. 97.

[64] Sobiech, *Herz, Gott, Kreuz*, p. 345.

[65] Miniati, *Steno's Challenge*, p. 53–54, here 54: "public ethic". The term is not discussed by Miniati. On Drexel: Friedrich Wulf, "Drexel (Drexelius), Jeremias"; in: DHCJ 2 (2001), pp. 1146–1147.

[66] Stroppiana, "Lancisi", p. 5.

[67] Richard Toellner: "Heilkunde/Medizin II. Historisch"; in: TRE 14 (1985), pp. 743–752, here 749.

[68] VitTrans., p. 211, lines 9–15; cf. GlandOc., p. 82, line 7; VitHyd., p. 237, lines 2–3.

[69] Myol., p. 106, lines 8–9 = Kardel, *Steno on Muscles*, p. 222, lines 14–15.

great scientific awareness as well as his humility, and the statement represents a contemporary observation of what has been termed the Scientific Revolution (Révolution scientifique) since the 1930s.[70] It should be mentioned that Stensen's Parisian lecture on methodology *Discours sur l'anatomie du cerveau*, given in 1665, was well ahead of its time in brain research. The medical science of Stensen's studying years was determined by two "basic medical concepts"[71]: 'autopsy' on the one hand and 'authority' of received ancient medicine on the other. Medical practice was generally conducted independently of theory, although it was also not uncommon for practicing physicians to publish.

1.5.1   The 'Republic of Physicians'

Stensen lived at a time in which physicians – like all "learned men" (lit[t]erati) across the national borders – considered themselves a community of intellectuals, similar to the way we speak of the worldwide scientific community and its members today,[72] although the mobility of students as well as oral and written exchange of research results were facilitated greatly by the bonding force of the Latin language, which was a foreign language to *all* members.

The metaphor "res publica medica"[73] (*literally*: "medical republic"; *loosely*: "republic of physicians"[74]) can be found as early as the introductory letter (August 19, 1584) by Dr. med. Theodor Zwinger Sr. (1533–1588),[75] professor of theoretical medicine, for the treatise Παρατηρήσεων *sive observationum medicarum, rararum, novarum, admirabilium, & monstrosarum volumen in tomis septem* (Book of Medical, Rare, New, Wondrous and Monstrous Παρατηρήσεις or Observations in Seven Volumes; 7 vols., Basel and Freiburg in the Breisgau 1584–1597) written by the "Stadtphysikus" (city physician), ie the highest public health officer, of Freiburg

[70] By Dr. phil. Alexandre Koyré (1892–1964), see Hendrik Floris Cohen, "Wissenschaftliche Revolution"; in: ENZ 15 (2012), cols. 73–76, here 73–74; on Koyré: Charles C. Gillispie, "Koyré, Alexandre"; in: DSB 7 (1973), pp. 482–490, here 485 l. col., 487 r. col.

[71] The relationship between the two concepts hitherto seems insufficiently settled, see De Angelis, *Anthropologien*, pp. 13, 215–216.

[72] Cf. Storer, "Introduction", p. xxviii; a document from March 2012 can be found eg in White/Woods/Takai/Ishihara/Seki/Tilly, "Oocyte Formation", p. 419.

[73] Cited according to Pomata, "Praxis Historialis", pp. 134, 146n132.

[74] According to Delisle, "Controversy", pp. 163–164, 166–167, the "respublica medicorum" ("republic of physicians") is first mentioned by Dr. med. Pietro Andrea Mattioli (1501–1578) in a controversy during the years 1555–1565, here referring to a community of physicians with a Catholic and Italian background; on Mattioli: Cesare Preti, "Mattioli (Matthioli), Pietro Andrea"; in: DBI 72 (2008), pp. 308–312. On the "respublica medicorum" see Maclean, "Medical Republic of Letters", here pp. 15, 24–25 on the term.

[75] On Zwinger: Wilhelm Kühlmann, "Zwinger, Theodor, d. Ä."; in: Killy² 12 (2011), pp. 730–732.

in the Breisgau, Dr. med. Johannes Schenck von Grafenberg (1530–1598).[76] The term "respublica medica" represents a sub-entity of the so-called "republic of letters", the "civitas lit(t)eraria" or "respublica lit(t)eraria"[77] and can be found throughout medical publications of the early modern period, used eg by Stensen's academic mentor Dr. med. Thomas Bartholin (1616–1680),[78] who was a professor of anatomy at the University of Copenhagen from 1649 to 1656 and, following a hiatus due to illness, a *Professor honorarius* from 1661 onwards, as well as by Stensen himself.[79] This associative 'republic' of scholars, encompassing researchers as well as practitioners and variegated in the component 'republics' of the various scientific disciplines (eg medicine), was characterized by fundamental values which were exhibited in the lifestyles and activities of its members rather than being expressed in codified rules of behavior, and which in modern terminology would be subsumed in the term "ethos". Thus the 'republic of physicians' participated in the ethos of the early modern 'republic of letters' on the one hand, while on the other developing its own particular ethos and associated values pertaining to its respective areas of research and professional practice. Likewise, as a member of this community Stensen participated in its ethos while at the same time helping to shape it.

'Ethos' appears as a foreign word in German no earlier than the late nineteenth century "to denote the enduring attitude and mindset of an individual or a community, insofar as certain moral contexts are accentuated and certain values emphasized within it."[80] The term 'ethos', derived from Greek ἔθος (custom) or ἦθος (character) and indicating a process taking place out of habit within a group or community, or an individual human as a member thereof, describes "the 'morals' of a certain group, for example physicians, in terms of an unwritten or fixed code of ethics". This refers to the moral mindset of a person and the mutual permeation of professional and personal guiding values, with (in this case) the ideals of medicine and the histories of philosophy and medicine of the early modern period forming the backdrop.[81]

More precisely, "medical ethos" could be defined as "a sum of attitudes and mindset" distinguished from the more rule-based concept of medical ethics.[82] There exists no canon listing the constituents of this ethos. Corresponding to the objective definition is, on the subjective level, the individual's own mindset, his moral

---

[76] On Schenck von Grafenberg: Stephan, "Observationes", pp. 11–19.

[77] E.g. Geulincx, *De virtute*, fol. * 2ʳ–* 6ᵛ (letter of dedication from 07/271665), here * 2ᵛ, * 6ᵛ. – Other contemporary Latin variants also exist.

[78] E.g. in Bartholin, *Anatomes studiosis salutem*, line 33; AMPH 1 (1671/72), foreword "Th[omas] Bartholinus Fac[ultatis] Med[icae] decanus, lectori curioso s[alutem]" (unpaginated) from late December 1672, here the first page; E 2, p. 139, line 10. On Bartholin: HGK, s.v. "Bartholin, Thomas", pp. 70–71; Valdemar Meisen, "Bartholin, Thomas"; in: DBL³ 1 (1979), pp. 476–480.

[79] According to the only reference in GlandOr., p. 19, lines 19–20.

[80] Hans Reiner, "Ethos II"; in: HWPh. 2 (1972), cols. 812–815, here 815.

[81] Bergdolt, *Gewissen*, p. 18.

[82] Zwierlein, *Begegnung und Verantwortung*, p. 13.

character,[83] formed in part by inherent natural disposition and in part by conscious or unconscious shaping and training.[84] This subjective layer as well as the objective layer will be discussed in the course of this study.

Historical analysis of the professional ethics and the "medical 'self-perception'"[85] of physicians and anatomists, in particular those of the early modern period, constitutes a widespread desideratum for research which has hitherto merely been touched upon, eg for the period from the Renaissance to the middle of the seventeenth century, with a focus on certain areas of sexual ethics, by Winfried Schleiner in *Medical Ethics in the Renaissance* (Washington, D. C. 1995),[86] or by Klaus Bergdolt in *Das Gewissen der Medizin. Ärztliche Moral von der Antike bis heute* (Munich 2004)[87] for the entire early modern period. The few scientific writings dealing with the medical profession and explicitly mentioning 'ethos' in their title are for the most part insufficiently founded in theory,[88] resulting in rather diffuse descriptions of the 'ethos' inferred from the works of the respective authors examined therein.

In his emphatic rectoral inauguration address *Die Geburt des ärztlichen Ethos aus dem Geiste der Anatomie* (The Birth of the Anatomical Ethos from the Spirit of Anatomy) on November 20, 1948, Dr. med. Dr. phil. Gustav Sauser (1899–1968), professor of anatomy at the Leopold Franzens University of Innsbruck and described as a "Christian humanist" by his colleague Adolf Faller,[89] reminded his audience that 'ethos' (and therefore by extension medical ethos) implies a "position and mindset in the sense of a binding relationship", thereby already approaching "the broadest definition of 'religio' [in the sense of 'commitment']".[90] As anatomy represents not only an indispensable foundation of knowledge for a physician but also a form of education – a corpse illustrating not only the finiteness of human existence, but also "the physician's ethos of imperturbability in never relinquishing life in the face of death" –, the learning physician encounters "the first medical teacher whose hand will never truly leave him in clinic and profession" in the anatomist.[91] A love for life, intuitive art and conscientious craft, knowledge and scientific thinking, silence and *pietas* (reverence), Sauser summarized, are the "spectral colors" of medical ethos.[92] This ethos, shaped by Christian humanism and – 4 years after the end of World War II (1939–1945) – still affected by the horrible distortions effected

---

[83] Wolfgang Kluxen: "Ethos"; in: LThK³ 3 (1995), cosl. 939–940, here 939; Kluxen, "Ethos"; in: LBioeth. 1 (1998), pp. 693–694, here 693.

[84] Gerhard Funke: "Ethos I"; in: HWPh. 2 (1972), cols. 812–815, here 812.

[85] Maehle, "Werte und Normen", p. 339.

[86] Schleiner, *Renaissance*, p. vii.

[87] Bergdolt, *Gewissen*, esp. pp. 107–222.

[88] E.g. Korff, *Berufsethos*; cf. in contrast Schefer, "Berufsethos", pp. IX–XII.

[89] On Sauser: Adolf Faller, "In memoriam Prof. Dr. med. et phil., Magister pharm. Gustav Sauser 15.7.1899–17.6.1968"; in: Acta anat. 74 (1969), pp. 1–9, here 6.

[90] Sauser, *Ethos*, p. 9. Sauser refers to the etymology of the word from the Latin "religare".

[91] Ibid., pp. 10–12 with citations on pp. 10, 12.

[92] Ibid., p. 21.

on the professional ethics of medicine by totalitarianism, results in an anatomist's attitude towards the human subject heavily influenced by the utmost respect.

That Sauser's emphasis of the educational aspect of anatomy in the study of medicine is still valid today can be seen in a 1999 essay by Reinhard V. Putz, an anatomist from Innsbruck then teaching in Munich, which implicitly deals with the anatomist's ethos: Putz attributes "an important reflection on the foundations of one's own life"[93] to the gross anatomy practical due to the "student's personal involvement with the corpse" engendering "a confrontation with human mortality, and ultimately with one's own mortality and finiteness".[94] It appears meaningful to analyze the lasting significance of the ethos of the anatomist Stensen from this perspective as well.

1.5.2 Life as a Student and Researcher

What were the requirements for putting this ethos into practice in Stensen's time? Since the medical ethos as applied to medical science was linked to the overarching ethos of the 'republic of letters', the majority of anatomists in the second half of the seventeenth century, and likewise Stensen, were accustomed to discussing their findings with colleagues prior to publicizing them.[95] Scholars were connected in networks of correspondence[96]; observations and experiments were communicated and debated in letters to each other.

Stensen was introduced to science by Thomas Bartholin, his private instructor and study mentor. On August 5, 1662 st.v., in view of Stensen's discoveries, Bartholin encouraged his student to further engage in the study of anatomy, for which he appeared to be cut out and, according to Bartholin, seemed to "burn" for, "for the benefit of sick mortals".[97] In his answer sent from Leiden on August 26, 1662, Stensen stressed that he had been divested of an intermittent uncertainty whether to give up his studies of medicine in favor of geometry, his original educational interest, by occasionally quite harsh threats by highly famous men and some less than friendly public writings misinterpreting his views.[98] Stensen's interest in

---

[93] Putz, "Leichnam", p. 30.

[94] Ibid., p. 28.

[95] See in more detail in Miniati, *Steno's Challenge*, pp. 85–87, here 87.

[96] Detlef Döring, "Gelehrtenkorrespondenz"; in: ENZ 4 (2006), cols. 386–389; on Stensen's scientific networks during his time as anatomist and geologist in Florence, see Bek-Thomsen, "Diplomatic Skills", here p. 9, a statistical analysis of Stensen's correspondence from 1666 to 1672.

[97] E 8, p. 158, lines 8–10, here 9–10: "fervide [...] in mortalium aegrotantium commodum".

[98] AvCun., p. 115, lines 1–15 & p. 117, lines 3–5 = E 9, p. 158, line 42–p. 159, line 10 & p. 160, lines 22–24. The mentioned letters are not preserved. Descartes' recently published work *De homine* (Leiden 1662), which Stensen mentions at the end of the letter (see note 180, in Chap. 2 below), can likewise be presumed to have played a role, cf. Grell, "Conversions", p. 210.

science, presumably awakened at the early age of six in his father's goldsmith's shop,[99] is reflected in the topical diversity of his study diary.

As was customary at the time, Stensen embarked on the so-called *Peregrinatio academica*,[100] which starting in 1659, the end of his studies in Copenhagen, led him via Rostock to Amsterdam and Leiden. In his accountability report on his studies abroad, the *De musculis & glandulis observationum specimen. Cum epistolis duabus anatomicis* (Specimen of Studies on the Muscles and Glands. With Two Anatomical Letters; Copenhagen and Amsterdam 1664, Leiden 1683), containing Stensen's new discoveries on the muscular structure of the heart and the glandular system, he mentions that he could not praise his supporters enough for their "kindness" (humanitas) in providing him as a guest with the necessary implements in their hospitals[101]; this should be viewed against the background of a letter written to Bartholin by Stensen on January 9, 1662 from Leiden complaining about the difficulties in finding dissection material.[102] In the same letter, Stensen also informed Bartholin that Sylvius – the professor of anatomy in Leiden, Dr. med. Frans de le Boë (du Bois) Sylvius (1614–1672)[103] –, who was offering practical exercises in the current trimester, had given him the opportunity a few days earlier to examine to his heart's content a demonstration object left over after Sylvius' *Collegium medico-practicum* held in the small Anatomical Theater, ie the autopsy room expanded by Sylvius in the *St. Caecilia Gasthuis*, the municipal pauper's hospital in Leiden.[104] In his letters to his study mentor in Copenhagen, Stensen also writes of his "[financially] tight situation at home" imposing boundaries on his research.[105]

In contrast to most of his fellow students who became general practitioners, Stensen continued his wanderings – now as travels of a fully-fledged academic (Peregrinationes eruditae)[106] – beginning with a stay of about 9 months in France starting with his journey to Paris at the end of 1664. His precarious financial situation meant that he had to rely on 'discovery' during his travels by the members and scholars at the research locations of the 'republic of physicians', which at the time were often private academies or institutions associated with royal courts. Stensen

[99] In 1662 in NicPulv., p. 109, lines 6–9 = E 7, p. 153, lines 1–4, Stensen remembered this observation he had made no later than around the end of 1644; a further memory of this time by Stensen, which he made a note of during the final year of his studies in Copenhagen, can be found in Ziggelaar, *Chaos-Manuscript*, p. 290 (col. [121], fol. 60ʳ, N.171).

[100] According to Grell, "Students", pp. 173–174, a total of 52 Danish students of medicine, among them Stensen, studied abroad during the period from 1630 to 1660.

[101] MuscGland., p. 167, lines 12–14. –MuscGland. is the first treatise within the accountability report; the two other treatises are AnRaj. and VitTrans.

[102] SudOr., p. 102, line 9 = E 5, p. 147, lines 32–33.

[103] On Sylvius: DDPhS, s.v. "Sylvius, Franciscus dele Boë", cols. 1939–1943; Harmen Beukers: "Sylvius, Franciscus dele Boë"; in: DDPh. 2 (2003), pp. 973–975.

[104] SudOr., p. 101, line 23–p. 102, line 2 = E 5, p. 147, lines 22–25. On the dissection room cf. Huisman, F*inger of God*, pp. 94–96, 153–154, 160–161.

[105] DucSal., p. 7, lines 23–24 = E 1, p. 137, lines 19–20, here 19: "angusta domi res" and SudOr., p. 101, lines 11–12, here 11 = E 5, p. 147, lines 10–11, here 10: "angusta domi res".

[106] Simone Giese: "Peregrinatio academica"; in: ENZ 9 (2009), cols. 951–955, here 952.

apparently internalized this vagrant lifestyle so much that he continued to profit from it as a vicar apostolic whose sphere of influence extended over several countries. As Stensen was rarely able to take along more than the bare necessities (generally including his study diary documenting the period from March 8 to July 3, 1659 st.v., the so-called *Chaos* manuscript)[107] on his frequent and far-reaching journeys,[108] his research and publication activities were all the more dependent on his memory and extraordinary intuitive capabilities as well as on improvisation and organizational talent. In light of this fact Stensen's achievements, which in terms of their diversity can already only be explained as the accomplishments of a highly gifted person, must elicit even greater appreciation.

A well-founded education was necessary to be able to cultivate the ethos of the 'republic of physicians' within oneself. Even with the appropriate qualifications, however, the capability to engage in independent research as a student beyond the usual – which was rare – required substantial financial backing and time, something Stensen complained about the lack of.[109] He was aware that authoritative scientific studies "claim entirely a person who has nothing to do but this",[110] as he emphasized in the methodologically seminal Parisian *Discours* in 1665, alluding to a less than distinctive profile of medical research attributable to a lack of appreciation for autopsies. That Stensen was able to perform independent research and publish groundbreaking results as a student despite his lack of funds once again bespeaks his outstanding giftedness. Though able to actually deal with precious few of the things worthy of examination he encountered every day, as he writes in his unique modest style to Thomas Bartholin from Leiden on March 5, 1663, it was nevertheless necessary to tend to this, ie the research, until time and means had brought the meager fruits to a certain level of maturity.[111]

In his booklet *Epistolica dissertatio de genuina medicinam instituendi ratione* (Treatise by Letter on the Natural Way of Establishing Medicine; Amsterdam 1680)[112] – hitherto unnoticed in this regard – the surgeon Pieter Guenellon (1611–after 1680),[113] who practiced in Amsterdam from 1644 onwards, made reference to Stensen's admonition that authoritative scientific studies "claim entirely a person who has nothing to do but this". Guenellon's paper reports a conversation he had engaged in shortly before with Dr. med. Johannes Munnicks (1652–1711),[114] a

---

[107] Sobiech, *Herz, Gott, Kreuz*, p. 92.

[108] StenBrun. II, p. 506, lines 10–12 = E 388, p. 748, lines 19–21.

[109] See note 105 above. Cf. Grell, "Students", pp. 173, 177.

[110] Discours, p. 17, lines 6–9, 30–31, here 31: "un homme tout entier, qui n'ait que cela à faire."

[111] VesPul., p. 136, lines 17–20 = E 11, p. 172, lines 11–14.

[112] He was apparently familiar with Stensen's Parisian *Discours*, from which he cites: Guenellon, *Epistolica dissertatio*, p. 76, line 5–p. 77, line 5 (cf. Discours, p. 17, line 30–p. 18, lines 3, 13–20), here esp. p. 76, line 18–p. 77, line 5; Guenellon, *Epistolica dissertatio*, p. 24, line 17–p. 25, line 5 (cf. Discours, p. 16, lines 11–14).

[113] On Guenellon: DDPhS, s.v. "Guenellon Sen., Pierre (Pieter)", cols. 738–739.

[114] On Munnicks: DDPhS, s.v. "Munni(c)ks, Johannes", cols. 1387–1389.

professor of anatomy and botany at the University of Utrecht since 1677[115]; it mentions a manuscript – scheduled to be printed – of an anatomical textbook whose alleged "new pictures" of the human body in c. 80 panels were riddled with well-known and rather dated errors, as Guenellon claims. This half-baked work with its "chaotic contents" had been written by a "certain anatomical jackanapes" to whom most things remained cryptic and who conceitedly praised none but himself while scurrilously preying on others.[116] Appalled, Guenellon declaims: "If only he had chosen Stensen as his advisor!"[117] For besides "modesty",[118] the author could have learned the deleteriousness of publishing erroneous images of the parts of the body from Stensen – who was already a bishop at this time –, the "most famous and excellent promoter of anatomy".[119]

During his stay in Italy following that in France, Stensen was neither a practitioner nor the personal physician of the Florentine grand duke Ferdinando II, but instead an anatomist financially backed by the court of the Medici, who from the second half of 1666 until around the beginning of December 1667 conducted research at the Tuscan *Accademia del Cimento* (Experimental Academy), extant from 1657 to 1667 and working in the spirit of the Italian mathematician, physicist and philosopher Galileo Galilei (1564–1642), as well as at the Florentine hospital *Santa Maria Nuova*, which still exists today.[120] He was also one of the few scientists of his day concerned exclusively with research.[121] This relief from financial strains and freedom to engage in research, which facilitated Stensen's time-requiring precision, probably also abetted his publications, in which he reported his results succinctly and straightforwardly, to a certain degree.

---

[115] See details in Guenellon, *Epistolica dissertatio*, p. 3.

[116] Ibid., pp. 46–47: "nasutulum quendam [...] anatomicum [...] [47] [...] novis [...] iconibus [...] rerum chaos". It can be assumed that "novis iconibus" (or the synonymous "novis figuris") and "corpus humanum" occured in the title in question. A printed publication to which this applies cannot be traced; information can be found eg in the medical bibliography BMPh., published one year after Guenellon's booklet and covering the period from 1651 to 1681; cf. HGK, s.v. "Beughem, Cornelis van", p. 96.

[117] Guenellon, *Epistolica dissertatio*, pp. 47–48, here 48: "Utinam consultorem sibi elegisset Stenonium!"

[118] Ibid., p. 48: "modestiam".

[119] Ibid., p. 24: "nobilissimus & insignis anatomes promotor".

[120] Scherz, *Epistolae*, vol. 1, pp. 16–17; Kardel/Maquet, *Biography and Original Papers*, p. 157. On the hospital *Santa Maria Nuova*, see Ciuti, *Ospedale di Santa Maria Nuova di Firenze*.

[121] Scherz, "Anatomische Forschung", p. 251.

1.6   Medical Ethos in Its Relation to Theology Since the Early Modern Period

Similar to the professional ethics in medicine and the 'self-perception'[122] of physicians, early modern treatises on the relationship of medical professionals with religion as well as the relationship between medical ethics on the one side and Christian ethics and moral theology on the other represent a desideratum for research until today.[123] The relationship between anatomists – and anatomy as a science – and Christianity (and the Roman Catholic Church in particular) during the early modern period has likewise been subjected to precious little systematic examination.[124] Winfried Schleiner emphasizes that medical professionals as well as theologians wrote about medical ethics during the Renaissance era, and that knowledge about Catholic moral casuistry and the various papal pronouncements dealing with questions of health and medicine was not restricted to Catholic physicians.[125]

1.6.1   Between Experience-Relatedness and Reflection in Faith

The Toulousian professor Raymond of Sabunde (Latin: Sebundus) (†1436), a graduate of theology, the liberal arts and medicine who hailed from Catalonia and whose thinking revolved around experience, had already criticized the "errors of former heathen and unreliable philosophers" – and thus the faith placed in authority by the medicine of his time – in his *Liber creaturarum sive de homine* (Book on Creatures resp. on Man).[126] René Bernoulli puts him on a level with Niels Stensen, one of the physicians and natural scientists who, like Sebundus, "unconcerned with the respective prevailing doctrine dared to leap into the freedom of reflection and transcendence".[127]

At the University of Copenhagen, which during Stensen's time was in the domain of the Lutheran state church, faith played a fundamental role. This is illustrated by the fact that Dr. med. Dr. theol. Caspar Bartholin Sr. (1585–1629), a professor first

---

[122] See note 85 above.

[123] Maehle, "Werte und Normen", p. 339; Pott, *Medizin*, pp. 76, 231n146. Studies about the influence of confession on medical training are only slowly being begun for the sixteenth century, cf. the overview in Helm, "Religion and Medicine", pp. 51–54; on the impossibility of delimiting scientific progress in terms of confession (also in opposition to Robert K. Merton), see Greyerz, "Religion und Natur", pp. 55–57, here 57 on Stensen.

[124] Cf. the upcoming follow-on volumes on the seventeenth century of (1) Baldini/Spruit, *Catholic Church and Modern Science*, and (2) of the project "Römische Inquisition und Indexkongregation von 1542 bis 1966" headed by Hubert Wolf, under http://www.buchzensur.de.

[125] Schleiner, *Renaissance*, pp. vii–viii.

[126] Bernoulli, "Sebundus", pp. 10–11, 18–21; cited according to the translation on p. 19. On Sabunde: HGK, s.v. "Sebond, Raimond", pp. 598–599.

[127] Bernoulli, "Sebundus", p. 24.

of medicine and later of theology, opined the view in his study guide *De studio medico inchoando, continuando, & absolvendo consilium breve atque extemporaneum [...]* (Short and Extemporized Advice on the Beginning, Continuance and Completion of the Study of Medicine [...]; Copenhagen 1628) that medicine were only effective if it were based on the Lutheran faith and practiced by a pious physician, for otherwise the physician would be threatened twofold: in that he might fall prey to vices, or that he might be overcome by fearfulness in the face of particularly difficult cases or the failing of his medical art.[128] While medicine had not been secularized in the early modern period, its practice was certainly marked by a "dichotomy between faith and skepticism".[129] In the seventeenth century, the physician's profession was even viewed with "suspicion of atheism" by many Protestant apologists.[130] Indicative of this notion is a medieval saying still in common use during this time: "Ubi tres medici, duo athei" (Where [there are] three physicians, [there are] two atheists).[131] Thomas Keck quotes it in his "Annotations" supplementing the *Religio medici* (The Physician's Religion; London 1642) by Dr. med. Sir Thomas Browne (1605–1682) starting with the fourth edition 1656, claiming it to be "common speech (but onely [= only] amongst the unlearned sort)".[132] The Arminianist Browne structures his book, in which he develops a set of practical ethics with a view to death and the afterlife, along the lines of the theological virtues faith, hope and charity.[133]

An early work on medical ethics is the *Ventilabrum medico-theologicum [...]* (Medical Theological Winnowing Fan [...]; Antwerp 1666) by the Catholic physician Dr. med. Michiel Boudewijns (early seventeenth century–1681) of Antwerp, which deals with questions pertaining to medical practice, pharmacists, patients and

---

[128] Elkeles, "Aussagen zu ärztlichen Leitwerten", pp. 121–126; Grell, "Education", p. 79. On Bartholin: Hendrik Andreas Hens (Bjørn Kornerup/Valdemar Meisen), "Bartholin, Caspar"; in: DBL³ 1 (1979), pp. 470–472.

[129] Bergdolt, *Gewissen*, p. 179. On the term "orthotomy" and its religious implications see Danneberg, *Anatomie*, pp. 204–225, and on the wish by the Helmstedt professor Hermann Conring for Stensen in this regard see Sobiech, *Herz, Gott, Kreuz*, p. 91.

[130] Barth, *Atheismus und Orthodoxie*, pp. 79, 268–269.

[131] PSLMA 5 (1967), p. 445, no. 32070b.

[132] Browne, *Religio Medici* (Annotations upon Religio medici, The first Part, Sect. 1, Pag. 1), p. 187. Nothing further is known about Keck; see Hack-Molitor, *On Tiptoe in Heaven*, p. 32n70. *Religio Medici* was placed on the Roman Index in 1645; on Browne and his treatise: HGK, s.v. "Browne, Sir Thomas", pp. 133–134, here 134, incorrectly identified as indexed 'donec corrigatur', cf. by contrast ILP, s.v. "Browne, Thomas", p. 167. – Pope Pius VII ([* 1742] 1800–1823), who was staying in Paris for the occasion of the coronation of Napoléon Bonaparte (1769–1821), had the internist Dr. med. René-Théophile-Hyacinthe Laënnec (1781–1826) and four other young physicians, members of the French congregation "Sancta Maria, Auxilium Christianorum" (Holy Mary, Helper of Christians; est. 1801), introduced to him in the Grand Salon of the Louvre a few days after 12/18/1804; in an apparent soupçon of humor, the Pope said with a smile (Geoffroy de Grandmaison, *Congrégation*, pp. 50, 54–56, here 56): "Oh! A pious physician, an admirable thing!" (Oh! medicus pius, res miranda!). Cf. Kervran, *Laennec*, pp. 59, 75–76. On Laënnec: Henri Tribout de Morembert, "Laennec (René-Théophile-Hyacinthe)"; in: DBF 19 (2001), cols. 95–96.

[133] Andrew Cunningham: "Browne, Thomas"; in: DBPh. 1 (2000), pp. 130–132, here 130–131.

public health services from the point of view of moral theology.[134] Boudewijns, who had also studied theology, describes his opus as "something from a quasi mixed realm, which thus partakes in one as well as the other science", meaning theology as the science of the "immortal soul" (immortalis anima) and medicine as the science of the "mortal body" (mortale corpus).[135]

Biographical connections between the medical and priestly professions can be found frequently up to the present time, eg with the physician, priest and founder of the Barnabites Dr. med. Antonio Maria Zaccaria (1502/1503–1539),[136] who was canonized in 1897. Medical historian and psychiatrist Dr. med. Werner R. Leibbrand (1896–1974), who spent the final years of his life teaching in Munich, concerned himself with the theology of physicians, dealing (among others) with the seventeenth century and the "physician-priesthood of the Baroque".[137] In this context, mention should also be made of the Nuremberg physician Dr. med. Johann Georg Noesler (or Nößler) (* 1635), who repaired to the Benedictine abbey at Plankstetten in early 1667 and eventually converted there sometime before August 12, 1667.[138] Whether Stensen, whose interest in reports of conversions is well substantiated, knew about Noesler is unknown, however.

In the eighteenth century, Pope Benedict XIV ([* 1675] 1740–1758), who heavily advocated the development of anatomy as a science,[139] viewed anatomy as a means for exhibiting the perfection of creation.[140] He supported – not least in an attempt to question the widespread opinion that the female sex was inferior with regard to character – the academic careers of several gifted women, eg Anna Morandi Manzolini (1714–1774), who crafted anatomical wax models at the University of Bologna.[141]

In the twentieth century, Pope Paul VI ([* 1897] 1963–1978) emphasized in his final encyclical *Humanae vitae* (July 25, 1968) that he held in the utmost esteem all those physicians and assistants in the medical arts who strived in the practice of their profession to preserve what "the specific consideration of their Christian

---

[134] Boudewijns, *Ventilabrum*. On Boudewijns: DDPhS, s.v. "Boudewijns, Michiel", col. 236; De Nave/De Schepper, *Geneeskunde*, pp. 254–255 (no. 90); Ernst, *Boudewyns*, pp. 8–11.

[135] Boudewijns, *Ventilabrum*, foreword "Ad lectorem" (unpaginated), fol. ẽ 2ʳ–ẽ 4ᵛ, here ẽ 4ʳ: "materia [...] quasi mixti fori [...], tantum de una, quantum ex alia scientia participans".

[136] On Zaccaria: Egidio Caspani, "Antoine-Marie Zaccaria"; in: DSp. 1 (1937), cols. 720–723.

[137] Werner Leibbrand, "Priesterarzt"; in: KidW 2 (1949) 1st lot, no. 3, pp. 41–48; Leibbrand, *Äskulap*, pp. 13–23, 371n183; on Stensen: ibid., p. 210. On Leibbrand: Matthias M. Weber, "Werner Leibbrand"; in: PSexF, pp. 407–410.

[138] Noesler allegedly joined this Benedictine abbey; on Noesler: NGL 7 = 3rd suppl. vol., pp. 29–30; cf. Schmidt-Herrling, *Briefsammlung*, p. 435. According to information by telephone (08/02/2011) from Fr. Dr. Beda Maria Sonnenberg OSB, abbot of Plankstetten, Noesler is not mentioned in the relevant abbey-internal (unpublished) studies into the abbey at Plankstetten; it can therefore be assumed that he did not in fact join, but merely converted in the abbey.

[139] Messbarger, *Lady Anatomist*, p. 8.

[140] Enke, "Schönheit der Embryonen", pp. 207–208.

[141] Messbarger, *Lady Anatomist*, pp. 35, 108–109, 210n2.

vocation" – rendered as "a Christian professional ethos"[142] in the translation approved by the German bishops – demanded of them, rather than acting according to random human interest. They were to persistently pursue their intention to always give precedence to advice commensurate with "the faith as well as right reason".[143] This "right reason" (recta ratio) is defined in the tradition of Christian ethics as the rationality spurred by the conscience which "in evaluating a concrete situation is able to observe the rules of natural law".[144] The *Charter for Health Care Workers*, published in multiple languages by the *Pontificium Consilium de Apostolatu pro Valetudinis Administris* (*literally*: Pontifical Council for the Apostolate of Health Care Workers) states that "science and wisdom should go hand in hand" (cf. eg 1 Cor 2:6–9) and that "wisdom and conscience trace out for them [science and technology] the impassable limits of the human".[145]

It is interesting to consider against this background what the American physician Dr. med. Bernard N. Nathanson (1926–2011)[146] defined as the "minimal description of a doctor" at the end of the twentieth century: "a highly trained technician, daily exposed to exceptionally powerful material and spiritual temptations".[147] In his experience only those could endure without harm "the lure of the worldly temptations in the medical world: the uninterrupted flow of money, the drumfire of flattery, and the inebriating effects of special privilege" who possessed "an inflexible inner spiritual column" to support the immense weight of the obligations and responsibilities of a physician.[148] This seems to point to "wrong reason" contrasting the "right reason" mentioned by Pope Paul VI – namely reason biased purely towards one-sided, instrumental rationality and the pursuit of profit and reputation. As will be demonstrated in the following, Nathanson's warnings were topical for Stensen's times as well.

---

[142] Sekretariat der Deutschen Bischofskonferenz, *Acta*, p. 49. The German indefinite article "ein" apparently refers to the ethics of the different medical professions.

[143] AAS 60 (1968), pp. 499–503, here 50, no. 27: "praecipua christianae vocationis ratio […] et fidei et rectae rationi".

[144] Maarten J. F. M. Hoenen, "Recta ratio"; in: HWPh. 8 (1992), cols. 355–360, here 357 (Thomas Aquinas).

[145] Pontifical Council for Pastoral Assistance to Health Care Workers, *Charter*, p. 53, no. 45. On the foundation of this Pontifical Council in canon law, see AAS 77 (1985), pp. 457–461 (Motu Proprio *Dolentium Hominum*) in connection with AAS 80 (1988), pp. 841–930 (Apostolic Constitution *Pastor Bonus*), here pp. 900–901 with Art. 152–153.

[146] Born into Judaism, later atheist and abortionist. His emerging belief "that Someone had died for my sins and my evil two millennia ago" (see his autobiography Nathanson, *Hand of God*, p. 194) played a role in his path to conversion to the Roman Catholic faith on 12/08/1996, the Feast of the Immaculate Conception (ibid., pp. 197–203 ["Afterword" by C. John McCloskey III], here 199). On Nathanson: Alison Snyder, "Obituary Bernard Nathanson"; in: Lancet 377 (2011), p. 990.

[147] Nathanson, *Hand of God*, p. 109.

[148] Ibid.: "an inflexible inner spiritual column […] the lure of the wordly temptations in the medical world: the uninterrupted flow of money, the drumfire of flattery, and the inebriating effects of special privilege."

1.6.2   Niels Stensen, René Descartes and Faith

The medical ethos of the early modern period increasingly followed the philosopher and amateur anatomist René Descartes (1596–1650),[149] who in his anonymously published *Discours de la méthode pour bien conduire la raison, & chercher la ver-ité dans les sciences [...]* (Leiden 1637) taught methodical doubting as the path to insight and intended to avoid indecisiveness in his own actions while in the mean-time traveling this path by way of a "provisional morality".[150]

As expounded by the medical historian Gerrit A. Lindeboom (1905–1986), Descartes likens the human body in health and illness to a machine, eg an automatic clockwork. At times, Lindeboom writes, the French philosopher himself may have forgotten that this was but an analogy, making it somewhat understandable if some his contemporaries and followers thought the body *was* in fact a machine. His mechanical conception dispensed completely with any notion of teleology for the human organism.[151] For Stensen on the other hand, Descartes' philosophy was nei-ther beyond doubt nor did he expect it to provide information on the composition of reality; to him it represented a "methodological matrix"[152] which he adapted to the requirements of his research. In his own *Discours*, Stensen therefore lamented that Descartes' machine was interpreted by many as "a faithful report about that which is hidden deep in the recesses of the human body", resulting in major differences between the model thus absolutized and the reality encountered during dissections.[153]

In order to comprehend Stensen's experience-centered understanding of science, one of the first proponents of which was Raimund of Sabunde and which Stensen also applied to his examinations of the Christian confessions and religious convictions,[154] it is illuminating to cast a glance at the American sociologist Robert K. Merton (1910–2003) and his essay *Science and Democratic Social Structure*,

---

[149] On Descartes: HGK, s.v. "Descartes, René", pp. 222–223.

[150] This "provisional morality" is described in the third part of *Discours de la méthode*. Cf. Bergdolt, *Gewissen*, pp. 162–163.

[151] Lindeboom, *Descartes and Medicine*, pp. 58–61. On Lindeboom: Antonie M. Luyendijk-Elshout, "Gerrit Arie Lindeboom (1905–1986)"; in: Clio med. 21 (1987/1988), pp. 207–208.

[152] Iofrida, Stensen, pp. 892–895; cf. Olden-Jørgensen, "Steno and Descartes", pp. 149, 156, who describes Stensen as "a heuristic Cartesian".

[153] Discours, p. 7, line 38–p. 9, line 4, here p. 8, lines 33–34: "une rélation fidele, de ce qu'il y a de plus caché dans les ressors du corps humain". Cf. Lindeboom, *Descartes and Medicine*, pp. 60–61.

[154] This also appears to be Stensen's original contribution to controversial theology; for details see Sobiech, *Herz, Gott, Kreuz*, pp. 24, 148 (lost manuscript). Stensen's writings are characterized by their logical structure, even in terms of print layout: for example, in E 128, p. 344, lines 10–14 Stensen describes his habit of folding a sheet of paper down the middle and juxtaposing the con-trasting positions synoptically in the form of a table in order to achieve clarity also for the eyes, eg in AProd., p. 150, line 22–p. 151, line 21 to debunk Blasius' alleged duct (Fig. 2.2). The foundation for this method is already laid down in the *Chaos* manuscript, in which Stensen on 03/12/1659 st.v. demands distinct juxtaposing of position and counterposition for discussions (Ziggelaar, *Chaos-Manuscript*, p. 77 [col. 23, fol. 33ᵛ, N.8]).

published in 1942 and of great significance in the sociology of scientific knowledge.[155] In this essay, Merton paints a picture of science characterized, among other things, by "organized scepticism"; by this he meant "the suspension of judgment until 'the facts are at hand' and the detached scrutiny of beliefs in terms of empirical and logical criteria".[156]

This is precisely what Stensen practiced in the field of anatomy as a heuristic Cartesian as well as on his path of faith. According to Stanley Finger and Marco Piccolino, Stensen was an empiricist and opponent of theorization without evidence in concrete observations.[157] As a bishop, Stensen strongly objected in the written defense of his conversion *Defensio et plenior elucidatio epistolae de propria conversione* (Defense and deeper explanation of the letter regarding my own conversion; Hanover 1680) to allegations by Dr. theol. Johann Wilhelm Baier Sr. (1647–1695),[158] a Lutheran professor of church history at the University of Jena since 1675, that Descartes had been "the beginning of faith" for himself or that he sought adjudgement by the philosopher in questions of faith.[159] Instead, the decisive factor for Stensen's thusly prepared adoption of the Catholic faith had been his personal experiencing of the "inner light of God", ie God's mercy.[160] While he considered the Cartesian system disproved as a result of his heart dissections in 1662/1663, he nevertheless respected Descartes' critical mind. Stensen's incisive experience of 1662/1663 caused him to view medicine, in which – according to Richard Toellner – every theory must undergo the test of practice, as a "touchstone for the efficacy of philosophy in the early modern period"[161] with direct consequences for his own life. To him, Descartes represented a philosophical introduction to theology engendering increased existential depth.

As it happened, only a few months later – on November 20, 1663 – Descartes' philosophical writings, among them as part of the *Opera philosophica* (Philosophical Works; 2nd ed., Amsterdam 1650) the Latin translation of *Discours de la méthode*, entitled *Specimina philosophiae: seu dissertatio de methodo recte regendae rationis, & veritatis in scientiis investigandae [...]* (Philosophical specimens or treatise about the method of using reason correctly and exploring truth in the sciences [...]; Amsterdam 1644), were placed on the "Index of Prohibited Books" (Index librorum

---

[155] Merton, *Social Theory*, pp. 604–615. Merton's sociological theory is discussed in regard to bioethics in Jonathan R. Cole, "Science, Sociology of"; in: EBioeth. 4 (1978), pp. 1541–1546, here 1542. On Merton: Gerald Holton, "Robert K. Merton. 4 July 1910–23 February 2003"; in: PAPS 148 (2004), pp. 506–517.

[156] Merton, *Social Theory*, p. 614. On criticism of this approach cf. Barnes/Dolby, "Scientific Ethos", pp. 4, 10–11, in which the disruptive factors bias, emotional attachment, and polemics are pointed out.

[157] Finger/Piccolino, *Electric Fishes*, pp. 156, 186.

[158] On Baier: Ernst Koch, "Baier, Johann Wilhelm"; in: RGG[4] 1 (1998), cols. 1065–1066.

[159] DefConv., p. 390, lines 14–16, here 15: "fidei principium".

[160] Bambacari, *Signora Arnolfini*, pp. 37–38, here 38 = Sacra Congregatio pro Causis Sanctorum, *Positio*, p. 103: "interno lume di Dio".

[161] Richard Toellner, "Medizin III. Frühe Neuzeit"; in: HWPh. 5 (1980), cols. 984–992, here 986. On Stensen's heart dissections in 1662/1663, see Sobiech, *Herz, Gott, Kreuz*, p. 60.

prohibitorum) by the Sacred Congregation of the Index 'donec corrigatur' (until corrected), meaning until all questionable passages had been clearly designated as scientific hypotheses.[162] But even as a bishop, Stensen highlighted the usefulness of Cartesian philosophy for the exposing of prejudices in his letter to Baier Sr., emphasizing that everything human possessed a meritorious and an ignoble aspect.[163] Stensen was also self-critical enough to recognize the possibility that he might, as he wrote from Florence c. November 1675 to his former teacher at the Latin school Dr. med. Ole Borch (1626–1690),[164] a professor of philology, botany and chemistry in Copenhagen from 1660, in future be forced to abrogate something resulting from his adoption of "prejudices of modern philosophy".[165]

Helmut Thielicke (1908–1986), a Lutheran professor of systematic theology in Hamburg, states in his book *Being Human ... Becoming Human: An Essay in Christian Anthropology* (Garden City, NY 1984; German original version Munich 1976) that Stensen had repeatedly argued against the metaphysical, deductive analogy principle – which Descartes himself had still applied and thereby made incorrect deductions – in the name of modern empiricism. His refuting of errors amalgamated with metaphysical statements about God and the soul which had been generated in natural science due to the analogy principle caused Stensen to deduce, Thielicke claims, that the manner of belief in a creator would necessarily have to be measurable in empirically verifiable nature.[166] Thielicke used this as an example to make clear that theological discourse about humans had to stand up to the empirical findings of human science – including but not limited to anatomy –, something he himself seemed on occasion not to succeed in.[167] From Stensen's point of view, caution was advised in the implementation of this postulate: in his letter from November 24, 1671 to his friend Dr. med. Marcello Malpighi (1628–1694),[168] a professor of medicine, anatomist and later personal physician (archiater) to Pope Innocent XII

---

[162] ILP, s.v. "Descartes, René", pp. 281–282, here 281. Descartes' posthumous work *De homine* (Leiden 1662), which Stensen read immediately after its publication (see note 180, in Chap. 2 below), was not placed on the Index.

[163] For details see DefConv., p. 388, lines 12–26.

[164] On Borch: HGK, s.v. "Borrichius, Olaus", p. 122.

[165] E 108, p. 309, lines 6–13, here 11–12: "praejudicia [...] philosophiae modernae".

[166] Thielicke, *Anthropology*, pp. 179–182, 480. On Thielicke: Arnulf v. Scheliha, "Thielicke, Helmut"; in: RGG⁴ 8 (2005), cols. 363–364.

[167] Reis, *Lebensrecht*, p. 42n180 mentions critically a lecture by Thielicke in 1983 in which he claimed that the beginning of human life could "apparently not" be determined by applying medical criteria. In addition, Werner Leibbrand had in 1949 accused Thielicke of being "still too biased by psychotherapeutic methods which in effect are already a part of history" (Werner Leibbrand, "Priesterarzt"; in: KidW 2 [1949] 1st lot, no. 3, pp. 41–48, here 46–47 with citation on p. 47, l. col.).

[168] On Malpighi: Cesare Preti, "Malpighi, Marcello"; in: DBI 68 (2007), pp. 271–276. During his sojourn in Rome in May/June 1666, Stensen had often participated as a guest at the vespertine academy meetings of Giovanni Guglielmo Riva (1621–1677), surgeon at the Roman *Ospedale di Santa Maria della Consolazione* and later archiater to Pope Clement IX; there Malpighi had come to know and appreciate Stensen during the accompanying meals (Malpighi 1698, p. 43); cf. Scherz, *Epistolae*, vol. 1, p. 31. On Riva: Savio, "Riva", here p. 251n110 on Stensen.

([* 1615] 1691–1700), in which he discusses thoughts on the connection between soul and movement, Stensen claims to have "several friends in the Netherlands fully devoted to Cartesian philosophy, to the extent that they wish to make philosophy the judge over the experiencing of [God's] grace."[169]

Guido Giglioni claims this letter by Stensen to show "the anxieties of a mind, both religious and scientific, torn between the improvement of anatomical knowledge and the safeguarding of the reasons for faith."[170] Apologetic reasons, Giglioni continues, appeared to have guided Stensen's decisions even in regard to anatomy, which for him was not only a discipline capable of challenging everything related to divine grace, but also an activity which had cast a right light on the weak points and limitations of natural philosophy. While the latter assumption is certainly true, the former lacks any corroboration in the mentioned letter as well as in the entirety of Stensen's oeuvre. It is beyond question that Stensen was worried about endangerments of the Faith; however, during his lifetime he never saw a contradiction between the laws of nature and the Catholic faith to which they had led him. Stensen in fact referred to the danger of the undermining of scientific methodology – whose neglect already by Descartes himself he complained of[171] – by, in the words of the philosopher of science Georges Canguilhem (1904–1995),[172] "scientific ideologies" (idéologies scientifiques). According to Canguilhem, these are "explanatory systems whose theme is exaggerated in relation to the norm of the scientific character assigned to it by way of loan".[173] Stensen means the ability to accept criticism and the ability to differentiate in order to determine, from the perspective of faith, positive and negative aspects within "the Zeitgeist of the scientific revolution".[174]

Much uncertainty can be observed in literature regarding the question of the position of the "later" Stensen towards the relationship between faith and science; this is most likely due to (1) Stensen's writings being read only fragmentarily, and

---

[169] E 65, p. 249, lines 1–26, here 24–26 = Adelmann, *Correspondence*, vol. 2, no. 271, pp. 597–598: "certi amici in Ollanda dati tutti alla filosofia Cartesiana à segno tale che di volere fare la filosofia judice delle notizie della grazia." For details on these friends see Sobiech, *Herz, Gott, Kreuz*, pp. 56–57.

[170] Giglioni, "Machines", pp. 150–151, here 150. – This would lend itself to comparison with the Cartesianism debate of 1676/1677 moderated by Cardinal de Retz (1613–1679), cf. Malina Stefanovska, "Retz, Jean-François-Paul de Gondi, cardinal de"; in: DFPh. 2 (2008), pp. 1077–1083, here 1082.

[171] DefConv., p. 388, line 34–p. 389, line 4; ibid., p. 390, lines 5–6; Sobiech, *Herz, Gott, Kreuz*, pp. 60–61.

[172] Canguilhem also attained a degree in medicine; on Canguilhem: Borck/Hess/Schmidgen, *Introduction*, pp. 12–26, here 14.

[173] Canguilhem, Idéologie, p. 44: "des systèmes explicatifs dont l'objet est hyperbolique, relativement à la norme de scientificité qui lui est appliquée par emprunt." These are to be differentiated from the so-called "ideologies of scientists" (idéologies de scientifiques), created by scientists within their scientific discourse in order to verbalize methods of examination and their practical implementation, as well as to establish the position of science within their culture (ibid., pp. 43–44). On placement in Canguilhem's biography see Borck/Hess/Schmidgen, *Introduction*, pp. 25–26.

[174] Finger/Piccolino, *Electric Fishes*, p. 154.

(2) that Stensen's motivation for his change of profession, ie becoming a priest, is not properly understood.

As stated accurately by Andrew Jamison, the point of fact is that for Stensen – as for the mathematician Blaise Pascal (1623–1662) –, the faith he was searching for could not be found in the natural sciences. He reacted to the exclusion of God from the theories of nature not with new theory, but by attempting to live the Faith, turning his own life into a paradigm and viewing it as an experiment of sorts, which in Jamison's opinion constitutes something typically Danish. Stensen, Jamison states, was dissatisfied with natural science; for while it brought him closer to understanding the glory of God, it was not close enough.[175] This observation by Jamison is worthy of emphasis, as it is reflected throughout Stensen's entire oeuvre.[176]

Stensen's wish to point out the narrowness of human knowledge is expressed in his letter written on January 25, 1678 to the scholar Melchisédech Thévenot (c. 1622–1692), at whose Parisian private academy Stensen had delivered his lecture *Discours sur l'anatomie du cerveau [...]* (Lecture on the Anatomy of the Brain [...]; Paris 1669; Latin translation Leiden 1671) sometime around February 1665.[177] In the letter penned 13 years later, he declares (presumably alluding to 1 Cor 13:12): "Oh Lord, all the sights of the world are nothing but vanity, and that which is found there as certainty is so little in comparison with that which we will see in our first glance cast at the divine being!"[178] Stensen wrote this letter during his time as Vicar Apostolic, when Gottfried Wilhelm Leibniz (1646–1716) in Hanover was unavailingly attempting to get him to return to natural science.[179] Even though the association of curiosity with vanity possesses a long tradition – particularly in theology –,[180] this statement by Stensen is no mere topos. His intent, according to Neil Kenny, is by no means an "urging" of Thévenot to "abandon naturalist endeavour".[181] Rather, Stensen is making a metaphysical and furthermore very personal statement about his stimulus for becoming a priest and thereafter occupying himself with natural science only with regard to the salvation of souls; it is only in this context that we can understand how he could consider his previous research a waste of time on the path to his conversion.[182]

---

[175] Jamison, *National Components*, pp. 219–220.

[176] For further details from a biographical point of view see Sobiech, *Herz, Gott, Kreuz*, pp. 68–79.

[177] Discours; cf. Kardel/Maquet, *Biography and Original Papers*, pp. 125–126, 130. On Thévenot: Nicholas Dew, "Thévenot, Nicolas Melchisédech"; in: DFPh. 2 (2008), pp. 1214–1216, here 1214 on Stensen.

[178] E 146, p. 372, lines 10–13: "Helas Monsieur, que toutes les curiositez du monde ne sont que vanitez, et ce qu'il y a de solide est si peu à l'égard de ce que nous verrons à la première oeillade, que nous donnerons à la divine essence!"

[179] See in further detail in Sobiech, *Herz, Gott, Kreuz*, p. 330. On Leibniz: HGK, s.v. "Leibniz, Georg Wilhelm Friedrich", pp. 397–399.

[180] André Cabassut, "Curiosité"; in: DSp. 15 (1991), cols. 2654–2661, here 2660 on the work *De imitatione Christi* by Thomas von Kempen CRV (1379/1380–1471).

[181] Kenny, *Curiosity*, p. 101. Kenny uses Dew, *The Pursuit of Oriental Learning* only as a secondary source for the basis of his argumentation.

[182] Sobiech, *Herz, Gott, Kreuz*, pp. 48–49.

Similarly to Stensen's motto "Pulchra sunt, quae videntur, pulchriora, quae sciuntur, longe pulcherrima, quae ignorantur" (Beautiful is that which is seen, more beautiful that which is known, most beautiful by far that which is not known)[183] from his Copenhagen inaugural address, which exhibits a similar hierarchy, his words directed at Thévenot thus bespeak a deep veneration for the natural sciences. Stensen's final anatomical notes on the brain and nerves were recorded in Hamburg in 1684,[184] at a time when the unedifying controversies in the local Catholic mission were distressing him. It may rightly be assumed that he attempted to compensate these woes by temporarily reengaging in anatomical studies.

The assertion by Nicholas Dew that Stensen followed the theological tradition of regarding curiosity as a vice, owing to its ability to engender excessive faith in human knowledge which in turn results in a fall into the mortal sin of pride, and fearing that it might incite the unwise reader to heresy, can therefore not be unconditionally endorsed.[185] Bearing in mind that on April 5, 1673 st.v. he informed Heinrich Meibom Jr. that God by no means abhorred the researcher's diligence,[186] one must recognize that Stensen, like Thomas Aquinas OP (1225–1274) does not intend to devalue or condemn curiosity as a sign of human intelligence, but to criticize it only where it crosses over into "distraction during the observation of intramundane things".[187]

1.6.3 Niels Stensen and the Legacy of the Time of the So-Called Enlightenment

With reference to the figures of medical authority, the "Ancients" (veteres), Stensen wrote the following in one of his publications during his studies in Leiden: "They ignited the flame; it is our task to keep it burning and brighten it as it goes forth."[188] With this allusion to the legend of the stealing of the fire by Prometheus, Stensen takes up the metaphor of a burning torch to be passed along among researchers, as used by the philosopher of science Francis Bacon (1561–1626),[189] who considered the continuous and inexorable progress of science to be real and technology to be *the* driving force of history.[190]

---

[183] See note 241, in Chap. 2 below.

[184] Add. 24, pp. 949–951; cf. Sobiech, *Herz, Gott, Kreuz*, pp. 148–149.

[185] Dew, *Orientalism*, pp. 129–130. He is referring to ibid., p. 129 without analyzing Stensen's motivation for his warning towards Thévenot being "probably a sincere one".

[186] See note 184, in Chap. 2 below.

[187] Groh, *Schöpfung*, pp. 398–403, here 403.

[188] GlandOr., p. 19, lines 2–7, here 6–7: "Lumen accenderunt illi; nostrum est, ut accensum conservetur, & luculentius eundo inardescat."

[189] On Bacon: Graham Rees, "Bacon, Francis"; in: DBPh. 1 (2000), pp. 39–46, here 39–40.

[190] Spedding/Ellis/Heath, *Works of Francis Bacon* (Chapter "Prometheus, sive status hominis"), pp. 675–676.

In the same vein are Stensen's programmatic words "To overthrow what is false, to lend steadfastness to what is true, to cast a bright light on what is unclear, to advance the unexamined" from his geological *Prodromus* composed after his conversion.[191] During his studies in Copenhagen – as documented by his study diary *Chaos* – he had occupied himself with Bacon's method and read the translation *De dignitate & augmentis scientiarum [...]* (Leiden 1645) of Bacon's *The Two Bookes [= Books] [...] of the Proficience and Advancement of Learning, Divine and Humane [= Human]* (London 1605), which delivered a scathing criticism of contemporary scientific methodology.[192] On March 23, 1659 st.v. Stensen decided to conduct an investigation following Bacon's as well as Descartes' methods.[193] In his reading during his studies he did not limit himself to the field of medicine, instead traversing into various subject areas.

The light metaphor used by Stensen – and later also by representatives of the so-called Enlightenment who claimed the metaphor for themselves – invites comparison with the Spanish professor of medicine Dr. Juan de Cabriada (c. 1665–after 1714),[194] author of the *Carta filosofica, medico-chymica [...]* (Philosophical, medico-chemical letter [...]; Madrid 1686 [corr. 1687]) denouncing the Galenic medicine practiced in Spain and its Aristotelic-philosophical foundations, and calling for freedom of choice in selecting a philosophical foundation for medicine. De Cabriada cites Stensen as reference for this request.[195] It would surely be informative to examine to what extent De Cabriada's "light of truth" and "rays of reason"[196] differ from Stensen's words about the natural light of the intellect.[197] At any rate, De Cabriada's book, whose publication received the ecclesiastical imprimatur, describes Francis Bacon's principles as entirely compatible with the Faith.[198]

Does this mean Stensen can be considered a precursor to the Age of Enlightenment? His Parisian *Discours sur l'anatomie du cerveau* is sometimes

---

[191] See note 42 above.

[192] Ziggelaar, *Chaos-Manuscript*, pp. 81 (col. 24, fol. 33^v), 142 (col. 48, fol. 41^v).

[193] Ibid., p. 124 (col. 38, fol. 39^r, N.41). – The citation of Francis Bacon mentioned in Iofrida, "Stensen", p. 895 as being "on the title page"of Stensen's geological *Prodromus* is actually only extant on the title page (recto) of the 2nd edition Pistoia 1763. On Bacon's *De dignitate [...]* (in its Latin translation London 1623), placed on the Roman Index 'donec corrigatur' in 1668, cf. ILP, s.v. "Bacon, Francis", p. 95.

[194] On De Cabriada: José M. López Piñero, "Cabriada, Juan de"; in: DHCME 1 (1983), pp. 149–152, here 149–151 on his magnum opus *Carta filosofica*.

[195] Israel, *Radical Enlightenment*, pp. 529–530. On further reception of Stensen in Spanish medicine of the early modern era, cf. López Piñero, *Medicina*, pp. 122, 131.

[196] De Cabriada, *Carta filosofica, medico-chymica*, dedication (unpaginated), here the next to last page: "Luz de la Verdad" and "Rayos de la Razon".

[197] E 65, p. 249, line 27; Bruun, "Fem nyfundne Niels Stensen-breve", p. 150, line 20 (darkness from which God shall lead anatomy with the help of man); Add. 24, p. 950, line 43; supplementarily VesPul., p. 134, lines 12–13 = E 11, p. 170, lines 9–10; Prooemium, p. 254, lines 21–22. See also Sobiech, *Herz, Gott, Kreuz*, pp. 72–73, 149, 171–172. On the discussion of the Christian apologists about the 'Lumen naturale' see Barth, *Atheismus und Orthodoxie*, pp. 201–205.

[198] Israel, *Radical Enlightenment*, p. 530.

viewed as a step on the path towards the "mechanization of the mind".[199] On the other hand – as has recently been emphasized from a theological perspective in an overview of the effects of the empiricism and rationalism arising in the seventeenth century –, it is precisely this *Discours* with which Stensen had fundamentally criticized Cartesian philosophy, whose influence lingered on in the shape of Cartesianism and Spinozism.[200] Meanwhile, Matthew Cobb interprets Stensen's discovery that human reproduction is no different than that of every other animal as a step towards a materialist conception of the actual origins of all life, and surmises: "Steno would have been horrified had he realized the profound implications of his idea."[201] Whether Cobb's assessment is correct has yet to be examined.

With reference to the scientific ethos as an "ethos of epistemological rationality", its foundations laid in Stensen's time and its applicability to scientific research generally accepted to this day, the philosopher Julian Nida-Rümelin requests the following for the ongoing debates on scientific ethics: "The handed-down scientific ethos of the early modern era and the European Enlightenment is in need of complementation with the dimension of responsible scientific practice."[202] Could Stensen – albeit himself a member of the early modern 'republic of physicians' – contribute to this complementation?

---

[199] Overmann, *Materialismus*, p. 174.

[200] Vilar, "Sinnfrage", p. 245.

[201] Cobb, *Egg & Sperm Race*, p. 100.

[202] Nida-Rümelin, "Wissenschaftsethik", pp. 847, 854, 857–858, here 857–858.

Chapter 2
Ethos and Bioethics of the Anatomist Niels Stensen

The term *bioethics* was introduced to the scientific debate – initially in the USA – in 1970/1971 by the freethinking American biochemist and professor of oncology Dr. phil. Van Rensselaer Potter (1911–2001)[1] in his essay *Bioethics: the Science of Survival* (1970) and his book *Bioethics: Bridge to the Future* (Englewood Cliffs, NJ 1971).[2] Potter's evolutionistic concept, employing the key phrases 'man's survival' and 'improvement in the quality of life', is part of a secularization process taking place at this time.[3] Owing to the author's mechanistic view of man, it exhibits considerable shortcomings from a philosophical as well as from a theological perspective, including a number of grave misconclusions.[4] By citing 1 Thess 5:21, "Prove all things; hold fast that which is good", however – intending it to be applied to

---

[1] On Potter: James E. Trosko/Henry C. Pitot, "In Memoriam: Professor Emeritus Van Rensselaer Potter II (1911–2001)"; in: CanRes. 63 (2003), p. 1724; Peter J. Whitehouse, "The Rebirth of Bioethics: A Tribute to Van Rensselaer Potter"; in: GlobBioeth. 14 (2001), pp. 37–45, here 39.

[2] Potter, *Bioethics* (front cover: photograph of Earth from space; no source reference); the first chapter (ibid., pp. 1–29) is a reprint of Potter's essay *Bioethics, the science of survival* (1970) = Potter, "Science of Survival", with an unreproduced introduction ibid., p. 127. This essay is not cited as a preliminary publication in Potter, *Bioethics*, pp. xi–xiii (Acknowledgments). On the Italian translation *Bioetica: ponte verso il futuro*. Messina: Sicania, 2000 cf. the review by Joseph Joblin in: Greg. 83 (2002), pp. 198–199.

[3] On placement in contemporary history see Chiodi, *Bioetica*, pp. 23–26. On the reception of Potter by Pierre Teilhard de Chardin SJ (1881–1955) and Hans Küng, cf. Marques Filho/Fabri dos Anjos, "Potter", pp. 429–431.

[4] In Potter, *Bioethics*, pp. 154–157 ("Biology for the Future") he predicts a political proposition which presumably unintentionally alludes to *Brave New World* by Aldous L. Huxley (1894–1963), namely "to add the antifertility chemicals to certain foods or to water supplies in large cities", from which he distances himself only humoristically (ibid., pp. 155–156 with citation on p. 155). Ibid., pp. 58–59 ("Rise of Unmanageable Knowledge"), here 59 he asserts that "Medically supervised abortions will undoubtedly become more widespread in the United States as the procedure becomes legalized, and will provide time for the development and adoption of more suitable methods of birth control." This likewise goes to show that for Potter the individual human being, for which – as for himself – he did not expect eternal life, was worth nothing, wherefore its medically

© Springer International Publishing Switzerland 2016

F. Sobiech, *Ethos, Bioethics, and Sexual Ethics in Work and Reception of the Anatomist Niels Stensen (1638-1686)*, Philosophy and Medicine 117, DOI 10.1007/978-3-319-32912-3_2

ethical traditions in bioethics[5] – Potter, who apparently possessed some knowledge of biblical texts,[6] provides his readers with an apt key phrase for criticism of his own book.

At the center of Potter's concept, which reached beyond the field of medicine and whose intent was to establish "bioethics" as a discipline that would "combine science and philosophy",[7] stands the term "humility with responsibility". Diametrically opposed to this was "arrogance",[8] which Potter defined as the ethical boundlessness of the natural sciences. The problematic issues raised by his concept will be pointed out in detail in a later section.

In the current bioethical debate, precious little significance appears to be attached to the "humility" of the scientist or medical practitioner.[9] From a philosophical and theological point of view, the term "humility" is seen in the context of the virtues and is thereby placed on a level with the other virtues,[10] eg love as the greatest of the three theological virtues (1 Cor 13:13). Potter merely mentions supporting the weak, the "concept of brotherly love" as one of several concepts to be balanced against each other.[11] Can Stensen's ethos, in which a humble attitude of the scientist and the physician – which he later interpreted theologically – played a major role since his studies in Copenhagen, lend new meaning to the term "humility with responsibility"?

As a result of his significant contributions to research into human reproduction, one of the focal points of Stensen's ethos was formed by questions concerning the beginning of human life which are among the subject matters of bioethics resp. biomedical ethics today. This justifies the retrospective use of the term "bioethics" as a superordinate concept for the pertinent thoughts of this seventeenth-century anatomist, further corroborated by the fact that Stensen and Potter share Descartes' philosophy, whose limited usefulness Stensen had defended as a bishop, as their starting point. Potter's thoughts will be discussed extensively in the following – in particular in the course of a comparison with Stensen's ethos conducted in the conclusion.

---

supervised killing would from his point of view afford 'science' more time to develop 'more effective' methods. The same applies to the identical opinion of Short, "Oocyte", p. 7, l. col.

[5] For example, in Potter, *Bioethics*, p. 90 he claims to have read to Book of Job "many times".

[6] See note 45, in Chap. 4 below.

[7] Potter, *Bioethics*, p. 110 and Potter, "Humility", p. 2297.

[8] Potter, "Humility", p. 2297.

[9] This is reflected eg in LBioeth. 3 (1998), pp. 823–894 (Index), where "Demut" (humility) is absent as a keyword.

[10] Cf. details in Weber, *Allgemeine Moraltheologie*, pp. 321–322; Austriaco, *Biomedicine and Beatitude*, pp. 41–42.

[11] Potter, *Bioethics*, pp. 112–115, here 115.

2.1 Research and Publishing in the 'Republic of Physicians'

With the beginning of his studies, Stensen joined the 'republic of physicians'. He studied for a total of 7 years: first at the University of Copenhagen (1656–1659), then, following a detour via Rostock and a brief stay at the university there (1659[?]/1660), during which time he met the professor of medicine Dr. med. Johannes Bacmeister Jr. (1624–1686),[12] in the Republic of the United Netherlands, first at the Amsterdam Athenaeum Illustre (1660), which he joined around March 20, 1660,[13] and finally at the University of Leiden (1660–1663).

Early on during his time in Amsterdam, a conflict arose which was to be formative for Stensen and in the course of which he contrasted the competence of Dr. med. Simon Paulli Jr. (1603–1680),[14] Thomas Bartholin, Bacmeister and even the company of his Copenhagen fellow students with the teachings of his Amsterdam lecturer Dr. med. Gerard Leonard Blasius (Blaës) (1624/1625–1692),[15] which he considered deficient.[16]

2.1.1 Conflicts of a Student: The Blasius and Deusing Episodes

As discussed by Edward G. Ruestow, Stensen was one of a group of young and highly talented anatomists associated with the Leiden school in the 1660s for whom the promise of fame, egoism and competition between individual researchers provided strong incentives for the "ethos of a persistent search" for previously unknown structures at the subtler levels of gross anatomy.[17] To what degree this schematic characterization applies to Stensen and his actions and behavior as a human being, a student and a scientist respectively will be examined in detail in the following.

Stensen had his early anatomical findings published a few years after the end of his studies by his Copenhagen study mentor Thomas Bartholin in the latter's edition *Epistolarum medicinalium centuria I.–IV. [...]* (Medical letters' first to fourth group of a hundred [...]; 4 vols., Copenhagen 1663–1667),[18] which was read far beyond

---

[12] On Bacmeister: Schumacher/Wischhusen, *Anatomia Rostochiensis*, pp. 57–59.

[13] GlandOr., p. 23, lines 8–10.

[14] On Paulli: Egill Snorrason/Anne Fox Maule, "Paulli, Simon"; in: DBL³ 11 (1982), pp. 181–183.

[15] On Blasius: DDPhS, s.v. "Blasius (Leonh. fil.), Gerardus", cols. 156–158.

[16] AProd., p. 147, lines 30–34.

[17] Ruestow, *Microscope*, pp. 45–46, here 46.

[18] Kragh, "New Science", p. 59; Valdemar Meisen, "Bartholin, Thomas"; in: DBL³ 1 (1979), pp. 476–480, here 479.

Denmark. These scientific epistles mirror the human and scientific environment which Stensen was a part of. Bartholin had fostered Stensen already in the years before: in a letter written in Leiden on May 21, 1662 and published in the mentioned edition in 1667, Stensen thanks Bartholin for making his name known throughout "the entire learned world", meaning his peers and the Danish king.[19]

2.1.1.1 Development and Progression of the Conflicts

In his treatise *Apologiae prodromus [...]* (Leiden 1663), which brought the Amsterdam Blasius episode to a close, Stensen indicates "with what means" and "artifice" Blasius had made him a guest in his private home around Easter, March 28, 1660, and claims to have learned nothing but useless chemical experiments on this occasion.[20] Blasius had been permitted to host private anatomical and chemical exercises for the students of the Amsterdam Athenaeum Illustre in his home since May 1658, and had been given an unpaid post as lecturer on October 7, 1659.[21] It was customary in Amsterdam in the seventeenth century for students to be put up by professors, and Blasius was known to take in foreign students.[22] The frontispiece of his study handbook *Medicina generalis [...]* (Amsterdam 1661) shows him as an academic teacher in a seminar scene in his private house on the Verversgracht canal with four students, one of which may theoretically represent Stensen (Fig. 2.1). Stensen had, in fact, dedicated his *Disputatio physica de thermis [...]* (Physical disputation on hot springs [...]; Amsterdam 1660), held on July 8, 1660 at the Athenaeum Illustre with the professor of philosophy Dr. phil. Arnold Senguerd (1610–1667) in chair, to his "host" Blasius among others – and was not the only Amsterdam student to do so.[23]

What constituted the "artifice" criticized by Stensen can likely not be reconstructed: there was presumably more than one letter of recommendation involved, however, as Stensen alludes to the contents of a "first commendatory letter" which his Copenhagen study mentor Thomas Bartholin had addressed to Blasius in Stensen's name.[24] This letter seems to have appeared important to Stensen in retrospect, since it is the only one he refers to and at the same time emphasizes.

---

[19] Bartholin, *Epistolarum centuria IV* (Epist[ola] I. Cur nicotinae pulvis oculos clariores reddat/De lactea gelatina observatio), pp. 1–10, here 2 = NicPulv., p. 107, line 24–p. 108, line 1, here p. 107, line 25 = E 7, p. 151, lines 31–33, here 32–33: "universo litterato orbi".

[20] AProd., p. 147, lines 27–30, here 27: "quibus [...] artibus".

[21] DDPhS, s.v. "Blasius (Leonh. fil.), Gerardus", cols. 156–158, here 157; Miert, Amsterdam, pp. 92, 319.

[22] Miert, Amsterdam, p. 118.

[23] Scherz 1969, p. 50, lines 19–22, here 22: "Hospiti"; Miert, *Amsterdam Athenaeum*, p. 118n11. On Senguerd, whom Stensen does not mention in his other writings: Gerhard Berthold Wiesenfeldt, "Senguerd, Arnold"; in: DDPh. 2 (2003), pp. 909–911.

[24] AProd., p. 147, lines 26–27: "commendatitiis litteris primis".

Fig. 2.1 Gerard Leonard Blasius with students of the Amsterdam Athenaeum Illustre in his private home on the Verversgracht (*today*: Groenburgwal) on the frontispiece of his *Medicina generalis [...]* (Amsterdam 1661)
SUB Göttingen, 8 THER 711, first page recto before the title page

Blasius on the other hand, in a letter to Bartholin from Amsterdam on July 16, 1661, only mentions a letter of recommendation by Bartholin which Stensen carried on his person.[25]

Being the keen student he was, Stensen closely observed not only his objects of study but also his host; in a letter to Bartholin from Leiden on August 26, 1662, he notes that Blasius had once damaged his own private interests with a thoughtless comment uttered in excitement, as he, Stensen, could prove if he were so inclined.[26]

Three weeks after his arrival in Amsterdam, as Stensen writes to Bartholin from Leiden on April 22, 1661, he saw an opportunity for anatomical sections following the 6th session of his anatomy course at Blasius' house and asked his host for permission to purchase suitable animal objects and dissect them with his own hands. After Blasius had given his permission, Stensen proceeded to visualize and mimic what he had seen done by various men well-versed in anatomy. His luck had been so good that in the first sheep's head which he had bought on April 7, 1660 and begun to work on alone in his dissection room, he came upon a duct which, to the best of his knowledge, no-one had yet described.[27] Stensen's room was located directly above Blasius' living room (see Fig. 2.1).[28] Puzzled by this finding, Stensen called Blasius to ask his opinion and showed him the half-dissected duct. Blasius initially accused his student of the use of force before seeking refuge in the explanation – quite common at the time – that nature often "fooled about" and finally referring to a discovery made by Thomas Wharton (1614–1673) which he thought might be what Stensen had found.[29]

While Stensen was already on his so-called Peregrinatio erudita following the conclusion of his studies, the mathematician Giovanni Alfonso Borelli (1608–1679)[30] wrote to Malpighi from Florence on August 29, 1665, looking back on the above-mentioned episode and stating that his reading of a printed commentary by Blasius, which he claimed any ordinary printer could have written just as well, had given him the opinion that Blasius was "a very superficial person".[31] The work in question is Blasius' Appendix to *Exercitatio anatomica de structura & usu renum [...]* (Anatomical exercise on the structure and function of the kidneys [...]; Amsterdam 1665) by the Pisan professor of anatomy Dr. med. Lorenzo Bellini

---

[25] Bartholin, *Epistolarum centuria III* (Epistola XLIII. Examen disputationis Stenonianae de inventione ductus salivalis exterioris), pp. 158–184, here 160.

[26] AvCun., p. 115, lines 16–22 = E 9, p. 159, lines 11–17.

[27] Described by Stensen in DucSal., p. 4, lines 18–25 = E 1, p. 134, lines 15–22; GlandOr., p. 23, lines 5–20; AProd., p. 146, lines 6–8.

[28] This hitherto unknown fact is documented in Buder, *Nützliche Sammlung*, pp. 662–717 (XXXVII. Extract eines Reise Iournals eines gelehrten Medici), here 668.

[29] DucSal., p. 4, lines 8–10, 30–33 = E 1, p. 134, lines 5–7, 27–29; GlandOr., p. 23, lines 20–23. On the so-called *Ductus Whartonianus* cf. E 1, p. 137n5.

[30] On Borelli: Ugo Baldini, "Borelli, Giovanni Alfonso"; in: DBI 12 (1970), pp. 543–551.

[31] Adelmann, "Supplement", no. 142A, p. 57: "un uomo assai superficiale".

(1643–1704).[32] Blasius, who himself admitted to Thomas Bartholin that his *Medicina generalis* was scientifically shallow,[33] nevertheless became a didactically quite successful professor – probably due in part to the influence of gifted students like Stensen.[34]

Following this rebuff by Blasius, Stensen waited and kept his silence; for he knew that something similar had already been discovered and he could not be sure that others had not already observed what he had seen. Seven months after his transition to the University of Leiden, c. January/February 1661, in the *Collegium medico-practicum* – which was always held in winter due to the cold which was useful for dissections, and had begun in November 1660 –,[35] he finally seized an opportunity to ask his academic teachers, first Sylvius and then the professor of anatomy and surgery Dr. med. Johannes van Horne (1621–1670),[36] for advice.[37] He later continued this practice of not exclusively trusting his own eyes with regard to his discoveries, but instead to repeat the dissection under the eyes of competent witnesses.[38] In this respect he acted like the majority of anatomists in the second half of the seventeenth century, who would have their discoveries reviewed by peers prior to publishing.[39]

Frans de le Boë Sylvius, who taught in Leiden from 1638 to 1641 and as professor of practical medicine from 1658, and under whom Thomas Bartholin had studied,[40] was renowned throughout Europe for his excellent teaching. He held practical exercises twice a week within the framework of the *Collegium medico-practicum* at the *St. Caecilia Gasthuis*,[41] with some of the dissections being performed by Dr. med. Johannes Antonides van der Linden (1609–1664), professor of medicine and editor of the medical bibliography *De scriptis medicis libri duo* (Two books about medical writings; Amsterdam 1662)[42] published for the last time in a revised edition while retaining the name "Lindenius" in the year of Stensen's death. Van Horne was occasionally present at these sessions as well,[43]

---

[32] Bellini/Blasius, *Exercitatio anatomica*, pp. 97–132. On Bellini and his *Exercitatio anatomica*: ibid., p. 57n6; Giulio Coari/Claudio Mutini, "Bellini, Lorenzo"; in: DBI 7 (1965), pp. 713–716, here 713–714.

[33] In a letter on 07/16/1661 from Amsterdam, cf. Miert, *Amsterdam Athenaeum*, p. 93n239.

[34] Scherz, "Niels Stensen's First Dissertation", p. 251n13; Miert, *Amsterdam Athenaeum*, pp. 147, 164, 324, 327.

[35] Moe, "Glands", p. 57.

[36] On Van Horne: DDPhS, s.v. "Horne (Hornius, Hoorn), Johannes van", cols. 908–910.

[37] DucSal., p. 5, lines 2–5 = E 1, p. 134, line 37–p. 135, line 1; AProd., p. 146, lines 14–17.

[38] GlandOr., p. 42, line 14; VitTrans., p. 212, lines 27–32; HepRed., p. 62, lines 22–31.

[39] See notes 95 & 96, in Chap.1 above.

[40] HGK, s.v. "Bartholin, Thomas", pp. 70–71, here 70; Huisman, *Finger of God*, p. 145.

[41] DDPhS, s.v. "Sylvius, Franciscus dele Boë", cols. 1939–1943, here 1939–1941; details on Sylvius' teaching activities: Huisman, *Finger of God*, pp. 144–154.

[42] On Van der Linden: DDPhS, s.v. "Linden (Lindanus), Johannes Antonides van der", cols. 1200–1203.

[43] Huisman, *Finger of God*, p. 95.

and Stensen mentions that other interested persons – presumably students registered for other study programs – were also allowed to attend.[44] In addition, Sylvius
and Van der Linden had been teaching directly at bedside daily in 3-month alter-
nating shifts since 1659.[45]

Following Stensen's indication of the duct he had found, Sylvius proceeded to
the small autopsy room next door to the *St. Caecilia* hospital, which he had expanded
into a small anatomical theater, and investigated his student's claims on a human
corpse. It turned out Stensen had discovered the hitherto unknown excretory duct of
the parotid gland, the *Ductus parotideus* or *Ductus salivaris exterior*, which Sylvius
was able to demonstrate reliably several times. The same was achieved "twice
publicly"[46] shortly thereafter by Van Horne, in the larger anatomical theater built in
1593 for which he was responsible.[47] Stensen notes that Van Horne used his,
Stensen's, name "certainly several times" in connection with the newly discovered
duct "before a large circle of scholars of all fields" (ie not only physicians) during
these demonstrations[48]; it was owed only to their "charitableness" (humanitas),
Stensen adds, that both professors had attributed the discovery to him.[49] These public ascriptions of the duct to Stensen marked the beginning of the eponym *Ductus
Stenonianus* or *Ductus Stenonis*.[50]

It was a few weeks thereafter[51] that Blasius, who had been informed of Van
Horne's demonstrations,[52] began his literary offensive against Stensen. This was
because Blasius, who had been made the first extraordinary professor of medicine
and city physician in Amsterdam on September 4, 1660,[53] had already claimed
Stensen's discovery for himself by this time, as Ole Borch wrote to Thomas
Bartholin from Leiden on March 20, 1661. Borch added that Stensen, though quite
modest and somewhat inexperienced with regard to ambition up to this point, had
stated very clearly that he was prepared to defend himself against Blasius if matters

---

[44] GlandOr., p. 22, lines 1–4 & p. 37, lines 31–32.

[45] DDPhS, s.v. "Linden (Lindanus), Johannes Antonides van der", cols. 1200–1203, here 1202;
Huisman, *Finger of God*, p. 152.

[46] As Ole Borch wrote to Thomas Bartholin from Leiden on 03/03/1661, cf. Bartholin, *Epistolarum
centuria III* (Epist[ola] LXXXV. De fovea septi pellucidi [...]), pp. 360–369, here 362: "bis
publice".

[47] Huisman, *Finger of God*, p. 83.

[48] DucSal., p. 4, lines 4–6 & p. 5, lines 5–6, 26–28, here 26–27=E 1, p. 134, lines 1–3 & p. 135,
lines 1–3, 22–25, here 23: "in tanta omnis generis eruditorum corona [...] quidem iterato".

[49] AProd., p. 146, lines 17–21 on Sylvius and Van Horne.

[50] FachwA, s.v. "Steno(nius) (dänisch Stensen), Niels", pp. 223–224, here 224: "ductus
Stenonis=ductus parotideus [1661]".

[51] DucSal., p. 4, lines 6–7=E 1, p. 134, lines 3–4.

[52] DucSal., p. 5, lines 12–13=E 1, p. 135, lines 8–9.

[53] DDPhS, s.v. "Blasius (Leonh. fil.), Gerardus", cols. 156–158, here 157; Miert, Amsterdam,
pp. 92, 320.

came to a head.[54] However, Borch also wrote "I will not make this cause my own",[55] indicating that Stensen would stand alone in his defense. Apparently, Blasius had been of the opinion that Stensen was unable to make proper sense of his discovery,[56] and now saw himself forced to defend against Stensen's claims:

In the "Preface" (Praefatio) to his *Medicina generalis*, which appeared before Easter 1661 and was presumably undergoing final editing or already being printed during these weeks (the letter of dedication being dated March 22, 1661),[57] Blasius states that his book offered "hardly anything new worth mentioning"[58] with regard to the state of the art. However, he follows this assertion with an offhand remark about seeing himself forced to remind the reader that "my diligent student" Stensen had seized a convenient opportunity a few months earlier to demonstrate to spectators at the Leiden hospital in a human head the duct that he, Blasius, had discovered during his own anatomical exercises in the spring of 1660 in a *calf's* head (note the divergence from Stensen's own – correct – description).[59] How did Stensen react to this false account?

Thomas Bartholin had encouraged Stensen in a letter[60] to describe his discovery. In his answer to Bartholin, penned after Easter from Leiden on April 22, 1661 and representing the first preserved letter to his former Copenhagen mentor, Stensen calls it a "small discovery" (inventiuncula). He also felt compelled to briefly mention, Stensen continues, the envy it had brought upon him; he wished to do so only in order to repudiate the "spiteful allegation of plagiarism", however, as sadly his only alternatives were to either make waves about such a "matter of not great importance" or, because modest silence would be interpreted as a guilty conscience, to allow an ignominious badge of shame to be forced upon himself.[61] Stensen admits to having initially been disconcerted after reading the "Praefatio" to the *Medicina generalis*, as he had assumed Blasius to have described the duct correctly, which of course would have been disadvantageous for Stensen's defense. Upon reading the respective section in the book, however, he recognized that Blasius had been incapable of doing so. In fact, Stensen was able to prove Blasius' complete ignorance even of Wharton's discovery as well as relevant glands due to the fact, according to Stensen, that Blasius had described neither the origin nor the orifice of the duct cor-

---

[54] Bartholin, *Epistolarum centuria III* (Epist[ola] LXXXVII. […]), pp. 374–377, here 376.

[55] Ibid.: "Ego litem hanc non facio meam." – In his letter from Leiden on 03/03/1661, Borch had not been aware of this: cf. note 46, in Chap. 2 above.

[56] According to Moe, "Glands", p. 55.

[57] Blasius, *Medicina generalis* (Dedicatio), fol. * 4ᵛ.

[58] Ibid. (Praefatio), fol. * 5ᵛ = DucSal. (Notes), p. 223 ad "P. 4. l. 12–13 from top", here l. col.: "vix aliquid novi, quod considerationem mereatur."

[59] Blasius, *Medicina generalis* (Praefatio), fol. * 5ᵛ = DucSal. (Notes), p. 223 ad "P. 4. l. 12–13 from top", here r. col.: "discipulo meo industrio". Cf. Stensen's factual description of Blasius' actions against him in AProd., p. 146, lines 21–25; ibid., p. 151, lines 24–25.

[60] The letter (date unknown) is not preserved.

[61] DucSal., p. 3, line 17–p. 4, line 4, here p. 3, lines 19, 21–23 & p. 5, lines 29–30 = E 1, p. 133, line 20–p. 134, line 1, here p. 133, lines 22, 24–26 & p. 135, lines 25–26: "inventiuncula […] invidiosum plagii crimen […] re non ita magni ponderis".

rectly as well as ascribing such minuscule function to the gland from which it originated that if he, Stensen, were not certain of having shown Blasius the duct on
several occasions, he would have to assume that the professor had never actually
seen it.[62] Furthermore, Stensen expressed his amazement to Bartholin that Blasius
could appear so clueless "in such an insignificant matter" as this duct.[63]

On May 10, 1661 st.v., Bartholin, who was friends with Blasius, wrote to Stensen
that both he and Blasius, "with intent to further the interests of the republic of physicians", should take care not to let their rivalry turn to envy.[64] The writings of learned
men should be free of bile, as it burdened rather than benefited the reader if an
acceptable level was surpassed; although he was aware it was not necessary to
admonish Stensen in this regard as he knew his "prudence and calm character".[65]
Bartholin could rely on Stensen, whose defense always remained factual in contrast
with Blasius' hurtful statements; at one particular point, however, Stensen nevertheless was unable to refrain from saying that the duct described by Blasius could at
best be found in the inhabitants of the moon or the intermundia stipulated by the
philosopher Epicurus (342/341 B.C.–271/270 B.C.).[66] Blasius never publicly admitted his misconduct.

On May 21, 1662, Stensen reported to Bartholin in regard to Blasius, openly
admitting to being filled with a deep sadness – not because of the factual case, as
even any judge chosen by Blasius could not have ruled it in the professor's favor
with a clear conscience, but because of Blasius himself: nothing went against
Stensen's character more than to raise his quill against him.[67] In closing, Stensen
writes: "But neither his currish letters nor the cartloads of nasty insults seemed to
me so important as to use them for a reason to overstep the boundaries of discretion."[68]
He did not think, Stensen stated almost a year later, on April 30, 1663, that massive
folios often contained as many errors as Blasius had made in defense "of his [–
alleged –] discovery",[69] and added: "I feel sorry for Blasius, who could not restrain

---

[62] DucSal., p. 5, lines 31–36 = E 1, p. 135, lines 28–33. Cf. GlandOr., p. 19, line 21–p. 20, line 2;
AProd., p. 146, line 38–p. 147, line 8; ibid., p. 148, lines 11–16.

[63] GlandOr., p. 20, line 2: "in re adeo exigua".

[64] E 2, p. 139, lines 9–11, here 9–10: "in promovendis reipublicae medicae commodis intentus".

[65] Ibid., lines 11–14, here 13–14: "modestiam [...] et sedatam mentem".

[66] MuscGland., p. 183, lines 26–29. On Epicurus' "intermediate worlds" (Greek μετακόσμια; Latin
intermundia), cf. Usener, *Epicurea*, pp. 240–241, here Fragmentum 359. – In the year 1970, a
moon crater was named "Steno" after Stensen, cf. *Gazetteer of Planetary Nomenclature* of the
International Astronomical Union at http://planetarynames.wr.usgs.gov, here especially http://
planetarynames.wr.usgs.gov/Feature/5696.

[67] NicPulv., p. 108, lines 9–16 = E 7, p. 152, lines 4–11; cf. AvCun., p. 115, lines 19–25 = E 9,
p. 159, lines 14–20.

[68] NicPulv., p. 108, lines 21–23 = E 7, p. 152, lines 16–18: "Mihi vero nec mordaces ejus literae, nec
convitiorum plaustra tanti visa sunt, ut eorum nomine modestiae limites transgrederer."

[69] NovMusc., p. 160, lines 25–26, here 26 = E 13, p. 178, lines 35–36, here 36: "proprii inventi".

his tongue nor hand until he had personally in his writings displayed publicly a temper scarcely befitting an honorable person."[70]

Stensen also had to deal with Dr. med. Anton Deusing (1612–1666), a professor of anatomy and philosophy at the University of Groningen, public health officer and personal physician.[71] Deusing had little experience in dissection and had not taught anatomy since 1654, but excelled at producing large numbers of technically mediocre writings of usually polemical content. His cantankerousness was well-known; Sylvius, target of many of Deusing's polemical attacks, defended against the latter's "needling and predations"[72] and his "truly diabolical manner of writing"[73] in 1664. In his hastily composed and unobjective polemic pamphlet *Appendix ad dissertationem de hepatis officio seu vindiciae hepatis redivivi, leni correctione tangentes sequiorem interpretationem clarissimi viri D[omini] J[oannis] van Horne [...]* (Appendix to the Treatise on the Service of the Liver or Legal Claim of the Resurrected Liver, Touching with Gentle Correction on the Inferior Interpretation by Mr. Johannes van Horne [...]; Groningen 1661),[74] in which Deusing attempted to repudiate "with a hasty hand" Stensen's printed edition of his *Disputatio anatomica de glandulis oris, & nuper observatis inde prodeuntibus vasis [...]* (Anatomical Disputation on the Glands of the Mouth and the Recently Observed Ducts Emanating from Them; 2 parts, Leiden 1661; defended in Leiden on July 6 and 9)[75] in Groningen on July 19, he addressed the presider over the disputation Van Horne as its author[76] – a circumstance which will be revisited in the following. In the preface to his written response *Responsio ad vindicias hepatis redivivi [...]* (Response to the Legal Claim of the Resurrected Liver [...]), dated November 28, 1661, Stensen replied: "I have heard from others that your [medical] practice takes much of your time; [your] writings, however, which appear daily, demonstrate that you busy yourself with reading and writing various things: It is obvious to everyone, however, that a mind exhausted from various concerns and tasks proves less astute in many activities, and that its albeit very ingenious perspicacity becomes dulled

---

[70] NovMusc., p. 160, lines 27–29 = E 13, p. 178, line 37–p. 179, line 2: "Miseret me Blasii, qui linguam ante manumque non potuit frenare, quam suis ipse scriptis suam publico manifestam reddidisset indolem honesto homini parum convenientem."

[71] On Deusing: DDPhS, s.v. "Deusing, Anton (Antonius)", cols. 432–434; Gert-Jan C. Lokhorst, "Deusing, Antonius"; in: DDPh. 1 (2003), pp. 265–266.

[72] Sylvius, *Epistola apologetica*, title page (recto): "cavillationes atque calumnias".

[73] Ibid., pp. 5–6: "vere Diabolico [...] scribendi genere".

[74] Maar had been unable to find a copy of this treatise by Deusing, cf. HepRed. (Notes) preliminary remarks, p. 237 r. col.

[75] Stensen, *Disputatio anatomica 6. Iulii* (pt. 1); Stensen, *Disputatio anatomica 9. Iulii* (pt. 2). Cf. the title entry in Maar, *Opera philosophica*, vol. 2, p. 35, no. 1,I/II (not edited by Maar). Both parts were corrected, extended, and – with their titles changed to *Observationes anatomicae de glandulis oris, & novis inde prodeuntibus salivae vasis* (GlandOr.) – republished as the first of the four treatises of Stensen, *Observationes anatomicae*. The three other treatises are HepRed., GlandOc. and NarVas.

[76] Deusing, *Appendix*, p. 5 (with an explicit reference to the second part of Stensen's disputation, see previous note), 40: "celeri manu".

from ceaseless speculation about sundries."[77] With regard to Deusing's emotional diction, which he criticized as such in his response, Stensen noted that mutually escalating agitated tempers formed the worst conceivable premise for a scientific controversy, for they deprived the "matter itself" of its due appreciation.[78] It is characteristical of Stensen's orientation towards factuality and request for scientific precision that he ends his reply to Deusing's *Vindiciae* by commenting that instead of continuing to press the written controversy, Deusing should rather occupy himself with the anatomical facts, the "matter itself"; he, Stensen, would not oppose Deusing's arguments if they were methodical and verified by results, while Deusing was to expect resistance from Stensen if his arguments continued to lack such validation.[79]

Stensen concluded his *Apologiae prodromus, quo demonstratur, judicem Blasianum & rei anatomicae imperitum esse, & affectuum suorum servum* (Precursor to an Apology by Which Will be Shown that the Blasian Arbiter is Ignorant of Anatomy and a Servant to His Fervor; Leiden 1663) in the same fashion as with Deusing, therewith also ending his controversy with Blasius (Fig. 2.2). "Arbiter" refers to Dr. med. Nicolaas Hoboken (1632–c. 1678),[80] a friend of Blasius with a medical practice in Utrecht who had supported Blasius and created an offensive anagram of Stensen's name.[81] By virtue of the nonexistence of the duct described by Blasius, demonstrated in the final section of the *Apologiae prodromus*, Stensen portrayed Hoboken as a man betrayed by his gullibility and blind loyalty to a friend.[82] Furthermore, after beginning his response by citing the vituperations bestowed on him by Blasius, Stensen continued with the words: "See the honorable words of Blasius! Who would have expected this from a professor, or from any honorable man in a matter not yet proven?"[83] It had sufficed for him to show how much credence Blasius deserved[84]; he did not wish to reciprocate a single one of Blasius' hurtful cusses, Stensen wrote, in order to prove how much he valued the standing of a professor of the Amsterdam Athenaeum, professorial dignity per se and especially the pursuit of discretion.[85] The dispute was not continued by Blasius, who by this

---

[77] HepRed., p. 61, lines 11–16: "Ab aliis didici, praxin tibi tuam multum temporis eripere; ex scriptis vero, quae luci quotidie exponuntur, te variis tum legendis, tum scribendis occupari abunde liquet: omnibus autem manifestum est, defatigatum aliis curis & laboribus animum quibusdam in negotiis minus observari lynceum, aciemque ejus vel acutissimam non interrupta variorum speculatione hebetari."

[78] Ibid., p. 64, line 36; ibid., p. 65, lines 2–6, here 3: "res ipsa"; cf. AProd., p. 154, lines 12–13.

[79] HepRed., p. 61, line 23: "res ipsas", & p. 73, lines 21–25, here 21: "ipsas res".

[80] Appointed as professor at the Gymnasium Illustre in Steinfurt in 1663 and as Professor of Medicine and as adjunct Professor of Mathematics at the University of Harderwijk in 1669/1670; on Hoboken: DDPhS, s.v. "Hoboken (Hobokenus), Nicolaus", cols. 871–872.

[81] AProd., p. 146, lines 2–4.

[82] Ibid., p. 148, lines 21–26; ibid., p. 153, line 36–p. 154, line 7.

[83] Ibid., p. 145, lines 14–20, here 19–20: "Ecce honesta Blasii verba! Quis haec a professore, imo quis ab ullo viro honesto in causa nondum probata expectasset?"

[84] Ibid., p. 153, lines 28–30.

[85] Ibid., lines 30–35. Cf. SudOr., p. 102, line 37–p. 103, line 3 = E 5, p. 148, lines 20–25.

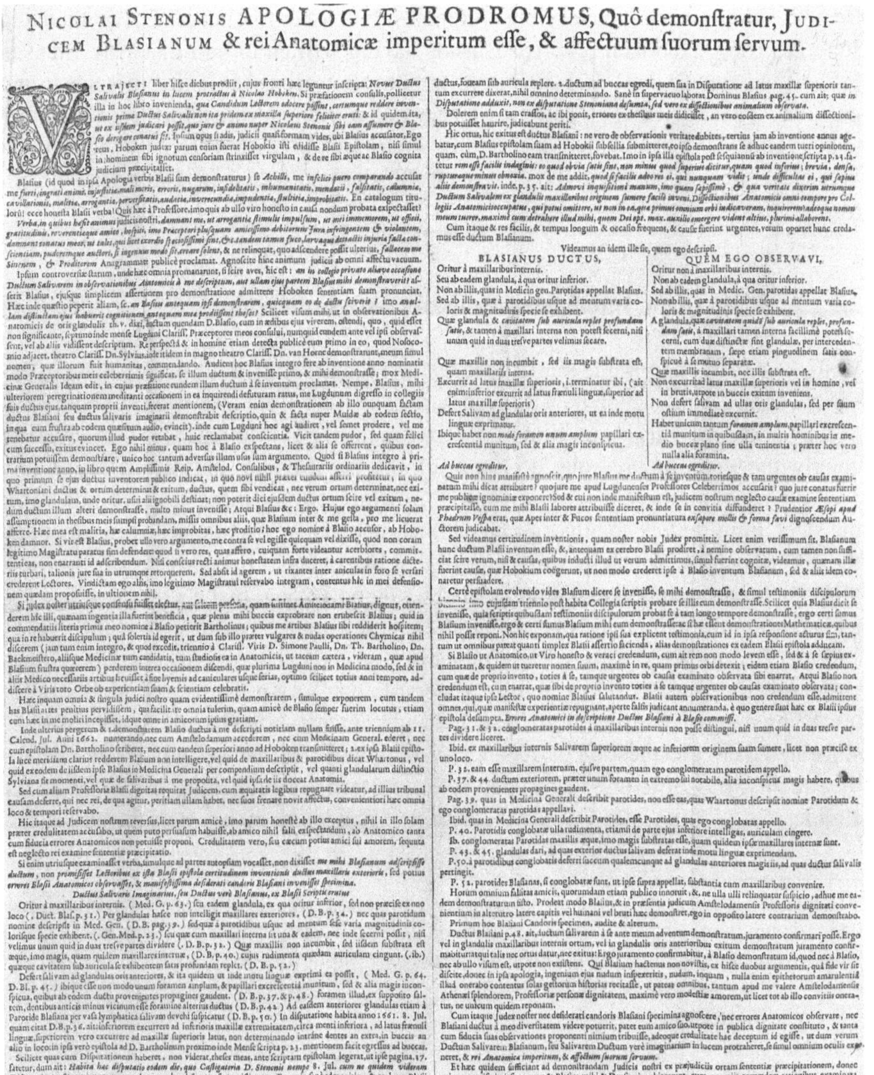

Fig. 2.2 Niels Stensen's *Apologiae prodromus [...]* (Leiden 1663) consisting of a foldable sheet printed on one side, perhaps designed for shipping by post
Concerning the synoptic table in the 2nd col. see note 154, in Chap. 1 above
BL, General Reference Collection 548.e.32.(2.)

time was hard-pressed for feasible arguments. In a letter to Thomas Bartholin dated June 26, 1663, Hoboken reported Blasius having written to him that "Mr. Stensen had visited our Blasius and displayed nothing but great benignity in speech and gestures." He added that he thought Stensen had refrained from a further publication against Blasius of his own volition[86]; this assumption is corroborated by the fact that Stensen had entitled his treatise against Blasius published in 1663 "Precursor" (prodromus).

2.1.1.2   Plagiarism, Intellectual Property and Authorship

With the words "honorable man", Stensen had reminded Blasius of his own professorial ethos against which he was necessarily measured. He also spoke explicitly of a "plagiarism".[87] This accusation of plagiarism must be distinguished – which is not clearly the case in previous accounts of the controversy between Blasius and Stensen – from a mere priority dispute,[88] in which two or more scientists discover the same thing *independently* of one another, usually with some offset in time, and the question arises to whom the honor of first discovery is due.

What Stensen thought about plagiarists is demonstrated by his comment in a letter to Thévenot, written in 1666 within the first few months of his arrival in Tuscany and attached to his Myology. Here he calls it an indicator of a deceitful and malicious spirit to present someone else's work as one's own. He would gladly relinquish any first discovery honors in such disagreements; however, these could not be separated from the question of decency, so that the more one was guided by love for that quality, the more one had to endeavor to defend oneself so as not to allow the slightest suspicion to arise against oneself.[89] That Stensen was serious about sur-

---

[86] Hoboken, *Anatomia* (Spicilegium epistolarum, rem potissimum generatoriam referentium, cum dd. [= dominorum] ad quos scriptae, responsionibus [pp. 67–219]), pp. 76–98 (letter no. IV), here 91–93 with citation on p. 93: "d[ominum] Stenonium Blasium nostrum invisisse, nihilque nisi summam modestiam ore & gestibus ostendisse."

[87] See note 61, in Chap. 2 above.

[88] Miniati, *Steno's Challenge*, p. 79 claims that Thomas Bartholin should have had no real doubt about Stensen's "primacy" regarding the discovery of the *Ductus parotideus*, as Borch had informed him of Stensen's discovery in his letter from Leiden on 03/03/1661 (Bartholin, *Epistolarum centuria III* [Epist(ola) LXXXV. (...)], pp. 360–369, here 362 [Miniati, *Steno's Challenge*, p. 79n34 cites an unfounded source]), ie before Blasius made his claim to priority. The case of Stensen and Blasius, however, is not one of two researchers independently making the same discovery, but in fact a case of plagiarism, which is also not clarified in Baumann, *Mehrfachentdeckungen*, pp. 24–25, no. 12. In addition, Stensen had mentioned his discovery to Jakob Henrik Paulli (1637–1704) (on Paulli: Carl O. Bøggild-Andersen, "v. Rosenschild, Jacob Henrik Paulli"; in: DBL³ 12 [1982], pp. 373–374) in a (presumably lost) letter in April 1660 and enclosed a description (DucSal., p. 4, line 36–p. 5, line 2 = E 1, p. 134, lines 33–36).

[89] Myol., p. 95, line 1–p. 106, line 38, here p. 95, lines 29–34 = Kardel, *Steno on Muscles*, p. 184, line 8–p. 224, line 24, here p. 186, lines 18–24; on temporal placement of the letter see Kardel, *Steno on Muscles*, pp. 22–23. For further examples of unjustified accusations of plagiarism see Merton, "Priorities", p. 474.

rendering the honor of discovery – so long as his reputation as an honest and hard-working scientist were not compromised – is demonstrated in 1670 by his even more accomodating behavior towards Reinier de Graaf, who was researching the anatomy of human reproduction in parallel with Stensen.

In his controversy with Blasius, as Stensen wrote to his Copenhagen mentor Thomas Bartholin on April 22, 1661 from Leiden, he would have been content to forgo "my right" to being identified as the discoverer of the *Ductus salivalis exterior* if the duct had not already been linked with his name by Van Horne, ie in public.[90] This means that had Blasius immediately intervened following Sylvius' demonstration – which apparently had been restricted to a group of students –, and had not Van Horne subsequently become active, Stensen would have viewed the relinquishment of his right as still feasible. Van Horne's public dissections, however, had changed the situation: now Stensen not only wanted to preserve his own scientific reputation, but also protect his professors in Leiden from accusations of spreading lies.[91] In view of these facts – going back to Ruestow's generalizing statement[92] – Stensen was not being egoistical, but was merely truthfully defending his reputation and would have been willing to rescind his claim of discovery for the sake of academic peace had this been a viable option.

That Stensen visited Blasius is commensurate with his unconfrontative character, as he himself stated that he was not fond of "duels".[93] In 1662, he humorously wrote that he nevertheless owed much to Blasius, namely that the latter had not only given him the opportunity to assert his claim to "what is mine", but thereby also to discover new things.[94] That Stensen, as a highly gifted anatomist, was the subject of hostility coming from his peers in later years as well is documented by two further statements he made: in the letter to Thévenot attached to his Myology, Stensen emphasizes that he did not wish to begin any "cantankerous quarrels" by defending against his critics, and that his remarks concerning the matter were primarily targeted not at the objections of his critics, but at the doubts of his friends so they may be assured that he had worked according to proper research practice, and so that the critics might over time become more lenient with him after recognizing the equitableness of his cause.[95] In a letter to his friend Dr. med. Heinrich Meibom Jr. (1638–1700),[96] the Lutheran personal physician to the dukes of Braunschweig-Wolfenbüttel and professor of medicine in Helmstedt since 1664, Stensen wrote on March 29, 1669 from Venice in regard to a discovery which Meibom had published in his *De*

---

[90] DucSal., p. 5, lines 26–29, here 28 = E 1, p. 135, lines 22–25, here 25: "jure meo".

[91] This is documented in Ole Borch's letter to Thomas Bartholin from Leiden on 03/20/1661, cf. Bartholin, *Epistolarum centuria III* (Epist[ola] LXXXVII. [...]), pp. 374–377, here 376.

[92] See note 17, in Chap. 2 above.

[93] As in his controversy with the Paderborn vicar general Laurentius von Dript OSB (1633–1686), see Sobiech, *Herz, Gott, Kreuz*, p. 313n281.

[94] GlandOr., p. 20, lines 3–5, here 4: "meum".

[95] Myol., p. 95, lines 21–26, here 21–22. = Kardel, *Steno on Muscles*, p. 186, lines 7–14, here 8–9: "litigiosas [...] controversias".

[96] On Meibom: Sabine Ahrens, "Meibom, Heinrich (d. J.)"; in: BBL, p. 487.

vasis palpebrarum novis epistola (Letter About New [Blood] Vessels of the Eyelids; Helmstedt 1666) that such a discovery, had it been made by himself, "would have given others an opportunity to vent their presumptuousness at me".[97] It can be assumed beyond reasonable doubt that Stensen was considered too intelligent by some of his peers, and that they perceived this to depreciate their own efforts.

That such calumniations are often difficult to dispose of once and for all is demonstrated in the 1780 volume of the periodical *Neues Magazin für Aerzte* by the Göttingen professor and clinic director Dr. med. Ernst Gottfried Baldinger (1738–1804).[98] In it can be found a review by Prof. Dr. med. Johann Friedrich Blumenbach (1752–1840)[99] of the *Bibliotheca anatomica [...]* (Anatomical Library [...]; Zurich 1774–1777) by the Swiss physician Dr. med. Albrecht von Haller (1708–1777), in which Blumenbach – unlike Haller[100] – refers to the autobiographical report on the study trip of Dr. med. Rudolph Wilhelm Krause (1642–1718)[101] partially published in 1735.[102] Krause, who had studied medicine in Leiden under Van Horne and Sylvius like Stensen, acquired his doctorate at the University of Padua c. 1667/1668 and was an associate professor of medicine at the University of Jena from 1671. Professionally, he was mostly a compiler. The hitherto unnoted printed original[103] of the equally unappreciated abridged publication of 1735 says about Krause's stay in Amsterdam in 1666/1667,[104] ie at a time when Stensen was already at the Medici court in Florence: "Having now heard that Dr. Gerard Blasius (namely the son of Leonard) was much praised for his teaching and anatomical works, and also that Mr. Swammerdam and Mr. Stensen had gained their knowledge in anatomy mostly from the helpful tuition of Mr. Blasius, I travelled back there. And there it was stated continually and explicitly that Mr. Blasius had first taught the *Ductus salivales* to his listeners in the *Collegium anatomicum*, among whom was also Stensen; that thereupon the latter, when he heard that Mr. van Horne was planning a dissection in

---

[97] Bruun, "Fem nyfundne Niels Stensen-breve.", p. 119, lines 10–14, here 13: "aliis ansam praebuisset insultandi mihi". On Meibom, *De vasis palpebrarum* cf. ibid., p. 123n11.

[98] On Baldinger: Hans H. Lauer, "Baldinger, Ernst Gottfried"; in: EMG 1 (2007), p. 134.

[99] NMAe. 2 (1780), pp. 33–39: "Des Herrn Professor Blumenbachs, zu Göttingen, Nachlese zu Herrn von Hallers Bibliotheca anatomica." On the physician and natural scientist Blumenbach: HGK, s.v. "Blumenbach, Johann Friedrich", pp. 106–107.

[100] Haller had mentioned the dispute between Stensen and Blasius without judgment (Haller, *Bibliotheca anatomica*, s.v. "§ 453. Nicolaus Stenonius", pp. 491–495, here 491). On Haller, who taught at the University of Göttingen: HGK, s.v. "Haller, Albrecht von", pp. 325–327.

[101] On Krause: Giese/Hagen, Jena, pp. 167–174; Mägdefrau, "Universität Jena" (essay), p. 158 and ibid. (Notes), p. 485.

[102] NMAe. 2 (1780), pp. 33–39 (Des Herrn Professor Blumenbachs [...]), here 38: "p. 491. Z. 6 v[om] E[nde] [...]". On this defamation's effects, which lasted until 1981, see supplementarily Gysel, "Conflit", pp. 545–546; a further example is Fischer-Defoy, "Studienreise", p. 325.

[103] Buder, *Nützliche Sammlung*, pp. 662–717 (XXXVII. Extract eines Reise Iournals eines gelehrten Medici), here 662 with explanatory "Anmerckung" (comment) followed by reproduction of the "Extract[s]".

[104] The terminus ante quem is the "Reise nach Italien Anno 1667" (journey to Italy in 1667; ibid., p. 672) following the sojourn in Amsterdam.

Leiden, had travelled posthaste to Leiden, entered the presentation, searched for the *Ductus salivales* in a calf's head and, with a stylus which he inserted into it, shown and taught it to Mr. van Horne when he presented the head, and because Stensen had not said that Mr. Blasius had discovered and taught him these [Ductus salivales], Mr. van Horne proclaimed loudly in the direction of the audience: See there, Stensen's duct; thus even afterwards, albeit wrongly, Stensen would have been considered the discoverer of the *Ductus salivales* at Mr. Blasius' expense."[105] After having thus spent "several days" in Amsterdam, Krause stayed with Blasius for some time, moving into the very same study chamber in which Stensen had lived.[106] There is no mention here of Blasius' conflict with Stensen; Krause leaves open the question of who was right. Blumenbach, however, seems to have had some faith in the story told by Krause.

From a sociological standpoint, the Blasius and Deusing cases occur within the relationship between a student who is as yet 'unfinished' and his socially higher-ranked academic teacher: Blasius, a not yet well-established lecturer,[107] attempted to seize the – from his point of view – propitious opportunity to appropriate and publish a new finding at the expense of its discoverer, the 'ingenuous' student. When Stensen unexpectedly – and against all accepted conventions – resisted this attempt, Blasius had to fear for his scientific reputation, which he subsequently tried to 'rescue' quite unscrupulously and with any available means. Against this background, Blasius' submissive gesture, the "humblest obsequiousness" of which he assures the authorities whose continued support of his own person he hoped for in the letter of dedication for his *Medicina generalis*,[108] gains an unsavory aftertaste. Nevertheless, the entire inglorious episode apparently did not damage him, as he was granted a full-fledged professorship in 1666.[109] During his later time as suffragan bishop in Münster, Stensen often witnessed how fear of public shame could lead

---

[105] Ibid., pp. 666–667: "Nachdem ich nun in Amsterdam den Herrn Doctorem Gerhardum (Leonhardi scilicet filium) Blasium sehr in in[667]formando und Anatomicis hatte rühmen hören, auch daß Herr D[ominus] Schwammerdam und Steno ihre Wißenschafft in Anatomicis meisten-theils aus des D[omini] Blasii treuer unterweisung erlernet hätten, reisete ich wieder dahin. Es wurde daselbst beständig und ausdrücklich gesagt, D[ominus] Blasius hätte die ductus Salivales zuerst im Collegio Anatomico seinen Zuhörern, darunter auch Steno war, gewiesen; worauf dieser, als Er vernommen, daß der Herr D[ominus] von Horne in Leiden eine Anatomie vor hätte, in aller Eil nach Leiden gefahren, in die demonstration gegangen, in einem Kalbskopfe die ductus Salivales gesucht, und mit einem stylo, den Er drein gethan, dem Herrn von Horne, als Er Caput demonstri-ret, præsentiret und gewiesen, und weil Steno nicht gesagt hatte, daß D[ominus] Blasius Ihm sol-che gewiesen und erfunden, hätte D[ominus] von Horne laut gegen die umstehenden Zuhörer gesagt: En ductum Stenonianum; dahero auch nachmals, obgleich fälschlich, Steno für dem Erfinder des ductus Salivalis, in fraudem D[omini] Blasii wäre gehalten worden."

[106] Ibid., p. 668.

[107] With similar examples Barber, "Resistance", pp. 550–552.

[108] Blasius, *Medicina generalis* (Dedicatio), fol. * 3ʳ–* 4ʳ, here * 3ᵛ–* 4ʳ: "humillimae [...] subjectionis".

[109] DDPhS, s.v. "Blasius (Leonh. fil.), Gerardus", cols. 156–158, here 157.

even clerics to dissemble an honorable outer appearance[110]; in the cases of occasionally even criminal behavior by individual clerics in Münster which he complained
of, his own distressful Amsterdam experiences may actually have proved useful.

Structural similarities with the above-mentioned incident exist for the discovery
of the spermatozoa by the Leiden student of medicine Johan Ham (1654–1725),
who in August 1677 informed the amateur researcher Antoni van Leeuwenhoek
(1632–1723) of his findings. Van Leeuwenhoek, who worked in Delft using self-
constructed microscopes, thereupon returned to his microscopic observations of
human sperm which he had begun 3 or 4 years previously, but eventually abandoned. There is no further data about Ham's medical training after his visit with Van
Leeuwenhoek; he was later employed in the Arnheim administration and as a diplo-
mat. Leeuwenhoek however, who even 20 years later praised Ham as a modest and
intelligent young man highly suited for the study of nature, always admitted that the
latter was the actual discoverer of the spermatozoa.[111] Such humility was not exhibited by everybody.

One must not forget that Stensen, though at the time still a student, was publishing independently like any full-fledged researcher which, with few exceptions, was
well outside the contemporary norm. In the letter of dedication for his *Medicina
generalis*, Blasius states that he had written the book "most of all to thank the academic youth which is entrusted to my and others' care".[112] It almost seems as though
he regarded this acknowledgement as a pretense for the right to appropriate discoveries made by his students. On the frontispiece of *Medicina generalis*, the depiction
of Blasius' seminar attendants on the Verversgracht, whose childlike physiognomy
and size are reminiscent rather of first-year high school students than of university
freshmen (Fig. 2.1), underscores their immaturity. Although Blasius also engaged in
the republishing of writings by deceased as well as living authors under their respective names – not under his own –,[113] he must still have been of the opinion that discoveries by his students appertained to himself,[114] presumably partly because he
thought he alone was capable of properly classifying and appreciating them.

---

[110]With a citation from *Opus imperfectum in Matthaeum* ParHAg., p. 14, lines 4–11, here 8–10 (=
Altoviti, *Obbligo* [Cap. III], p. 10, lines 3–11, here 8–11 = Attavanti, *Obbligo* [Cap. III], p. 3, line
31–p. 4, line 3, here p. 3, line 37–p. 4, line 3), cf. Migne, *Opera omnia* (In Matthaeum. Homilia xl.
ex capite xxj [a]), cols. 849–859, here 852. On this, see Joop van Banning/Franz Mali, "Opus
imperfectum in Matthaeum"; in: TRE 25 (1995), pp. 304–307.

[111]Committee of Dutch Scientists, *Collected Letters*, Letter No. 35 [22] (November 1677),
pp. 277–299, here 279n2 & pp. 281, 291; Lindeboom, "Leeuwenhoek", pp. 132–136; Lodewijk
C. Palm, "Leeuwenhoek, Antoni van"; in: DDPh. 2 (2003), pp. 599–603, here 600 l. col.; DDPhS,
s.v. "Ham, Johan(nes)", cols. 773–774; ibid., s.v. "Leeuwenhoek, Antoni van", cols. 1162–1165,
here 1163–1164.

[112]Blasius, *Medicina generalis* (Dedicatio), fol. * 4ʳ: "in gratiam maxime juventutis Academicae
curae meae post alios commissae, conscriptam".

[113]For the years 1659–1666 cf. Miert, *Amsterdam Athenaeum*, pp. 93, 319.

[114]Miniati, *Steno's Challenge*, p. 81 points out that Blasius also appropriated observations by Jan
Swammerdam in the year 1666.

Deusing on the other hand had ignored that Stensen had identified himself explicitly as "author and respondent"[115] of his disputation at the end of the acknowledgements on the rear of the title page of the disputation's first part, published on July 6, 1661, and had thus made clear that – although the opposite was customary – it had *not* been his presider Van Horne who had created the printed version. In his Amsterdam *Disputatio physica de thermis*, Stensen by contrast had not laid claims to authorship.[116] Such disputations held by students for purposes of practice, probably printed in small runs at the university's expense, were passed along to the student's own family and friends to demonstrate their study progress and generally represented a continuation of the professor's teaching activity[117]: Stensen's teacher Sylvius, for example, published ten disputations held by his students in Leiden between 1659 and 1663, introducing them as the "first returns of my academic efforts" in his preface of dedication to his own former preceptor, the Leiden professor of medicine and botany Dr. med. Adolph Vorstius (1597–1663),[118] on January 1, 1663.[119] In the preface to the reader held on January 11, Sylvius elaborates on this statement by describing himself as the author who had "written these medical disputations"[120]; it should be noted, however, that the students were identified as respondents, with their names and the dates of the disputations specified, in the table of contents preceding the disputations themselves.[121] According to contemporary customs, the publishing professor selected the disputations to be included and revised the material presented to him.[122] For students like Stensen to publish their own research results was an exception, which may in part explain the clashes between Stensen and Blasius together with Deusing.

Even the dissertation for the acquiring of an academic degree was usually a treatise written by the presider over the disputation (praeses), the 'doctoral adviser', who also chaired the dissertation's defense by the candidate.[123] By contrast, Stensen's doctoral thesis testifies to an unusually high level of independence for the

[115] Stensen, *Disputatio anatomica 6. Iulii*, title page (verso): "Auth[or] & Resp[ondens]". Cf. HepRed., p. 62, lines 9–14, where Stensen once again emphasizes that he is the author of the printed version of his Leiden disputation from 07/06 and 07/09/1661, then proceeds to specify the reasons for this fact.

[116] Cf. Scherz, *Geological Papers*, p. 50. According to Miert, *Amsterdam Athenaeum*, p. 154, other students did this as well even when they, as can be surmised, were the actual authors.

[117] Scherz, "First Dissertation", p. 251n12.

[118] On Vorstius: DDPhS, s.v. "Vorstius, Adolphus", cols. 2088–2089.

[119] Sylvius, *Disputationum medicarum*, fol. a2^r+^{v}, here a2^r: "laborum meorum academicorum primitias".

[120] Ibid., fol. a3^r+^{v}, here a3^r, no. 1: "has scripsi disputationes medicas". – On Blasius, *Institutionum medicarum compendium* (Amsterdam 1667) with 12 disputations by four students, published by Gerard Leonard Blasius under his own name, cf. Miert, Amsterdam, pp. 154, 322.

[121] Sylvius, *Disputationum medicarum*, s.v. "Elenchus contentorum in hoc opusculo." The pagination begins hereafter with the first disputation.

[122] Scherz, "First Dissertation", p. 251n12.

[123] Wollgast, *Promotionswesen*, p. 69.

period and, as can be inferred from the letter[124] accompanying his diploma, was extremely highly appraised.

The shock induced by Blasius continued to affect Stensen for a long time. It can justifiably be described as a 'trauma' for Stensen during his study years; this is expressed in his Florentine spiritual records, in which he contemplates the possibility that his writings on natural science might be destroyed or fall into the hands of evildoers who would use them against him or appropriate the discoveries detailed therein which he considered to be his.[125]

Stensen saw himself confronted with further hostility following the altercation with Blasius. In his posthumously published writings in which he mused about what constituted a "discoverer" (inventor), Malpighi made reference to a passage in Stensen's letter to Thévenot attached to the Myology in order to vindicate himself vis-à-vis various professors attacking him in publications and from their chairs with regard to his microscopic study *De pulmonibus observationes anatomicae* (Anatomical Observations on the Lungs; Bologna 1661), affirming that his discoveries were the result of his own research. Stensen had been in a similar position with regard to the yolk duct in birds, which he had discovered and published in the above-mentioned letter[126] dated June 12, 1664 st.v. – dedicated among others to the Amsterdam surgeon Dr. med. Paul Barbette (c. 1619–1665),[127] whom Thomas Bartholin held in high esteem, and part of Stensen's accountability report on his studies abroad –, when presumably worried friends had written him to call his attention to previous identical observations by other researchers. In this letter to Thévenot, Stensen mentions his predicament at the time, and Malpighi uses Stensen's justification published there to summarize his own: "By the way, as far as I [= Malpighi] am concerned, I wish to repeat with the extremely pleasant and conscientious Stensen in reference to his book on myology: Thus I am heartily pleased that I, even though it was not pointed out to me by them [= the other scientists], have [observed] the same thing that they have observed, and it disconcerts me not that I have seen it later than them, so long as I am not accused of having known this same thing from others."[128] Stensen's factual defense against unfair accusations was exemplary to Malpighi.

---

[124] E 17.

[125] Spir. 11, p. 130, lines 18–20. Stensen reflected on these and other experiences, as well as on imaginable occurrences, after his conversion.

[126] VitTrans.

[127] On Barbette: DDPhS, s.v. "Barbette, Paulus", cols. 69–71.

[128] Malpighi, *Opera posthuma*, pp. 9–140, here 10=Adelmann, *Embryology*, p. 820: "Caeterum, quod ad me spectat, cum amicissimo, & religiosissimo Stenone, lib[ro] Myologiae, repetam: Sic mihi gratulor, me non ab illis monitum eadem, quae illi observaverunt, nec me movet post alios me haec vidisse, modo ab aliis eadem habuisse non accuser." Stensen's justification cited by Malpighi can be found in Myol., p. 95, line 27–p. 96, line 10, here p. 96, lines 7–10=Kardel, *Steno on Muscles*, p. 186, lines 15–29 & p. 188, lines 1–11, here 8–11. In Myol., p. 95, lines 28–29=Kardel, *Steno on Muscles*, p. 186, lines 17–18, Stensen refers to his letter VitTrans., cf. VitTrans. (Notes) preliminary remarks, p. 264; the "letters by his friends" (Amicorum litterae) which he mentions in Myol., p. 96, line 6=Kardel, *Steno on Muscles*, p. 188, lines 5–6 are not preserved. On Malpighi, *De pulmonibus* see Cesare Preti, "Malpighi, Marcello"; in: DBI 68 (2007), pp. 271–276, here 272.

2.1.2   As Student and Researcher: Personality and Research Ethos

That Stensen defended himself as a student to prevent himself being called a plagiarist and his Leiden professors being tagged as purveyors of lies, while at the same time not giving back as he was given and even declaring his willingness to relinquish his rights, indicates a remarkably conscientious personality structure. What precisely characterized his research ethos as a student?

Four basic attitudes distinguishing Stensen's life and research form a 'fundamental ethos' of sorts, determining his methods of scientific examination – which consisted primarily of autopsy and a continuous practice of dissection – and turning them into tools in the hands of a concrete researcher defined by his character and talent.

2.1.2.1   Intuition and Reflection

Already in Amsterdam, at the beginning of his Peregrinatio academica, Stensen benefitted from his agile and mercurial intellect which was his source of intuition and inspiration: he discovered the *Ductus salivalis exterior* only because he had followed a casual idea to examine the vessels running through the mouth in a sheep's head prior to dissecting its brain.[129] Even mistakes by others offered him the opportunity of discovering something new while refuting their errors, as when he stated in a letter to Thomas Bartholin on April 22, 1661 that he owed it to Blasius' less than equitable accusation – namely that during the dissection of a calf's head under the Blasius' guidance he, Stensen, had not found even a shadow of the vessel Blasius claimed to have observed – that he had discovered several new lymphatic vessels there instead.[130] The contrast between the instructor Blasius and the independently researching student Stensen could not be demonstrated more strikingly.

Stensen's study diary for the year 1659 during his time in Copenhagen proves quite informative; its title *Chaos* characterizes it as a disordered collection of excerpts and notes.[131] Stensen presumably borrowed the term "Chaos" from his study literature, probably also with allusion to Gen 1:2.[132] Of the private notes interspersed among his excerpts, the evening jottings summarily evaluating the respective days are of particular interest.

The mental agility which facilitated his discoveries also entailed certain disadvantages from Stensen's point of view: on Saturday, March 19, 1659 st.v., he expresses his wish for God to free him of his mind's plague of mercurially occupying

---

[129] DucSal., p. 4, lines 25–27 = E 1, p. 134, lines 22–24.

[130] DucSal., p. 6, lines 4–9 = E 1, p. 136, lines 1–5. MuscGland., p. 167, lines 16–18 states that a dissection had – against all expectations – brought with it the exploration of several other things.

[131] On the term "chaos" with a negative connotation, see by contrast note 116, in Chap. 1 above.

[132] Ziggelaar, *Chaos-Manuscript*, p. 15.

itself with various things, so that instead he might do only *one*.[133] On Sunday, March 27 st.v., he asks God to allow him to command himself to abstain from opinions and in particular from precipitous and ill-deliberated verdicts on things *more or less* known to him, unless he knew them very precisely.[134] On Saturday, May 14 st.v., he notes down that vicissitude and overhaste are highly detrimental, adding that he hoped to be making excellent progress in his studies. In this note Stensen also subdivides his daily study schedule into time for reading and for excerpting – perhaps at Ole Borch's suggestion –, with any time left over assigned to visits.[135]

Stensen also meticulously records times not spent studying, putting down the following in his diary: on Thursday, March 10, 1659 st.v., that he had spent the entire day studying[136]; on Saturday, March 12 st.v., that 1 h before and 5 h after midday were lost[137]; on Monday, March 14 st.v., that he had been unable to do anything persistently due to the cold[138]; on Wednesday, March 16 st.v., that nothing besides medicine must be done before midday and that he had been unable to do anything persistently due to the cold[139]; on Thursday, March 17 st.v., that he had done too little[140]; on Friday, March 18 st.v., that he had not done anything that day[141]; on Monday, March 21 st.v., that he had done little and, after midday until 3 p.m., had taken a walk and engaged in a conversation which had nevertheless yielded something useful[142]; on Tuesday, March 22 st.v., that he had done too little, and therefore asked God to let him work properly and consistently[143]; on Friday, May 6 st.v., can be found the curt appraisal that he had done nothing for his studies the entire day.[144] The note from March 21 st.v. bespeaks Stensen's internal disquiet, growing throughout the day and finally mollified only in the evening, when he jotted down the entry, by the retrospectively legitimating consideration in regard to the walk and interlocution originally perceived as 'idleness'. It is here in particular that the constant self-reflection which accompanied Stensen's daily routines, as well as his researcher's spirit that wished to let never an hour go to waste, become

---

[133] Ibid., p. 95 (col. 30, fol. 37^r, N.26); Sobiech, *Herz, Gott, Kreuz*, p. 144.

[134] Ziggelaar, *Chaos-Manuscript*, p. 142 (col. 48, fol. 41^v, N.45).

[135] Ibid., p. 281 (col. 114, fol. 58^r, N.150) and ibid., p. 308n110. Further deliberations on Stensen's study schedule can be found ibid., p. 95 (col. 30, fol. 37^r, N.27); ibid., pp. 96–97 (col. 30, fol. 37^r, N.29); ibid., p. 100 (col. 32, fol. 37^v, N.33). On Borch's mentorship during Stensen's Copenhagen study days cf. Schepelern, *Olai Borrichii Itinerarium*, pp. vii–xliii (introduction), here esp. pp. xxxiii–xxxix.

[136] Ziggelaar, *Chaos-Manuscript*, p. 50 (col. 12, fol. 30^v, N.6).

[137] Ibid., p. 76 (col. 23, fol. 33^v, N.8).

[138] Ibid., p. 81 (col. 24, fol. 33^v).

[139] Ibid., p. 89 (col. 27, fol. 36^v, N.20 and N.20a).

[140] Ibid., p. 91 (col. 28, fol. 36^v).

[141] Ibid., p. 94 (col. 29, fol. 37^r, N.25).

[142] Ibid., p. 113 (col. 33, fol. 38^r, N.34).

[143] Ibid., p. 123 (col. 38, fol. 39^r, N.40).

[144] Ibid., p. 271 (col. 108, fol. 56^v, N.134).

apparent. To not be engrossed in his studies – for whatever reason – appears to have weighed heavily on Stensen.

During his studies in Leiden, Stensen would often dissect in the evenings or even at night, as then he was able to do so undisturbed; this prompted him to include the humorous comment in a letter to Thomas Bartholin on March 5, 1663 that his obser-vations on the brains of animals still had "much nocturnal quality" about them, and were riddled with gaps due to the regular study regimen.[145] These evening dissec-tions also attracted spectators since Stensen kept close professional ties with his peers.[146] One fellow student in Leiden for example, Dr. med. Sir Robert Sibbald (1641–1722), who became the first professor of medicine in Edinburgh in 1685, mentions in his autobiography how Stensen, "who became famous afterwards for his writings", had sometimes dissected in his, Sibbald's, study room.[147] One would certainly not be incorrect to assert that Stensen pursued his studies with far above average conscientiousness.

2.1.2.2   Empathy and Caution

In Amsterdam, as he reported to Thomas Bartholin on September 12, 1661, Stensen had conducted a 3 h vivisection of a dog in order to perform an experiment by the amateur anatomist Lodewijk de Bils (1624–1669).[148] The animal, Stensen claimed, could have survived the entire day under this torture; and as a single attempt would not suffice, he would repeat the procedure as soon as an opportunity presented itself in order to be able to draw reliable conclusions, though he had to admit it was not without horror that he tormented the creatures designated for these vivisections with such extended anguish. This leads Stensen to a pertinent question: "The Cartesians pride themselves keenly with the certainty of their philosophy; I wish they could convince me so definitely as they themselves are convinced that the animals possess no soul and that it makes no difference whether one touches, cuts or burns the nerves of a living animal or the strings of a machine driven by motion; for I would then examine the viscera and vessels of living creatures several hours longer and with more pleasure, as I see much that is worthy of investigation which seems oth-erwise impossible to hope for."[149] Here Stensen found himself in a conflict between

---

[145] VesPul., p. 136, lines 13–17, here 15 = E 11, p. 172, lines 7–11, here 9: "multum […] noctis". Cf. AvCun., p. 119, lines 34–35 = E 9, p. 163, line 13, where Stensen reports having opened the third of four rabbits for dissection at around 6 p.m.

[146] For details see Kardel/Maquet, *Biography and Original Papers*, p. 74.

[147] Cited according to Wright-St Clair, "Bedside", p. 445. On Sibbald: Charles W. J. Withers, "Sibbald, Sir Robert"; in: ODNB² 50 (2004), pp. 483–485, here 483 on Stensen.

[148] On De Bils: DDPhS, s.v. "Bils, Lodewijk (Louis) de", cols. 148–150, here 149; for details on De Bils' experiment see Maehle, *Tierversuch*, p. 87n300.

[149] VarOb., p. 57, lines 6–18, here 12–18 = E 3, p. 142, lines 12–24, here 18–24: "De philosophiae certitudine multum gloriantur Cartesiani; vellem, ita certo mihi persuaderent, ac ipsi sunt persuasi, nullam esse brutis animam nec differre, utrum bruti vivi nervos an automati, quod actu movetur, chordas tangas, disseces, uras; vivi enim animalis viscera et vasa aliquot horis saepius et libentius

his compassion and the necessities of research.[150] When Stensen published De Bils' experiment, which he had repeated on two more dogs in the presence of friends in Leiden, as a part of his response to Deusing, he abstained from this lament – presumably owing to the scientific imperative.[151] As a result of his own practical experience, however, he considered the Cartesian conception disproved.[152] In this situation, Stensen demonstrated caution and empathy in that he respected the animals' lives and did not wish to subject them to unnecessary suffering.

The caution which Stensen exercised in his research, as applied to humans, is clearly evidenced by his answer to a question by his students in Copenhagen pertaining to the experiments with blood transfusions conducted in England, France and Italy, namely "whether a blood transfusion can be done without risk? We do not yet know", Stensen replied, "the nature of blood itself and even less the distinguishing attributes of the varied nature of blood, and therefore a blood transfusion cannot be done without risk, for if the nature of the variants of blood to be unified is not known, one would likely submit the patient to a thousand dangers; but every direct injection into the blood is subject to danger as well as being foreign to the rational order of nature that has encased the blood especially well."[153] For Remacle Rome OSB (1893–1974), curator of the Museum of Geology at the Catholic University of Leuven,[154] this "judicious response" demonstrates – with a view to the foolhardy and sometimes reprehensible contemporary transfusion experiments advertised in the *Philosophical Transactions* of the *Royal Society* (est. 1660) published by Henry Oldenburg (c. 1618–1677)[155] – "a characteristic of Stensen's scientific spirit".[156] And indeed this wording by Stensen subsumes the scientific ethos oriented towards the patient's well-being that defined him. The *Exercitia academica* by his Danish

---

rimarer, cum multa videam inquirenda, quae alia ratione non licet exspectare." – That Stensen did not frequently engage in vivisection is evidenced by DucSal., p. 7, lines 22–23 = E 1, p. 137, lines 18–19, where he mentions that he would have conducted an experiment had he been more versed in the dissection of live animals.

[150] Maehle, *Tierversuch*, pp. 87–88. On vivisection in the early modern era and its contemporary evaluation see Bergdolt, *Gewissen*, pp. 253–255.

[151] HepRed., p. 64, lines 17–20.

[152] During his time in Florence, Stensen enunciates his respect for animals – be they 'useful' or 'harmful' – as creations of God in Spir. 5, p. 92, lines 35–36.

[153] ExAc., p. 292, lines 23–29: "Num tuto possit exerceri transfusio sanguinis? Necdum ipsius sanguinis naturam, multo minus signa diversae naturae sanguinis distinguentia cognoscimus, adeoque nec transfusionem sanguinis tuto exercere licet, cum, nisi nota fuerit jungendorum sanguinum natura, mille periculis aeger exponatur; sed & omnis immediata in sanguinem injectio periculo exposita est & a naturae instituto aliena, quae sanguinem adeo bene clausit."

[154] Later active on the newly founded campus of Louvain-la-Neuve; on Rome: De Deckker, "Rome", pp. 7–8, on his writings about Stensen ibid., pp. 8–9.

[155] On Oldenburg: Sarah Hutton, "Oldenburg, Henry"; in: DBPh. 2 (2000), pp. 617–618, here 617 on Stensen.

[156] Rome, "Sténon", p. 530: "une caractéristique de l'esprit scientifique de Sténon [...] réponse judicieuse". On the first blood transfusions beginning in 1665 and their dangers see Meli, *Mechanism, Experiment, Disease*, pp. 138–141, here 141, and Savio, "Riva", pp. 252–254 on the three blood transfusions (December 1667) by the surgeon Riva, whose academy meetings Stensen had attended in 1666 (see note 168, in Chap. 1 above).

student Holger Jakobsen (1650–1701) – who undertook many a dissection under Stensen's instruction in Copenhagen from 1672 to 1674 –, containing lecture notes taken during and after Stensen's sections as "Anatomicus regius",[157] attest to the latter being a commendable and innovative teacher.[158] That Stensen habitually spoke to his students while facing them over his dissection object is also confirmed by André Graindorge in 1665.

The four interrelated stances described above, which formed Stensen's 'fundamental ethos', constitute essential aspects – in the field of medical research – of the "dimension of responsible scientific practice" called for by Nida-Rümelin. They will be revisited in the course of this study as applicable to concrete situations in Stensen's life.

2.1.3 The Human Brain as an Instrument of the Researcher

Stensen, who held his *Discours* on the anatomy of the brain – the significance of which was to extend far beyond the seventeenth century – in Paris in 1665, was unusually gifted. The letter written by the university senate to accompany his doctoral diploma, and stating that he had received the best possible grade for the final exam which he presumably passed in autumn 1663,[159] emphasizes that Stensen had repeatedly proven "with the highest distinction" his "extraordinary learning" in public disputations and his publications.[160] At Stensen's request – brought forward by Sylvius – the Leiden University senate, after hearing the medical faculty's verdict and recommendations by other professors, decided to exempt Stensen from the as yet outstanding oral defense of his doctoral thesis in Leiden on the grounds that his case was "exceptional", and mailed his diploma to him in Paris in 1664 in absentia. The reason for this was that Stensen, due to his journeys to Copenhagen and, having been passed over in the appointment of new chairs at the University of Copenhagen, subsequently to Paris, was unavailable owing to "serious reasons".[161] The handling of Stensen's case therefore represented an extremely rare exception.

---

[157] ExAc., pp. 287–310 with ibid. (Notes) preliminary remarks, pp. 347–348; supplementarily to ExAc., pp. 297–310 cf. Add. 12, pp. 927–928. A complete edition of Holger Jakobsen's notes has hitherto not been published. On Jakobsen: Gordon Norrie, "Jacobæus, Holger"; in: DBL³ 7 (1981), pp. 209–210.

[158] Cf. the compilation in Scherz, *Pionier der Wissenschaft*, pp. 258–271.

[159] E 17, p. 183, lines 7–11; Sobiech, *Herz, Gott, Kreuz*, p. 35. – The doctoral diploma is not preserved.

[160] E 17, p. 183, lines 11–13: "maxima cum laude [...] eruditionem singularem".

[161] Ibid., lines 13–20, here 14, 19: "graves causas".

2.1.3.1   Between Intellectual Heights and Self-Ignorance

In his *De sapientia veterum liber* (Book About the Wisdom of the Ancients; London 1609), Francis Bacon had underlined the preeminent meaning of the legend of the theft of fire by Prometheus – to which Stensen made reference as a student – as being "that Man, if one looks to final causes, may be regarded as the center of the world; insomuch that if man were taken away from reality, the rest would seem to roam about and wave up and down without intention, and would not head for a goal like a besom without a binding, as the saying is. For all comes to man's aid; and it is man whoderives use and fruit from all particular things. Certainly even the revolutions and courses of the stars serve both for the differences of the seasons and the division of the areas of the world, and the meteors for signsof weather conditions, and the winds now for sailing and now for the mills and machines; and plants and animals of any species are measured on the basis either of the abodes of man and his places of refuge or hisclothing or his subsistence or medicine, or lessening his labors, or finally his entertainment and his comfort: insomuch that all things seem not to run their own course but that of Man."[162]

In the proem of his anatomical observations on the glands of the mouth, published 1662 in Leiden, Stensen declares it worthy of admiration that divine foreordination had provided the human mind with the ability to envision, at any time one desired, images of things one had previously perceived through the senses, and to realistically view absent things in all details through such envisioning as though they were present. He cites the example of an aged, but still spry natural scientist who had spent his life conducting experiments and who made one wonder how he, within the sphere of his cranium – being as it were of such small size – could traverse the immeasurable universe, expanded to nearly boundless spaces, and all its celestial objects in his mind: "Here [= in the cranium], climbing up to the stars, he will narrate to us the unchanging order of the fixed stars, the undeceiving aberrations of the planets [ie their orbits] and the wanderings of the comets which follow no laws; returning from there in an instant he will hasten through the air and describe the pleasing multitudes of colors, the amazing shapes of the firebrands in those regions in which they appear at intervals; from there descending back to Earth he will explain the meticulously devised works of nature and the no less artful replications of them to be found there: finally he will penetrate into the core of the Earth

---

[162] Bacon, De sapientia veterum, pp. 94–95 = Spedding/Ellis/Heath, *Works of Francis Bacon* (chapter "Prometheus, sive status hominis"), pp. 670–671: "quod homo veluti centrum mundi sit, quatenus ad causas finales; adeo ut sublato e rebus homine, reliqua vagari sine proposito videantur et fluctuari, atque quod aiunt scopae dissolutae esse, nec finem petere. Omnia enim subserviunt homini, isque usum et fructum ex singulis elicit et capit. Etenim astrorum conversiones et periodi et ad disctinctiones temporum et ad plagarum mundi distributionem faciunt; et meteora ad praesagia tempestatum; et venti tum ad navigandum, tum ad molas et machinas; et plantae atque animalia cujuscunque generis, aut ad domicilia hominis et latebras, aut ad vestes, aut ad victum, aut ad medicinam, aut ad levandos labores, aut denique ad delectationem et solatium referuntur: adeo ut omnia prorsus non suam rem agere videantur, sed hominis."

and reveal the hidden secrets of the minerals. He lets all of these images obey his own will, as if the macrocosm were enclosed in the microcosm."[163]

Thus Stensen, who had delved into astronomy from a technical point of view during his studies in Copenhagen,[164] drafts a similar notion of the natural sciences to that of the philosopher of science Bacon, whose works he had read as a student in 1659, although Stensen is able to formulate in a psychologically more delicate fashion. One may also notice that Stensen describes not man as the center of the world, but rather man's ability to mentally appropriate his environment in order to conduct research. On February 3, 1673 st.v., he spoke as "Anatomicus regius" in the course of his public dissections about how it was beyond the greatest possible admiration that man could see such distant bodies as the stars with the help of his eyes.[165] The image of the aged natural scientist represents the great fascination which the process of human thought provided for Stensen, and furthermore appears to mirror Stensen's self-reflection.

The pioneer of neurophysiology Dr. med. Marie-Jean-Pierre Flourens (1794–1867), who taught as a professor at the Parisian *Collège de France* and *Muséum national d'histoire naturelle*,[166] wrote the following in the January edition of the Milanese *Annali universali di medicina*: "The great anatomist Stensen, late bishop and vicar apostolic of the Pope, ingeniously said that the soul, which knows the outside world so well, and everything that lies outside of itself, once it has entered its own house no longer knows where it lives."[167] Flourens was alluding to the proem to the Parisian *Discours*, in which Stensen had pointed out the ignorance of thinking and researching man with regard to the structure of his own brain, the "most important organ of our soul". For the soul, Stensen thought, believed to have penetrated – with the help of the brain with which it accomplished admirable things – everything around it so deeply that nothing in the world could limit its insight. "However", he added, "once it has returned into its own house, it is incapable of describing it and cannot recognize itself there."[168] According to Manlio Iofrida, Stensen's empiricism

---

[163] GlandOr., p. 13, lines 1–21, here 13–21: "Hic ad astra evolans constantem fixorum ordinem, fallere nescios planetarum errores, omni lege carentes cometarum excursus nobis evolvet; inde momento relapsus aërem pervagabitur, & jucundas colorum varietates, stupendas ignium formas in illis regionibus se per intervalla ostendentes depinget; hinc in terram descendens, varia, quae se ibi offerunt, accuratissime elaborata Naturae opera, & illis vix cedentia artis imitamina exponet: tandem in terrae viscera penetrabit, & abdita mineralium mysteria revelabit. Has ille ideas omnes nutui suo habet obsequentes, ac si macrocosmus microcosmo lateret inclusus."

[164] Scherz, *Anatomy of the Brain*, pp. 77–78; Kardel/Maquet, *Biography and Original Papers*, pp. 52–54; Ziggelaar, *Chaos-Manuscript*, pp. 474–475.

[165] ExAc., p. 304, lines 38–41.

[166] On Flourens: Stéphane Le Tourneur, "Flourens (Marie-Jean-Pierre)"; in: DBF 14 (1979), cols. 140–142.

[167] Flourens, "Cervello", p. 173: "Il grande anatomico *Stenone*, morto vescovo e vicario apostolico del Papa, diceva con molto spirito 'che l'anima, che conosce sì bene il mondo esterno, e tutto ciò che è fuori di essa, rientrata che sia nella propria casa, non sa più dove abita'."

[168] Discours, p. 4, lines 2–7, here 2–3, 6–7: "le principal organe de nostre ame […] cependant, quand elle est rentrée dans sa propre maison, elle ne la sçauroit décrire, & ne s'y connoist plus elle-mesme."

here connects with the topic of the "intransparency of the mind to itself" typical for philosophy in the latter half of the seventeenth century.[169]

Stensen was also aware that the human senses could deceive: if the human brain, as he wrote while studying in Leiden, wished to concentrate on a small detail of the section, it found this difficult to do, for the human mind sought entertainment in multiplicity and was therefore powerless precisely in its strength, so that it had difficulty reaching a decision and could not free itself of ancillary thoughts to a degree that it might be consistently free for "the one thing alone". The reason for this was that the multitude of delicate parts connected into the totality of the human body was so great that it deceived even the most attentive observer and played games with him.[170] Only an undistracted mind that knew what could and had to be considered, Stensen stated in Copenhagen in 1664, would be free for what was to be deliberated; thus every substantial section, given the delicacy and multifariousness of the object of examination which made even the slightest observation important, required undivided and circumspect attention so that a vagrant mind propelled by swirling thoughts would not distract the all too swift eyes from their current examination, thereby becoming susceptible to error.[171]

In his Copenhagen inaugural address, Stensen indicates a spiritual solution to this dilemma by placing the aforementioned "one thing alone" in the greater context of the supernatural ultimate goal of man: the senses' task is not to portray things the way they are, nor to render judgment about them – as stated by Descartes –[172]; their function instead is to provide the mind with those requirements of things for examination sufficient to obtain a knowledge of things which enables man to achieve his goal.[173] The purpose of sensory perception is to reach the goal set for man by God.

2.1.3.2 Precipitance and Pride as Obstacles

In his letter dedicated to Barbette, Stensen warned that any anatomist who practiced dissections even only irregularly would not deny occasionally having persuaded himself through a precipitous judgment or an overeager exclamation of "Eureka!" (εὕρηκα) when observing something that at first glance seemed unfamiliar, and either confirmed a preconceived opinion or prompted a new notion, that he had seen something which in truth he had *not* seen, and taken no time for a closer examination. Therefore, Stensen added, the "utmost care" (maxima cautio) was required so

---

[169] Iofrida, "Stensen", p. 893.

[170] GlandOr., p. 13, line 21–p. 14, line 2, here p. 13, lines 27–28: "uni soli".

[171] VitTrans., p. 212, lines 15–24.

[172] Koch, "Dänemark", p. 1254; cf. Kardel, *Steno*, p. 77.

[173] Prooemium, p. 254, lines 3–8.

as not to be deceived by such initial appearances.[174] Van Rensselaer Potter, who identified his own term "bioethics" as the result of a 'eureka' moment,[175] highlighted the "inherent possibility of error" during such experiences as an example concretizing his concept of "humility with responsibility".[176] It was for this reason, Stensen claimed, that many a book purported outrageous anatomical propositions which other researchers could not retrace in their dissections no matter how hard they tried – because they simply did not exist.[177] Robert K. Merton points to Stensen's discovery of the pineal gland in animals, which disproved Descartes' assertion that it existed only in humans, the latter having prematurely provided an explanation for something he had not yet verified.[178] Descartes had delineated his brain theory in his work *De homine* [...] (Treatise on Man [...]; Leiden 1662), written in 1632/1633 and published posthumously by Dr. phil. Florentius Schuyl (1619–1669),[179] professor of philosophy at the Athenaeum in's-Hertogenbosch. Stensen, who was independently conducting dissections of animal brains at the time, dryly noted in his letter to Thomas Bartholin from Leiden on August 26, 1662 with regard to Descartes' recently published work that in it "one can observe not inelegant illustrations which surely arose from a brilliant brain [ie that of Schuyl]; whether they can be found inside any brain, however, I rather doubt."[180]

Stensen's teacher Sylvius spoke out similarly, saying that no one possessed the privilege to pass off any fabrication emanating from their own mind as the truth without having subjected it to sufficient examination by unbiased sensory percep-

---

[174] VitTrans., p. 212, lines 2–6, 11–14, 24–27 with citation in lines 24–25. In VesPul., p. 134, lines 10–22 = E 11, p. 170, lines 7–19 Stensen writes in reference to his investigations of the chyliferous vessels that descriptions by various authors, who all claimed experience as their witness but nevertheless did not sufficiently reveal their objects to the reader, cast doubt on the credibility of the experiment, and that the reader eventually assumed to have seen that which he had initially observed with the "eyes of the mind" (mentis oculi), ie that which was described in the publication, likewise with the "eyes of the body" (corporis oculi), namely as putatively commensurate with reality.

[175] Ten Have, "Bioethics", p. 61.

[176] Potter, "Humility", p. 2297; in more detail see ibid., pp. 2302–2304.

[177] For more details see VitTrans., p. 212, lines 15–18.

[178] Merton uses this to exemplify that before "social facts" can be "'explained'" it must be certain that they actually exist (Merton, "Notes", p. xiii). On Stensen's discovery of the pineal gland in animals: VesPul., p. 136, lines 3–12 = E 11, p. 171, line 35–p. 172, line 6.

[179] Descartes, *De homine*. Schuyl attained the additional degree of Dr. med. at Leiden University in 1664 and was appointed Professor of Medicine there in the same year; on Schuyl: DDPhS, s.v. "Schuyl, Florentius", cols. 1791–1794, here 1791–1792.

[180] AvCun., p. 120, lines 13–17, here 13, 15–17 = E 9, p. 163, lines 28–33, here 28, 31–32: "hisce diebus [...] figurae conspiciuntur non inelegantes, quas ex ingenioso cerebro prodiisse certum est; an vero tales in ullo cerebro conspiciendae, valde dubitarem." Cf. the similar comment in AProd., p. 151, lines 33–35 on Blasius' nonexistent duct.

tion or reliable consideration by reason. Sylvius, like Stensen, was a heuristic Cartesian.[181]

Stensen observed how discoveries were made without plan: it often happened, he writes to Heinrich Meibom Jr. on April 5, 1673 from Copenhagen, that God left one in one's original ignorance when one searched for something with the utmost fervor, whereas many a time when one was not searching, or even occupying oneself with something else, rich insights beyond all expectations would manifest themselves.[182] Robert K. Merton later described this phenomenon as the "serendipity pattern".[183] This did not occur, Stensen goes on, because God detested the researcher's diligence, but instead in order to prevent pride in the discoverer. He had experienced both himself: that keenest efforts remained unsuccessful as well as that discoveries of exceeding significance fell to him even though he was practically idle and would hardly have even dreamed of them.[184]

Stensen later elucidated the "pride" of the researcher mentioned here, which according to him caused one to love what one believed oneself to be,[185] in a theological manner against the background of the doctrine of original sin. Already in the *Chaos* manuscript, on June 23, 1659 st.v., he had excerpted from a letter by the Parisian physician Samuel Sorbière (1615–1670),[186] who had converted to Catholicism in 1653, to his colleague Dr. med. Jean Pecquet (1622–1674),[187] printed in Pecquet's *Experimenta nova anatomica [...]* (New Anatomical Experiments [...]; 2nd ed., Paris 1654), that one should not brag about the award of a doctorate and the

---

[181] Cook, "Low Countries", pp. 125–126; Theo Verbeek, "Descartes, René"; in: DDPh. 1 (2003), pp. 254–260, here 259 l. col. – Lester S. King (1908–2002) states dryly in King, *Medical Enlightenment*, pp. 93–98 that Sylvius, primarily oriented towards clincial practice and, in the sense of an extreme skepticism, ill-disposed towards the teachings of any medical authority as well as – at least in terms of pretension – towards his own, had not thought in any detail about the process of sensorial perception, nor about criteria of truth or a scientific method; the lack of evidence and the dogmatic phrasing of Sylvius' medical assertions conflicted with his own premises. On King: Charles G. Roland, "Lester Snow King (1908–2002)"; in: JHMAS 58 (2003), pp. 362–366. Stensen, by contrast, in 1667 in Florence (Myol., p. 103, lines 8–16, here 9, 15 = Kardel, *Steno on Muscles*, p. 212, lines 6–17, here 7, 14–15) described Sylvius as "my very famous teacher Sylvius" (celeberrimus praeceptor meus Sylvius) who, so as not to create the pretense of valuing his personal honor more highly than the public good, had impressed upon his students daily that he could not achieve everything, and that therefore in those things which he believed he had not yet fully comprehended himself he presented his notions merely as "opinions and assumptions" (opiniones suspicionesque), and had encouraged his students to undertake their own examinations by offering them the material.

[182] Bruun, "Fem nyfundne Niels Stensen-breve", p. 150, lines 20–23.

[183] Merton, *Social Theory*, pp. 157–162, here 159, according to which this pattern contains "the unanticipated, anomalous and strategic datum which exerts pressure upon the investigator for a new direction of inquiry which extends theory".

[184] Bruun, "Fem nyfundne Niels Stensen-breve", p. 150, lines 23–26.

[185] For details see Sobiech, *Herz, Gott, Kreuz*, pp. 129–138.

[186] On Sorbière: Franck Lessay, "Sorbière, Samuel"; in: DFPh. 2 (2008), pp. 1186–1190.

[187] On Pecquet: Sylvie Taussig, "Pecquet, Jean"; in: DFPh. 2 (2008), pp. 969–971. Stensen presumably made Pecquet's acquaintance in Paris in 1664.

privileges of the academies.[188] In another excerpt created on June 28, 1659 st.v., Stensen underlined a sentence stating that everyone wished to become famous by telling others something new.[189] In the proem to his Parisian *Discours*, he argued that if one was to describe in detail the interconnection of the nerves in the brain, one would arrive at a point at which one would have to admit one's ignorance if one did not wish to increase the ranks of those who would rather be admired by the public than be honest.[190] Finally, in one of his last anatomical publications released in the year of his ordination as priest, Stensen writes that the illusion of honor in regard to one's own discovery amounts to vanity.[191]

2.2 Medicine, Patient and Creator

Stensen's ethos – in particular following his conversion – exhibits a 'horizontal' and a 'vertical' dimension: the researcher in the 'horizontal' network of the 'republic of physicians' and in his 'vertical' relationship with God, the Creator. There are similarities to be found with the physician, in whose 'horizontal' dimension the relationship with the patient comes to the fore.

2.2.1 Physician-Patient-Relationship

The precise and detailed pharmaceutical and practical medical notes in the *Chaos* manuscript show that in 1659, Stensen was still studying with a view to becoming a practicing physician.[192] Although he never was a medical practitioner, he

---

[188] Ziggelaar, *Chaos-Manuscript*, p. 364 (col. 153, fol. 68ʳ) and Pecquet, *Experimenta*, pp. 164–180 (letter by Sorbière to Pecquet, Paris, 08/13/1654), here 172.

[189] Ziggelaar, *Chaos-Manuscript*, p. 436 (col. 179, fol. 74ᵛ). This is Stensen's excerpt or his copy of an excerpt by Ole Borch of Gassendi, *Animadversiones in decimum librum Diogenis Laertii*, a philological reconstruction of the atomistic philosophy of Epicurus by Dr. theol. Pierre Gassendi (1592–1655); on Gassendi: Sylvie Taussig, "Gassendi, Pierre"; in: DFPh. 1 (2008), pp. 526–533, here 529.

[190] Discours, p. 4, lines 18–22. Otherwise, Stensen writes, one deceived first oneself and subsequently others (ibid., p. 6, lines 13–15).

[191] OvaViv. II, p. 178, line 30. Stensen's criticism (Op. 9, p. 505, lines 6–8) of his own earlier joy as a researcher at making a name for oneself with one's anatomical publications shows that he was by no means insusceptible to the honor which a discovery brought with it; it should be taken into consideration that his sensitive character caused him to be more severe in his judgment concerning himself.

[192] Kardel/Maquet, *Biography and Original Papers*, pp. 48–49; advice on memorizing medications and their properties in Ziggelaar, *Chaos-Manuscript*, p. 326 (col. 135, fol. 63ᵛ, N.212). Notes eg on the awakening of hope in patients presumably trace back to Ole Borch as a practicing physician (ibid., pp. 315–316 [col. 128, fol. 61ᵛ, N.206] & p. 377n7).

nevertheless regularly expressed himself with regard to practiced medicine as the
area of application of his research from his study days on.

2.2.1.1 Criticism of Contemporary Medical Craft

Already as a student Stensen was aware of the deep inadequacy and limitations of
the medicine of his time resulting from a want of fundamental research. On June 23,
1659 he excerpted from a letter by the astronomer Adrien Auzout (1622–1691)[193] to
Jean Pecquet, printed in Pecquet's aforementioned *Experimenta nova anatomica*,
that once the correct medical theory had been found, previously hidden causes for
diseases could subsequently sometimes be discovered, and diagnosis and medication therefor occur on a secure basis.[194] On the same day, Stensen twice excerpted
with emphasis from the letter to Pecquet by Samuel Sorbière that metaphors were
to be rejected in medicine; not the meaning of words, but the "things" were to be
examined.[195] Already on March 11, 1659 st.v. he had excerpted from *Historiarum,
et observationum medicophysicarum centuriae IV [...]* (Four Groups of a Hundred
Medico-Physical Histories and Observations [...]; Paris 1656) by the Parisian physician Dr. med. Pierre Borel (c. 1620–1671)[196] the "invectives" against French medical practitioners who, with their applications of ice and pompous Latin lingo, were
leading many people down to "Orcus".[197]

Finally, on the previously mentioned 23rd of June 1659 st.v. Stensen excerpted
the following from Sorbière's letter to Pecquet: "Definition of medicine: In considering this, I do fear that someone might define medicine as 1. the art of babbling
buffoonery at the patient all day long with a grave eyebrow and applying uncertain
remedies so that his heart's grief might be somewhat assuaged, and so one calmly
awaits the restoration of good health, in which nature plays its part, or death by way
of fate running its course. Or one might say this in reference to [medical] practice:
2. a source for asserting foolish things and stabbing at the air with sesquipedalians
that sound like something big although they mean nothing; the boldness to prescribe
very unsafe medications as though they were highly effective; the audacity to gloat
over the boldness that caused no harm; fending off reproval of harmful incidents and
especially the skill of attributing [such incidents] to others. With such great insis-

---

[193] On Auzout: Philippe Hamou, "Auzout, Adrien"; in: DFPh. 1 (2008), pp. 62–64.

[194] Ziggelaar, *Chaos-Manuscript*, pp. 361–362 (col. 152, fol. 67ᵛ) and Pecquet, *Experimenta*,
pp. 155–163 (letter by Auzout to Pecquet, Paris, 03/01/1651), here 160.

[195] Ziggelaar, *Chaos-Manuscript*, pp. 362–363 (col. 153, fol. 68ʳ), here 363: "Res [...] res" and
Pecquet, *Experimenta*, pp. 164–180 (letter by Sorbière to Pecquet, Paris, 08/13/1654), here
167–168.

[196] On Borel: Philippe Hamou, "Borel, Pierre"; in: DFPh. 1 (2008), pp. 173–176.

[197] Ziggelaar, *Chaos-Manuscript*, p. 52 (col. 13, fol. 31ʳ): "invectivam [...] orcum" and Borel,
Historiarum (Centuria III, De glacie. Observatio XI), pp. 203–205, here 203, where Borel com-
plains of the pain agonizing him when he sees that Nature, originally an aide to medicine, had
become its enemy and that it was certain that the number of patients cured by physicians in France
did not approach the number of patients led down to Orcus by them with pompous Latin lingo.

tence is such great buffoonery put forward."[198] Stensen considered this criticism a fitting analysis of his times and an incentive for science.

In keeping with this assessment, Stensen wrote to Malpighi from Vienna on October 27, 1669 with regard to the latter's recurring fever that he wished for God to return him to full health "as we find our art so lacking that little can be expected from it."[199] The theory of so-called Antiperistasis, dating back to Aristotle (384 B.C.– 322 B.C.), was used at the time for example to explain fever in cases of pneumonia; Stensen disproved this theory in 1671.[200]

Stensen had also lamented contemporary medicine in his Myology: "How well would our ancestors have taken care of us, how well of all humanity, if only those who spent their entire lives engaged in anatomical exercises had bequeathed to posterity solely what is certain. Our knowledge would be less broad, but also less dangerous: and if medicine, based on those certain principles, could not eliminate a sick person's suffering, it would not add any more. We possess only very extensive volumes on anatomy and medicine: yet we drag the poor soul through a thousand miseries, through a thousand torments we guide them in the direction of an albeit dry [= non-bloody] death; and, which is our greatest squalor, often when we believe to be helping we are in fact doing the most harm. While I, by lamenting the daily misery, present the source of it, I myself do not promise a cure; I hope it can be found in time by others. And why should it not be allowed to hope for great things, if anatomy were guided back to where experience could find its joy only in certainty, and reason in that which is proven, that is, if anatomy were to swear an oath to [follow] the words of mathematics?"[201] With the term "non-bloody death", a

---

[198] Ziggelaar, *Chaos-Manuscript*, pp. 363–364 (col. 153, fol. 68ʳ, N′.222): "*Definit[io] med[icinae]* haec animadvertens verebor ne quis definiret medicinam. 1. Artem nugarum gravi supercilio coram aegroto effutiendarum, et remediorum incertorum adhibendorum, ut animi moeror aliquantulum deliniatur, et placide bonae valetudinis restauratio natura partes suas agente, vel mors fatis properantibus exspectetur. vel ratione praxeos diceret. 2. Fontem proferendorum deliramentorum, ventilandorumque verborum sesquipedalium ut magnum quid sonantium, cum nihil plane significent; temeritatem remediorum incertissimorum, tanquam praesentissima essent, praescribendorum. Superbiendi ex non infausta temeritate audaciam. declinandae ex infaustis eventibus vituperationis, atque aliis authoribus imputandae solertiam. tam magno nisu tam magnae nugae proferuntur." On this, see Pecquet, *Experimenta*, pp. 164–180 (letter by Sorbière to Pecquet, Paris, 08/13/1654), here 171–172.

[199] E 45, p. 212, lines 3–8, here 7–8 = Adelmann, *Correspondence*, vol. 1, no. 212, p. 430: "già che noi troviamo tanto imperfetta la nostra arte che poco da essa si può sperare."

[200] More information in Sobiech, "Experience", p. 182.

[201] Myol., p. 64, lines 21–34 = Kardel, *Steno on Muscles*, p. 84, line 24 & p. 86, lines 1–24: "Quam bene nobis, quam bene toti humano generi consuluissent majores nostri, si, qui totam aetatem in exercitiis anatomicis contrivere, non nisi sola certa posteritati tradidissent. Minus ampla esset cognitio nostra, sed & minus periculosa: & si certis hisce principiis innixa medicina dolores aegris non tolleret, non adderet illis novos. Modo vastissima habemus anatomes & medicinae volumina: nihilominus inter mille cruciatus miseram animam trahimus, per mille tormenta ad mortem etiam siccam tendimus; &, quae summa nostra infelicitas est, saepe cum prodesse credimus, tum demum maxime nocemus. Dum communem miseriam deplorando causam ejus expono, remedium a me nullum promitto, ab aliis cum tempore inveniendum spero. Et quidni magna sperare liceret, si eo

quote from the Roman satirist D. Iunius Juvenal (c. 60–c. 127),[202] Stensen expresses his opinion that it was routine for the medicine of his time that medical treatment, though not a bloody execution by the executioner, often represented – despite being administered with the best intentions – a non-bloody execution by the physician. That these are the words of a highly gifted man analyzing the state of his times is shown by the fact that the theory of muscle contraction expounded in Stensen's Myology, which was generally considered his scientifically weakest work until well into the twentieth century, was eventually verified as late as the early 1990s.[203]

2.2.1.2   The 'Dialogic Triangle'

Even though Stensen had never been a practicing physician, he had nevertheless observed during his research trips to hospitals how the patient, who was "in the hands of the medical practitioners",[204] was dependent on their ability. The "hands" played a significant role in Stensen's spirituality – as the hands of the humans through which Christ acted on the one hand, and as the hands of God on the other.[205] He only put these observations down in writing, however, in his pastoral theological treatise *Parochorum hoc age*. In it he likens the role of the parish priest to that of the "chief physician" (praefectus medicus) in a hospital, whose duty it was to know the condition of each individual patient in order to mend their diseases; in view of the countless different afflictions and cures, he had to examine each patient thoroughly in regard to the signs of an illness, its causes and the appropriate remedies.[206] In religious conversations, Stensen writes as bishop in a record dating from his time in Hanover, one should deal with one's interlocutors like a physician with his patients, meaning: one should not talk about everything at once, not become angry at their impatience, but carefully examine various things while repeatedly saying prayer and strive to clarify the issue in a dialog strictly confined to what was necessary – ie the diagnosis to be tested on the patient – and without any digression.[207] This statement hints at Stensen's criticism of contemporary medicine that it missed the heart of the matter, namely the patient's well-being, in its use of a great many learned words.

---

reduceretur anatome, ut in solis certis experientia, in solis demonstratis ratio acquiesceret, id est, in matheseos verba anatome juraret?"

[202] Adamietz, *Juvenal*, p. 210 (Satura X,113): "sicca morte".

[203] By computer calculation; see details in Kardel, *Steno on Muscles*, pp. 51–55.

[204] Letter to Bartolomeo Hortensio Mauro (1634–1725), secretary to Ferdinand II, on 12/04/1680 about the prince-bishop's illness in E 209, p. 456, line 9: "tra le mani de' medici". On Mauro: Hugo Thielen, "Mauro, Bartolomeo Ortensio (Hortensio)"; in: HBL, p. 247.

[205] See details in Sobiech, *Herz, Gott, Kreuz*, pp. 50, 173, 182–185.

[206] ParHAg., p. 28, lines 12–13, 16 (= Altoviti, *Obbligo* [Cap. XVIII], p. 31, line 34–p. 32, lines 1, 5 = Attavanti, *Obbligo* [Cap. XVIII], p. 24, lines 4–6, 10); ibid., p. 32, lines 8–17 (= Altoviti, *Obbligo* [Cap. XXIII], p. 37, line 24–p. 38, line 3 = Attavanti, *Obbligo* [Cap. XXIII], p. 29, lines 17–31).

[207] Op. 9, p. 484, lines 20–29.

To be found in the notes by the suffragan bishop Stensen on visiting the sick are various reflections which may recall the bedside teaching by Sylvius which he underwent in the *St. Caecilia Gasthuis* in Leiden: in order to prescribe the appropriate diet and medication for a patient, Stensen writes, one had to know his condition.[208] One should further support him eg with prayers and encouragement by people who have influence on him by way of reputation or favor, and in addition a priest should be brought to the patient to ask him about his illness and "suffer vicariously with him" so that the patient might obtain God's grace to confess.[209] If one appreciated the patient's own opinion and that of the people around him, the uncertainty of the situation and other things simultaneously, one's mind would become scattered by the manifold impressions; but if one instead imagined Jesus as a sufferer or Jesus as a physician, one could make better use of one's intellect by virtue of being strengthened by simple-mindedness, ie concentration on the essential.[210] Stensen subsequently brings up the point that one should not overestimate oneself in one's wise calculations, in the prescribing of medication and in medical advice, as many things were not dependent on one's own will; rather one would recognize from the circumstances presenting themselves, and sent by God, what was to be done or avoided – if one thought about God, who was present, with a "simple eye" (simplex oculus) (Matt 6:22 [Vul.]; Luke 11:34 [Vul.]) and asked him for the solution to the problem.[211] It can be presumed that Stensen, while making house calls eg as parish priest of the collegiate church of St. Ludgeri in Münster in 1680/1681, did not act only pastorally but would also apply his medical knowledge in cases of illness.

His records suggest a sort of 'dialogic triangle' featuring the relationship planes physician-patient, physician-God, and patient-God, ie one 'horizontal' and two 'vertical' planes. If one projects them onto the situation of a purely medical examination, they express what the philosopher Eduard Zwierlein identifies as "reverential self-moderation" in light of the "limit of comprehension" of the unknowable within the patient and the reason why "humility" constitutes a category of the physician's ethos.[212]

By emphasizing the interpersonal and spiritual dimension for the establishment of a diagnosis, Stensen addresses a notion appealed for by the American neurologist George L. Engel (1913–1999), namely that the classic seventeenth century biomedical paradigm prevalent in clinical practice to this day with its "reductionist, mechanistic statements" in the spirit of René Descartes and Sir Isaac Newton (1643–1727) should be supplemented with a paradigm following Albert Einstein (1879–1955) and Werner Karl Heisenberg (1901–1976) which collects additional information

---

[208] Op. 3, p. 424, lines 8–9.

[209] Op. 1, p. 408, lines 28–31, here 31: "compatitur".

[210] Op. 9, p. 492, lines 9–12. On the "simple-mindedness" of the senses "in Jesus" resp. on the theological tradition harkening back to Johannes Tauler OP (after 1300–1361), which was formative for Stensen's later records, cf. Sobiech, *Herz, Gott, Kreuz*, pp. 123, 167–168.

[211] More details in Op. 9, p. 492, lines 14–20. According to Beda Venerabilis (c. 672–735) the "simplex oculus" means security, cf. Schleusener-Eichholz, *Auge*, vol. 2, pp. 691–693, here 692.

[212] Zwierlein, *Begegnung und Verantwortung*, pp. 64–65.

attainable only through "interpersonal exchange" in conversation with the patient. Only if both of these paradigms were meshed together and complemented one another would situationally adequate scientific medical action and reliable diagnoses become possible.[213] This was a characteristic of the scientific and simultaneously "caring (merciful, pastoral) role of the physician."[214]

2.2.2   On the Metaphysics of Anatomy

The practice of anatomy in the early modern period occurred in a safe zone characterized by sacralization. Particular importance was attached to distancing the handling of the corpses to be dissected from common everyday activities in order to promote appropriate behavior by the anatomist and, in case of public dissections, by the audience. Stensen was introduced to this tradition at the beginning of his studies in 1656.

2.2.2.1   The Hand of the Anatomist and the Tradition of the "Haec Sacra"

The first volume of the *Acta Hafniensia*, which features reports on the years 1671/1672, is preceded by a prayer to the "Triune God" in which Thomas Bartholin praises Him as "the only creator of nature and medicine" who "supplies His physicians with beneficial remedies" and "impels the better talents and encourages them to fruition" – by which he means medical science. As for the goal of anatomical research, Bartholin informs the reader that "the experiments with nature and the strict examinations of our works are owed to only God alone".[215] From such a metaphysical perspective, science becomes a divine service.

This is also reflected in the vocabulary used: for example, in his letter to Stensen on February 14, 1662 st.v., Bartholin mentions the "haec sacra" (*literally*: "these holies"), ie the dissections which Stensen was engaging in.[216] In another letter to Stensen on July 25, 1663 st.v., he writes of the "inquisitive hand of the scholars", ie the anatomists, whose results we, meaning the members of the 'republic of physicians', eagerly awaited.[217] The hitherto disregarded term "haec sacra" refers demon-

---

[213] By drawing on concrete cases experienced by George L. Engel: Engel, "Wissenschaft", esp. pp. 5–7, 10–11. On Engel: A. Scott Dowling, "George Engel, M.D. (1913–1999)"; in: AJP 162 (2005), p. 2039.

[214] Engel, "Wissenschaft", p. 8, r. col.

[215] AMPH 1 (1671/1672), prayer (unpaginated), here first page, lines 1, 12–13, 15, 20–21: "Deo TRIUNI/[…]/Solus naturae omnis autor/& medicinae/[…]/Salutaria remedia medicis suis suggerit/[…]/Ingenia meliora excitat/Ad frugem animat" and second page, lines 11–14: "Cui soli Deo/ Experimenta naturae debentur/Et/Actorum nostrorum censurae".

[216] E 6, p. 150, line 21.

[217] E 14, p. 181, lines 19–21, here 20: "curiosa doctorum manus".

stratively and deictically to something beyond one's self; it can also be found in an essay by Bartholin in which he underscores the necessity of the previously enacted Danish Medicinal Ordinance of 1672. This ordinance will be revisited in the following. Bartholin writes that one could regularly meet midwives who plunged "in haec sacra" (into these holies) – in this case the practice of midwifery – quasi with unwashed hands, that is without a medical degree.[218] This motif of 'unwashed' or 'unclean' hands, referring to the practice of medicine without any formal medical education, was a common theme in anatomy from the second half of the sixteenth century onward.[219] In the *Acta Hafniensia*, Bartholin describes how his son Caspar Bartholin Jr. (1655–1738), who studied medicine in Copenhagen under Stensen (among others) from 1672 to 1674 and received his doctorate in 1678,[220] reported to him from Copenhagen on February 15, 1674 on the dissection of a fox, during which Caspar had been "supported by the help of the most excellent anatomist Mr. Niels Stensen", stating that in the presence of Stensen – who had lent his "fine hand" to the success – "these holies" (haec sacra) had been undertaken.[221] The hand of the scientifically active anatomist thus acted in a sacralized safe room.

Sacralization of the public dissection was common until at least the middle of the eighteenth century,[222] with the anatomist usually stating the nature of the dissection and his expectations in regard to appropriate behavior by the audience in a leaflet announcing the procedure: Simon Paulli Jr. for example had betokened the "anatomical exercises" conducted in the course of the public sections as "holies [sacra] and secrets".[223] In his invitation to the dissection of a female corpse scheduled for Friday, February 28, 1673 st.v. in Helmstedt, Heinrich Meibom Jr. introduced anatomy as a "holy art".[224] Dr. med. Christoph Helwig Sr. (1642–1690),[225] a professor of medicine at the University of Greifswald since 1667, invited his students to the dissection of the genitals of a female corpse resulting from an execution,[226] to be conducted on Tuesday, April 3 1677 st.v. at 3 p.m. His program leaflet[227] as well as the entry badge, the so-called "Tessera anatomica" (anatomical badge), featured the

---

[218] AMPH 1 (1671/1672), pp. 286–292 (CXXXVIII. De obstetricis absentia querelae), here 290–291.

[219] Gadebusch Bondio, "Hand", pp. 98–99.

[220] On Bartholin: Valdemar Meisen, "Bartholin, Caspar"; in: DBL³ 1 (1979), pp. 472–474, here 473.

[221] AMPH 3 (1674/1675), pp. 32–35 (XXI. De caudae vulpinae odore violaceo, & ursi suctione. Caspari Bartholini Thom[ae] fil[ii]), here 32–33: "adjutus[…] opera summi anatomici d[omini] Nicolai Stenonii […] manum eru[33]ditam […] haec sacra".

[222] Bergmann, *Patient*, pp. 202, 206.

[223] Bartholin, *Anatomihuset*, p. 56, lines 17, 22: "sacris & mysteriis […] anatomica […] exercitia".

[224] Meibom, *Programma*, fol. A 4ʳ–A 4ᵛ, here A 4ʳ: "Sacrae Arti". The program has hitherto not been analyzed.

[225] On Helwig: GreifswaldK, s.v. "Christoph Helwig", pp. 94–95.

[226] Helwig, *B[enevolo] l[ectori] s[alutem]*, fol. 2ᵛ. On this, see Thümmel, *Greifswald*, p. 228 (without mention of the printed program). The program has hitherto not been analyzed.

[227] Helwig, *B[enevolo] l[ectori] s[alutem]*, fol. 2ʳ.

Fig. 2.3 "Tessera
anatomica" for the public
dissection by Christoph
Helwig Sr. on April 3,
1677 st.v. from 3 p.m. at
the University of
Greifswald
Size: c. 6 cm (height) ×
8 cm (width)
UB Greifswald, 520/Va 375
adn21, entry badge (printed
recto), bound between fol. 1ᵛ
& 2ʳ

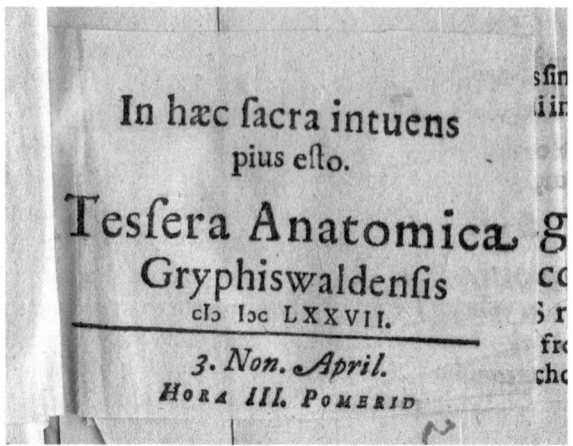

"haec sacra" motif. The anatomical badge even integrates it into a 'final monition'
to the students prior to the start of the proceedings in the motto "While observing
these holies [haec sacra] attentively, be pious" (Fig. 2.3).[228]

2.2.2.2 Technique "in the Hand of God" (in manu Dei) in Service to the Creator

The human hand, dissection of which was very popular in sixteenth and seventeenth
century anatomy due to its complexity and beauty as an indication of God – and thus
of the "haec sacra" – represented a "prominent vehicle for integrating sacred mystery with corporeal mechanism".[229] The dissecting hand of the anatomist had to be
well-trained in the appropriate sectioning techniques. Georges Canguilhem defined
technique as "the expression of an independent 'ability', creative in its depth, and
for which science, often in retrospect, compiles a development program or a codex
of precautionary measures".[230]

Stensen, in whose writings the term "haec sacra" for anatomy does not appear,[231]
was more prosaic in his diction, thereby providing an all the more clear view onto
the dissection object. In the inaugural address of his Copenhagen public dissection
he describes in simple terms the miracle reflected in the anatomy of the hand, and

---

[228] Ibid., entry badge (printed recto), bound between fol. 1ᵛ and 2ʳ: "In haec sacra intuens pius esto".

[229] Rowe, "God's Handy Worke", p. 287.

[230] Canguilhem, "Technique" (on "V. La morale et la pratique"), p. 77: "l'expression d'un 'pouvoir'
original, créateur en son fond, et pour lequel la science élabore[…], parfois à la suite, un pro-
gramme de développement ou un code de précautions". On this, see Borck/Hess/Schmidgen,
Introduction, pp. 14–15.

[231] Instead, he uses "sacra" (neuter plural) to refer to religious worship, cf. Sobiech, *Herz, Gott,
Kreuz*, p. 71.

speaks humbly of the "mistakes of his own hands" that might occur during the procedure.[232] Just like the visitors of a museum took no offense from the custodian's
pointer being of no artistic value, so was the case with the anatomist, who was
merely a "pointer in the hand of God" (radius seu virga in manu Dei) indicating the
curiosities of the human body as in a most exquisite museum.[233] Towards the end of
the inaugural address, Stensen once more speaks of his own deficiencies, namely
that the anatomist who, as a work of God with the help of God – who was not merely
watching, but actively participating –, was examining God's creation, must not arro-
gate anything in his research or his public dissections aside from shortcomings and
errors. Stensen therefore requests his audience to praise God's benevolence together
with him for everything that satisfied their expectations during the dissection, while
attributing any linguistic or manual mistakes made by him either to his impatience
or the pride hidden inside him, which might long for something more or greater than
the will of God, saying that even that which he otherwise would have achieved easily would then rightly be denied him by God in his section.[234]

Stensen's mention of the "eyes of the mind" (mentis oculi) with which the audience is to observe the dissected human body through the "eyes of the body" (oculi
corporis)[235] may also be in reference to the practical circumstances of the public
dissection, namely that "visibility is not without requirements" during an anatomical demonstration – ie that the audience was only able to understand the meaning
and context of what it was seeing with the help of the anatomist's explanations.[236]
Stensen's speech, however, has meaning beyond this; his "eyes of the mind" refer to
the natural light of reason of which the spectators were to avail themselves while
watching.[237]

Stensen had often expressed his awe regarding the natural beauty of the bodily
organs in his publications[238] – according to Plato (428/27 B.C.–348/47 B.C.), such
awe is a testament to appreciation for philosophy: namely "amazement" as the "pas-

---

[232] Prooemium, p. 251, line 19: "manuum lapsus".

[233] Ibid., lines 11–21, here 12, 14–15.

[234] Ibid., p. 255, lines 28–37, and see Sobiech, *Herz, Gott, Kreuz*, p. 134. On the "general concurrence" (concursus generalis) of God in the creation cf. ibid., p. 152.

[235] Prooemium, p. 254, lines 21–24.

[236] Rößler, "Blick", pp. 115–116, here 116.

[237] This is based on the window metaphor generally referring to the senses (usually the eyes) as
entrances to the souls; it is also mentioned by Augustine of Hippo. Cf. Eckart Scheerer, "Sinne,
die"; in: HWPh. 9 (1995), cols. 824–869, here 836; Sobiech, *Herz, Gott, Kreuz*, pp. 171–172.

[238] Cf. eg SudOr., p. 102, lines 30–32=E 5, p. 148, lines 15–17 with comparative progression. In
AvCun., p. 117, lines 9–10=E 9, p. 160, lines 28–29 Stensen admires the "quite elegant orifices"
(elegantes admodum […] insertiones) of vessels in a cormorant, in AvCun., p. 118, line 4=E 9,
p. 161, line 22 the "more elegant spectacle" (elegantius spectaculum) of the heartbeat of a young
chicken, and in AvCun., p. 118, lines 12–13=E 9, p. 161, lines 30–31 the "most appealing green"
(jucundissimam viriditatem) on the inside of the stomach. He reports to Thomas Bartholin on the
muscular fibers, "whose elegant structure I cannot admire enough" (elegantem earum mirari satis
nequeo structuram), in NovMusc., p. 157, line 14=E 13, p. 176, lines 1–2.

sion of the philosopher" and the only "beginning of philosophy".[239] This is further underscored by a statement by the anatomist Reinhard V. Putz, who emphasizes that students became aware of "aesthetic moments in many details of anatomical dissection" during their gross anatomy practicals, and many of them experienced "amazement at the fascinating perfection of the internal organization of the body".[240]

The motto of Stensen's Copenhagen inaugural address expresses this amazement thus: "Pulchra sunt, quae videntur, pulchriora, quae sciuntur, longe pulcherrima, quae ignorantur" (Beautiful is that which is seen, more beautiful that which is known, most beautiful by far that which is not known).[241] The professor of anatomy Dr. med. Jacques-Bénigne Winslow (Jacob Winsløw) (1669–1760), a grandnephew of Stensen who taught at the Parisian *Jardin du roi* from 1742 to 1758, was a member of the *Académie Royale des Sciences* from 1708, a professor of surgery from 1721, and "Docteur-Régent" at the Parisian medical faculty in 1728; he carried this motto forth to wider circles after 1732.[242] In contrast to his later sermons, Stensen does not differentiate in his inaugural address between the beauty of the soul originating from a merely natural likeness to God and a likeness made supernatural by His sanctifying grace[243]; this is connected to his practice oriented around natural law.

During the public dissection, Stensen relied less on verbose refutations of other researchers' errors and more on the overturning effect of the dissection itself: to underline this, he ended his inaugural address by citing Augustine of Hippo (354–430), stating that cognition of the truth, if it was only brought to light, was suitable for detecting everything incorrect and overthrow it.[244] For since his dissections of hearts conducted in 1662/1663, Stensen – as he informed Leibniz in autumn of 1677 – had made the experience that with a simple anatomical preparation, which he could have a 10-year-old boy complete within an hour, the most ingenious systems by "men who were practically idolized by all scholars" could be refuted "without a word by the visible result alone".[245] Stensen later accused "politicals" – by which he meant people viewing religion in terms of their own benefit and political agenda[246] – of foolishly taking the liberty of disseminating "oracles" on the nature of the soul even though they themselves could not even understand the admirable

---

[239] Staudacher, *Platon*, p. 44 (Theaitetos 155d): "φιλοσόφου [...] τὸ πάθος, τὸ θαυμάζειν [...] ἀρχὴ φιλοσοφίας".

[240] Putz, "Leichnam", p. 29.

[241] Prooemium, p. 254, lines 19–20. For details see Sobiech, *Herz, Gott, Kreuz*, pp. 153–156, here 154.

[242] Sobiech, *Herz, Gott, Kreuz.*, pp. 155–156. On Winslow: Otto Carl Aagaard, "Winsløw (Winslow), Jacob (Benignus)"; in: DBL³ 15 (1984), pp. 606–607, here 606 on Stensen.

[243] Sobiech, *Herz, Gott, Kreuz*, p. 155.

[244] Prooemium, p. 256, lines 23–26; cf. Daur, *Epistulae*, p. 120 (Epistula 118,12).

[245] E 143, p. 368, lines 1–5, here 3–4 = Leibniz-Forschungsstelle der Universität Münster, *Briefwechsel*, N. 160a, p. 577, lines 20–23 with citation in lines 20, 22: "Messieurs, que quasiment tous les sçavants adorent [...] sans aucune parole, la seule veue".

[246] Among them were eg the Cartesians and the Spinozists, see Sobiech, *Herz, Gott, Kreuz*, pp. 52–61.

works of God in a manual examination taking place only on a natural level.[247] Stensen thus viewed sectioning technique, as does Canguilhem, as the source of a "creative 'ability'" paving the way to the Creator.

2.3 Between Medicine and Theology

The historian and priest Dr. iur. Angelo Fabroni (1732–1803), who emphasized with appreciation the promotion of anatomy by the grand-ducal court of the Medici and the prestige the discipline enjoyed as a result,[248] reports as follows on the time during which Stensen worked at the court as a scientist: "Therefore all who came to Florence from various regions and were commended by their good scientific reputation were endowed with extraordinary honors and rewards. How generously Stensen was welcomed during this time, and for how many tasks was he consulted with! He proved his industriousness and acumen at the hospital *Santa Maria Nuova* together with Finch and [Stefano] Lorenzini by proving those things which were said at the time to be uncomprehended or incorrectly understood; with Redi he brought natural history and the dissection of animals to such perfection that he discovered much that was hitherto unknown; finally he let no day pass without approaching the [archducal] court with a gift of new things he had observed and compiled. He combined his unique scientific education with rare and difficult virtues, through which it happened that he was praised to the high heavens by everyone."[249] Fabroni's retrospect bespeaks a deep appreciation of Stensen's life and work.

From the second half of 1666 onwards, Stensen had indeed been able to engage in anatomical and surgical activity at the largest and most modern hospital in Florence at the time,[250] the *Santa Maria Nuova*, together with the English anatomist Dr. med. Sir John Finch (1626–1682)[251] and the physician and naturalist Stefano

---

[247] DefConv., p. 391, lines 6–8.

[248] Fabroni, *Historiae III*, (Caput XV. De anatomes magistris), p. 534. On Fabroni: Ugo Baldini, "Fabroni (Fabbroni), Angelo"; in: DBI 44 (1994), pp. 2–12, here 10 on Fabroni, *Historiae III*.

[249] Fabroni, *Historiae III*, (Caput XV. De anatomes magistris), p. 535: "Itaque quicunque ex diversibus regionibus Florentiam convenirent opinione doctrinae commendati, singularibus honoribus ac praemiis afficiebantur. Quam liberaliter hisce temporibus acceptus Stenonius, quantisque in rebus adhibitus! Is in Florentino S[anctae] Mariae Novae Nosocomio una cum Finchio, & Laurentio [*corr.*: Stephano; *cf.* Fabroni, *Vitae*, s.v. "Laurentius Lorenzinius", pp. 318–334, here 319–320] Lorenzinio documentum dedit singularis industriae atque acuminis in quaerendis rebus, quae tum aut incomprehensae, aut non bene perceptae dicebantur; is cum Redio zootomiam & naturalem historiam sic excoluit, ut multa antea incognita invenerit; is demum nullum abire diem sinebat, qui ad aulam accederet cum dono novarum rerum, quas observaverat atque collegerat. Adjungebat ad singularem doctrinam rariora atque difficiliora genera virtutis, quibus fiebat, ut summis laudibus ab omnibus ad caelum efferretur."

[250] E 97, p. 295, lines 30–31; E 290, p. 575, line 15.

[251] On Finch: Sarah Hutton, "Finch, Sir John"; in: ODNB[2] 19 (2004), pp. 577–578.

Lorenzini (1656–1731), who later spoke of "my beloved master Monsignor Niels Stensen, now bishop of Titiopolis and vicar apostolic in Hanover".[252]

The mathematician and pupil of Galilei Vincenzo Viviani (1622–1703),[253] who was employed at the Medici court, informed the chamberlain Bruto degl'Anibali de' Signori della Molara (†1685)[254] in a letter written on January 26, 1667 that Stensen had submitted plans to Grand Duke Ferdinando II de' Medici ([* 1610] 1621/1628–1670) for – among other things – a "thorough investigation of the fetus".[255] This was a focal point of Stensen's work in Florence, partly in collaboration with Dr. med. Francesco Redi (1626–1698),[256] personal physician to the grand dukes Ferdinando II and Cosimo III. Viviani also reported that Stensen had lauded the exceptionally broad support by Fernando II for observations and experiments of all sorts by way of providing various animals and especially the corpses from the hospital *Santa Maria Nuova*,[257] and that he, Stensen, could suffer no "greater shame" than "to be unable to show gratitude and reverence for this through my works".[258] This was the environment in which Stensen began to occupy himself increasingly with questions of human procreation, which in turn resulted in his establishing a new foundation for embryology.

2.3.1   The Beginning of Life: Research and Ethics

In his sermons, Stensen often made reference to the human body, eg to the artful design of its sensory organs, the movement of liquids and solids inside the body, the movements of the body itself and the fact that a person equipped with reason is able to take care of their own needs, well-being and happiness.[259] In this context, what

---

[252] Lorenzini, *Osservazioni intorno alle torpedini*, p. 4: "Monsign. Niccolò Stenone mio amatissimo Maestro, Vescovo oggi di Titopoli [sic!], e Vicario Apostolico in Hannover". On Lorenzini: DBioVolt., pp. 1089–1090; on his cooperation with Stensen cf. Altieri Biagi/Basile, *Scienziati del seicento*, s.v. "Stefano Lorenzini", pp. 791–822, here 793, 795–796.

[253] On Viviani: Altieri Biagi/Basile, *Scienziati del seicento*, s.v. "Vincenzo Viviani", pp. 331–372; on his cooperation with Stensen cf. Scherz, *Epistolae*, vol. 1, pp. 25–28.

[254] On Molara: Bernardi, *Paggio*, pp. 67–68 with p. 79n3; ibid., p. 120.

[255] BNCF, Ms. Gal. 252 [Div. 4ª, t. 142], fol. 109ᵛ: "una diligente esamine del feto". Cf. Scherz, *Epistolae*, vol. 1, p. 16; on the handwritten letter concept see CG-BNCF III 2.2, p. 41, no. 55. – According to Findlen, "Experiment", p. 50, Ferdinando II de' Medici was a patron following a tradition in which science was among the preferred leisure activities of sovereigns; he was greatly appreciated by the natural scientists of his time. Cf. also the comment in Fabroni, *Lettere*, pp. 24–38 (= E 36), here 24n1.

[256] He was also a poet; on Redi: John Henry, "Redi, Francesco"; in: DBSM 4 (1989), p. 26; Federico Luisetti, "Francesco Redi"; in: EILS 2 (2007), pp. 1557–1559.

[257] BNCF, Ms. Gal. 252 [Div. 4ª, t. 142], fol. 109ʳ.

[258] Ibid., fol. 109ᵛ: "magg[ior] rossore, che di non poter dimostrarne gratitudine, e ossequio con l'op[er]e".

[259] Sermo 12, p. 227, line 35–p. 228, line 3.

was his ethical position as researcher, priest and bishop on questions regarding the beginning of life?

The following Sects. 2.1.1.1, 2.1.1.2, 2.3.1.3 and 2.3.1.4 will deal with that which precedes the development of new life, and Sects. 2.3.1.5, 2.3.1.6 and 2.3.1.7 with life's early stages.

2.3.1.1 The Discovery of the Ovary and Scholastic Tradition

Stensen had presumably begun concerning himself with the female reproductive organs as early as 1662, for his study colleague, the entomologist Dr. med. Jan Swammerdam (1637–1680),[260] in his book *Miraculum naturae sive uteri muliebris fabrica [...]* (The Miracle of Nature or the Structure of the Female Uterus [...]; Leiden 1672, 2nd ed. 1679 and London 1680, 2nd ed. 1685) remembers having seen Stensen in the latter's study room in Leiden with "spermatic tubules of the [male] rat's testicles" in 1662.[261] Swammerdam furthermore asserts that as far as he knew Stensen was "the first" to have described the "spermatic parts of women" (partes seminales mulierum), and that he had "brought up" the term "ovary" in this context, ie introduced it into the scientific discourse.[262] It was eventually brought into common use as a technical term a few years later by a friend of Stensen's, the Delft anatomist Dr. med. Reinier de Graaf (1641–1673).[263] The *Miraculum naturae* was an attempt by Swammerdam to persuade the London *Royal Society* to rule in his favor in a priority dispute he was engaged in with De Graaf over the discovery of the ovarian follicles; however, on May 7, 1673 st.v. the Society awarded the priority to Stensen.[264] What had determined their decision?

Stensen, the actual discoverer of the so-called Graafian follicles, had not insisted on his rights as first discoverer but was instead prepared to let others continue his research on the follicles.[265] In his 1672 treatise on the female reproductive organs and his 1673 apology defending that treatise, De Graaf emphasizes that Stensen had trustfully complied with his, De Graaf's, request in 1670 to share his observations

---

[260] On Swammerdam: DDPhS, s.v. "Swammerdam, Jan (Johannes)", cols. 1923–1927.

[261] Swammerdam, *Miraculum naturae* (Caput III), p. 50: "tubulos seminales testiculorum muris majoris". On *Miraculum naturae* cf. DDPhS, s.v. "Swammerdam, Jan (Johannes)", cols. 1923–1927, here 1926.

[262] Swammerdam, *Miraculum naturae* (appendix), p. 54: "primus [...] partes seminales mulierum [...], ovarii mentione facta"; ibid., p. 55.

[263] Lindeboom, "Leeuwenhoek", p. 131. According to Redi, *Esperienze*, p. 50n74, Francesco Redi spoke of "testicoli o ovaie" (testicles or eggs) in his handwritten notes on 10/01/1660 and 12/27/1663. There is a difference between assumption and evidence, however. On De Graaf: DDPhS, s.v. "Graaf (1), Reinier (Regnerus) de", cols. 704–706; Han van Ruler, "Graaf, Reinier de"; in: DDPh. 1 (2003), pp. 348–349.

[264] Kardel, *Steno*, p. 46; Rome, "Royal Society", pp. 265–267.

[265] More details in Swammerdam, *Miraculum naturae* (Caput III), p. 51 = E 35, p. 205, lines 29–34 and Stensen's comment in OvaViv. I, p. 159, lines 1–6. Roger V. Short likewise advocates calling the ovarial follicles "Stensen's follicles" instead of "Graafian follicles" (Short, "Oocyte", p. 5).

on the female "testes" achieved through comparison.[266] Stensen's behavior is a striking example of magnanimity in the history of science.[267] As Matthew Cobb presumes,[268] it may have had to do with Stensen's new spiritual freedom due to his conversion amplifying the willingness to waive his rights previously demonstrated in connection with the *Ductus parotideus* in 1661.

Stensen put his discovery into words in the *Historia dissecti piscis ex canum genere* (History of a Dissected Dogfish),[269] the second appendix to his Myology probably composed in February 1667, thus falling into the period from 1666 to 1667 during which Stensen dissected many species of fish together with Francesco Redi.[270] A hand-written manuscript of the Myology including both appendices is preserved, though only the title page, dedication, corrections, margin notes and the short index are in Stensen's handwriting while the main text body was written by copyists (cf. Fig. 2.4).[271]

In contrast to the vague conclusions by analogy common in the medicine of his time, Stensen always emphasized the imperative of the "autopsy", meaning autoptic examination of the dissection material for the "anatomical arguments" to be gained therefrom, ie those confirmed by the results.[272] Already in his study diary and as early as April 15, 1659 st.v., he had rejected conclusions by analogy like the one in *Exercitationes de generatione animalium [...]* (Exercises on the Procreation of Animals [...]; London and Amsterdam 1651) by the English physician Dr. med. William Harvey (1578–1657)[273] stating that the fetus was connected to the uterus through the umbilical cord like the plants to the earth through their roots.[274] Using an anatomically verified functional analogy,[275] Stensen then proceeded to prove – under explicit correction of his own opinion published in the accountability report

---

[266] De Graaf, Opera omnia (Tractatus III. De mulierum organis generationi inservientibus [Leiden 1672], Cap. XII. De testibus muliebribus sive ovariis), pp. 300–301; ibid. (Tractatus IV. Partium genitalium defensio [Leiden 1673]), p. 478 ad "Pag. 51 [in Swammerdam, *Miraculum naturae*]". Cf. Roger, *Life Sciences*, p. 209.

[267] On similar examples see Merton, "Priorities", p. 450.

[268] Cobb, *Egg & Sperm Race*, pp. 107–108.

[269] Stensen, *Elementorum myologiae specimen*, pp. 111–119 (the title of the second appendix on p. 111 differs from the subtitle on the title page) = PiscCan.

[270] Guerrini, "Biografia Rediana", p. 59; Kardel/Maquet, *Biography and Original Papers*, pp. 149–151, 175.

[271] Kardel, *Steno on Muscles*, p. 231 (on form and content of the manuscript); see also Scherz, *Geological Papers*, p. 123n1; Scherz, "Danmarks Stensen-manuskript", p. 21 (title page), 28 (manuscript structure); Scherz, "Kostbares Stensenmanuskript"; Scherz, *Vom Wege*, pp. 70–71.

[272] NarVas., p. 93, lines 14–15: "argumenta anatomica [...] αὐτοψίαν".

[273] Harvey, *De generatione animalium* (Additamenta. De membranis), p. 541. On Harvey: Andrew Cunningham, "Harvey, William"; in: DBPh. 1 (2000), pp. 395–403, here 401–402 on *De generatione animalium*. Harvey was a proponent of Aristotle's research program, cf. ibid., pp. 397–398.

[274] Ziggelaar, *Chaos-Manuscript*, p. 249 (col. 96, fol. 53ᵛ, N.90); Sobiech, *Herz, Gott, Kreuz*, p. 292n149. – In GlandOr., p. 18, lines 26–31; VitTrans., p. 212, lines 2–6 Stensen explains the error of assimilation of observations as a result of the functionality of the human mind.

[275] For details see Lesky, "Säugetierovar", pp. 240–242; Lesky, "Zeugungsgeschehen", p. 11.

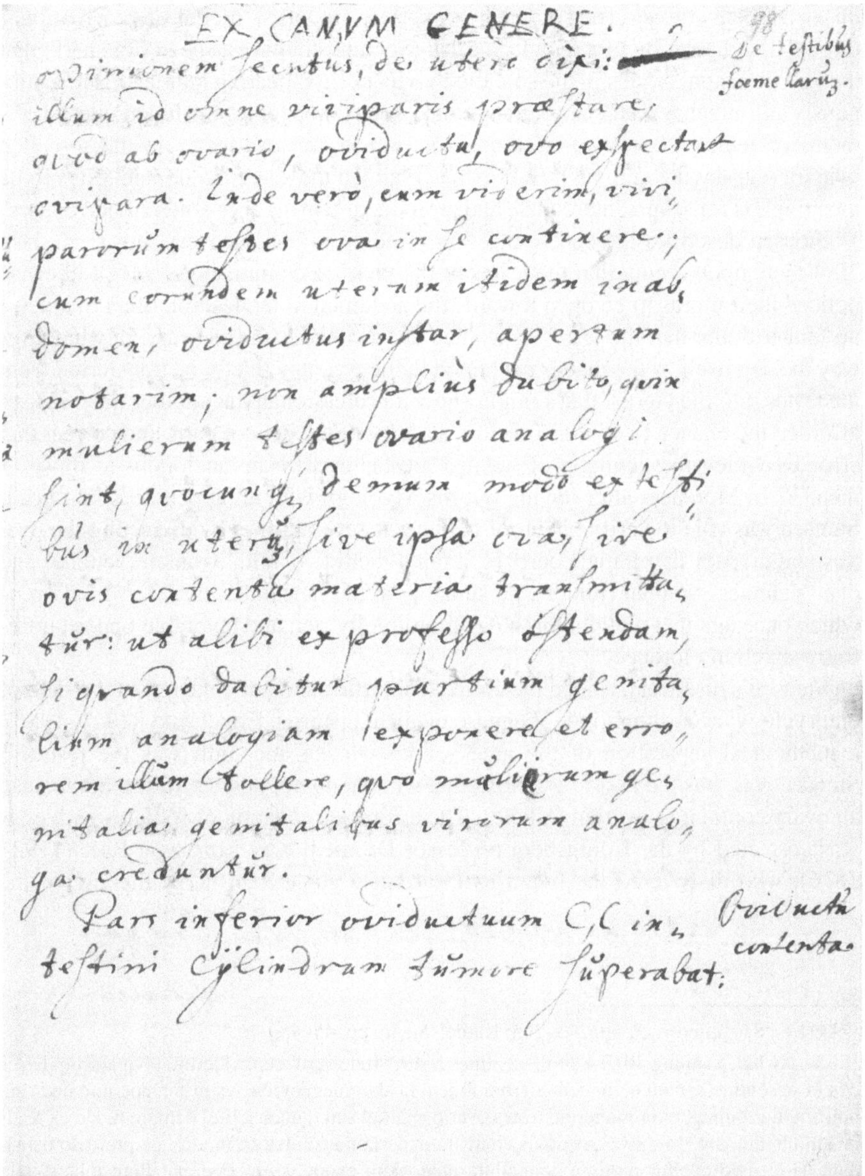

Fig. 2.4 Page from the handwritten manuscript (c. February 1667) of the second Appendix *Historia dissecti piscis ex canum genere* of Niels Stensen's *Elementorum myologiae specimen [...]* (Florence 1667)

Lines 4–18 are the decisive passage (transcription: see note 277, in Chap. 2 above) in which Stensen establishes that the so-called atrophied 'testes' of women are actually ovaries

KB, Ny klg. Samling 4019 4°, fol. 98ʳ

on his studies abroad (1664), which supported Aristotle's model of procreation –
through the dissection of a female dogfish (Scymnus lichia), a species of shark, that
the female sexual organs in the so-called vivipara (live-bearing animals), and analo-
gously in humans, neither produce female sperm nor represent functionless rudi-
ments of female 'testes' or 'testiculi', but are in fact ovaries as in the so-called
ovipara (egg-laying animals). Harvey had claimed the so-called female testes to be
meaningless for sexual intercourse and reproduction in his above-mentioned opus.[276]

Stensen describes the course of events leading to his discovery thus (Fig. 2.4):
"But then, upon seeing that the testes of the vivipara contain eggs, since I likewise
noticed their uterus to be open towards the abdomen in the fashion of an oviduct, I
no longer doubt that the testes of women are analogous to the ovary, in whichever
way the egg itself or the matter contained in the egg may finally be transferred from
the testes into the uterus: that I might show it professionally at another time, if I am
afforded the chance to demonstrate the analogy of the genital parts and correct that
error by which the genitalia of women are believed to be analogous to those of
men."[277] In Florence, after having put his Myology with all appendices into print,
Stensen was able to verify his proof in humans several times by dissecting the ovaries and uteri of five female corpses, among them two "old" women (vetulae) and
one "younger" woman (junior), presumably at the hospital *Santa Maria Nuova*,[278]
which once again casts light on the extraordinarily rich and favorable opportunities
for research in Florence.[279]

Stensen's discovery voided the entire scholastic canon of concepts in the field of
embryology; according to the Viennese medical historian Erna Lesky (1911–1986),
a methodical separation of the areas of knowledge and faith was the result.[280]
Stensen was thus the first to establish that women do not possess testes, but instead
an ovary containing ovarian follicles. The erroneous equation of follicles with eggs
was corrected by the Königsberg professor Dr. med. Karl Ernst von Baer (1792–
1876), who discovered the ovum as the female gamete in the epithelium of the

---

[276] Lesky, "Säugetierovar", pp. 243–245; Kardel, *Steno*, pp. 45–48.

[277] KB, Ny kgl. Samling 4019 4°, fol. 98ʳ, lines 5–19: "Inde vero, cum viderim, viviparorum testes
ova in se continere; cum eorundem uterum itidem in abdomen, oviductus instar, apertum notarim;
non amplius dubito, quin mulierum testes ovario analogi sint, quocunq[ue] demum modo ex testibus in uterum, sive ipsa ova, sive ovis contenta materia transmittatur: ut alibi ex professo osten-
dam, si quando dabitur partium genitalium analogiam exponere, et errorem illum tollere, quo
mulierum genitalia genitalibus virorum analoga creduntur." Cf. the transcription in PiscCan.,
p. 152, line 36–p. 153, line 7, here p. 152, line 38–p. 153, line 7. The more extensive treatise was
not realized by Stensen. – OvaViv. I and OvaViv. II show that the oviduct receives the 'eggs'
released by the ovary, cf. Lesky, "Säugetierovar", pp. 242–243.

[278] OvaViv. I, p. 165, line 22–p. 166, line 6, here p. 165, lines 27, 33. – Only published 1675 in the
Acta Hafniensia, cf. ibid. (Notes) preliminary remarks, p. 330.

[279] Scherz, *Vom Wege*, p. 78.

[280] Lesky, "Säugetierovar", pp. 248–249. On Lesky: Huldrych M. Koelbing, "Zur Erinnerung an
Erna Lesky (22. Mai 1911–17. November 1986)"; in: Gesnerus 44 (1987), pp. 3–5.

vesicular follicle.[281] By recognizing the ovarian follicles as egg containers, Stensen had furthermore – as emphasized by Adolf Faller – interpreted the so-called Graafian follicles "more correctly than De Graaf himself".[282] Stensen conducted his work without expectations in regard to results and did not cloud his vision with preconceived opinions.

The *Historia dissecti piscis ex canum genere* also received the necessary ecclesiastical imprimatur[283]: Benigno Bruni OMin. (1599/1600–1670),[284] consultor to the Holy Office, mentions explicitly in his censorial note of March 3, 1667, which honored Stensen highly, that he had read the "anatomical history of a dissected dogfish", thereby finding nothing disagreeing with the Faith or morality but instead seeing with pleasure that "everything was precisely described", wherefore he had decided that the work could go to print.[285] Stensen conveyed his knowledge on the ovary at various occasions in the following years: in the summer of 1671 he met the physician Dr. med. Paolo Maria Terzago (†1695) in Milan, dissecting for the latter during a presentation, among other things, the ovary of a heifer "including the extraction of an egg which he carefully stripped of its cortex".[286] As "Anatomicus regius" in Copenhagen, Stensen also taught his discovery to his students: Holger Jakobsen noted on Stensen's dissection of February 6, 1673 st.v., during which the latter discussed sexual intercourse and the female sexual organs using a woman's corpse which had been his object for public dissection since January 6 st.v., that women possessed an "ovary".[287] When Stensen ceased his research activities in mid-July 1674, Leeuwenhoek had already been sending letters to the *Royal Society* in London about the spermatozoa – which he referred to as "animalcula" (small creatures) for about a year.[288] Stensen himself had undertaken examinations on the path of the male semen by means of comparative anatomy together with Francesco

---

[281] Faller, "Heister", p. 64. – The natural philosopher Dr. med. Lorenz Oken (1779–1851) mentions Stensen in his habilitation thesis *Die Zeugung* (Bamberg and Würzburg 1805), completed in Würzburg, in connection with the female 'eggs' (Lorenz Oken, "Die Zeugung"; in: Bach/Breidbach/Engelhard, *Oken*, pp. 87–195, here 109; cf. ibid., p. XIII).

[282] Bierbaum/Faller, *Stensen*, p. 33 = Bierbaum/Faller/Traeger, *Stensen*, p. 33.

[283] Cf. the six censorial notes including the imprimatur relating to Myol. as well as its two appendices in Myol. (Notes) preliminary remarks, p. 319 r. col.; printed facsimile in Kardel, *Steno on Muscles*, p. 242.

[284] On Bruni: Tognocchi a Terrinca, *Theatrum* (Pars prima, Titulus II, Series V), p. 21.

[285] Myol. (Notes) preliminary remarks, p. 319 r. col. = Scherz, *Geological Papers*, p. 116: "dissecti[…] piscis ex canum genere historia anatomica […] omnia accurate descripta". The date can be found only in the handwritten copy; on this and a handwritten censorial note by Bruni to CanCap. on 02/23/1667 see Scherz, *Geological Papers*, p. 131n142 & p. 127n62.

[286] Corte, *Notizie*, s.v. "Paolo Maria Terzago", pp. 188–190, here 189: "coll'estrazione d'un uovo, che diligentemente spogliò della sua corteccia"; cf. Scherz, *Epistolae*, vol. 1, p. 34. On Terzago: Julius Pagel, "Terzago, Paolo Maria"; in: BLÄ 5 (1934), p. 534. Findlen, *Possessing Nature*, p. 220 incorrectly ascribes the dissection to Terzago.

[287] ExAc., p. 306, lines 11–12: "ovarium".

[288] DDPhS, s.v. "Leeuwenhoek, Antoni van", cols. 1162–1165, here 1163–1164.

Redi in autumn of 1667.[289] Factual knowledge on human reproduction began to be gradually pieced together from observations by various anatomists over these decades.

The official Dresden city physician Dr. med. Martin Schurig (1656–1733),[290] whose *Embryologia historico-medica [...]* (Historical-Medical Embryology [...]; Dresden and Leipzig 1732) coined the term "embryology",[291] took into consideration over 250 authors in the fields of gynecology, obstetrics, and sexology in his opus *Muliebria historico-medica, hoc est partium genitalium muliebrium consideratio physico-medico-forensis [...]* (The Female Genitalia in Medical History, Meaning a Physico-Medico-Forensic Consideration of the Female Genitalia [...]; Dresden and Leipzig 1729). Since he had already identified the "testiculi muliebris" as "ovaria", Schurig writes, he felt compelled to remind the reader with regard to the initial inventor of this designation that it was ascribed to Stensen "by some anatomists"; he also cites the respective sources like Swammerdam. However, Schurig follows this with accounts of discussions about other discoverers[292]; a result of, among others, Stensen's modesty and munificence.

The jurist and theologian Dr. theol. Dr. iur. utr. Francesco Emanuele Cangiamila (1702–1763),[293] provincial inquisitor to the Kingdom of Sicily from 1753, makes special mention of Stensen in his work *Embriologia sacra [...]* (Holy Embryology [...]; Palermo 1745): "Hence Stensen as the first and countless others have already brought to light that the sperm of the viviparous quadrupeds is not that which the Ancients thought it to be, which in truth was nothing but a simple lymph; instead it is an egg: that these [= the viviparous quadrupeds] already have their own ovary". He goes on to state that "all parts of the embryo [are] preconfigured in the egg, but they are as yet quasi wrapped into themselves".[294] Cangiamila's opus awakened the interest of theologians in the fetus.[295] The American moral theologian John

---

[289] For details see Goltz, "Uterus", pp. 256–260; Miniati, *Steno's Challenge*, p. 193 with n. 46.

[290] On Schurig: DresdÄ, p. 138.

[291] BRM, pp. 148–149 does not include a lemma "Embryologia".

[292] Schurig, *Muliebria* (Sectio III, Caput III, § 13), pp. 321–325, here 321: "a nonnullis anatomicis".

[293] On Cangiamila: Mario Condorelli, "Cangiamila, Francesco Emanuele"; in: DBI 18 (1975), pp. 72–74, here 73. The diagrams of fetuses were created by the Turin professor Dr. med. Giovanni Battista Bianchi (1681–1761), cf. Filippini, "Erste Geburt", p. 109. On Bianchi: Giuseppe Sperati, "Bianchi, Giovanni Battista"; in: DBI 10 (1968), pp. 118–120.

[294] Cangiamila, *Embriologia sacra* (Libro primo, capo IX), pp. 57–58, no. 2: "Stenone adunque il primo [...] ed altri senza numero ânno già messo in chiaro, che il seme delle Quadrupede vivipare non è già quello, che credevano gli Antichi, che in verità non era, se non una semplice Linfa; ma è un ovo: che elleno [= esse=Quadrupede vivipare] ânno ancora il loro ovario [...]. [...] parti tutte dell'Embrione [...] preesistono nell'uovo, ma sono ancora, per così dire, in se medesime involte." On the "lymph" (linfa) see details ibid. (Libro primo, capo IV), pp. 24–25, no. 6, here p. 24: "linfa delle Donne, volgarmente dagli Antichi creduta suo seme" (lymph of women, generally assumed by the Ancients to be their sperm).

[295] Mario Condorelli, "Cangiamila, Francesco Emanuele"; in: DBI 18 (1975), pp. 72–74, here 73; Filippini, "Erste Geburt", p. 112.

R. Connery SJ (1913–1987)[296] writes in *Abortion [...]* (Chicago 1977), a historical overview of the Roman Catholic view on induced abortion, that Cangiamila had been open to the notion of the 'egg' containing an embryo (the position of the so-called ovulists), but that this concept by no means made likely a simultaneous ensoulment of the embryo, ie creation and infusion of the soul at the moment of conception.[297]

Hans Reis on the other hand points out[298] that Aristotle's ideas on reproduction form the theoretical basis for his notion of the successive ensoulment of the human embryo – with the "vegetative soul" (anima vegetativa), the "sensitive soul" (anima sensitiva), and the "rational soul" (anima rationalis) following one another, each absorbing the previous, and together forming a single soul – which heavily influenced Thomas Aquinas and the scholastic theological tradition in his wake.[299] The Thomist philosopher Josef Pieper (1904–1997)[300] of Münster emphasizes in regard to Thomas' capacity as Doctor of the Church that, although "the whole of the truth is expressed in a unique, paradigmatic way" in his theological oeuvre he, Pieper, "had never even considered extending the authority of Saint Thomas to his biological doctrines". It was "relatively common opinion", Pieper states, that Thomas' natural philosophy was also the weakest element in his thinking. Hence those parts of Thomas' teachings as Doctor of the Church which were contingent on his times were not to be preserved.[301] Furthermore, as Paul Richter has recently elaborated, Thomas most probably did not wish to commit himself with regard to the biology of conception; rather he "offered up attempts at answers by trustworthy witnesses without fully adopting any one position himself".[302]

Stensen, whose conversion and theology were heavily influenced by his reading of the Bible – preferably in its original text – and the early Church Fathers,[303] avoided collisions between modern philosophy and theology by not considering the former as absolute, but instead using it as a heuristic instrument. He presumably thought similarly to Pieper: as early as March 30, 1659 st.v., while reading *Magnes [...]* (Magnet [...]; Cologne 1643) by the polyhistor Athanasius Kircher SJ (1601–1680),[304] a work which Stensen criticized for objectionable use of conclusions by

---

[296] In the years 1972/1973 and 1975/1976, Connery was a "Senior Research Scholar" at the *Kennedy Institute, Center for Bioethics*, Georgetown University, Washington, D.C., where he completed this work which outlines the Roman Catholic view on induced abortion from a moral theological perspective up to the mid-twentieth century, cf. Connery, *Abortion*, pp. 1–2, 5. On Connery: [editorial], "Medical-Moral Consultant Dies"; in: NJN 17,4 (1988), p. 13.

[297] Connery, *Abortion*, pp. 208, 224.

[298] Reis, *Lebensrecht*, p. 137n652.

[299] For detailed differentiation see Hack, *Beseelung*, pp. 316–333.

[300] On Pieper: Berthold Wald, "Pieper, Josef"; in: NDB 20 (2001), pp. 427–428.

[301] Pieper, "Thomas von Aquin", pp. 170–172 with citations on p. 171.

[302] Richter, *Beginn des Menschenlebens*, pp. 134–135, here 134.

[303] For details see Sobiech, *Herz, Gott, Kreuz*, pp. 116–120.

[304] On Kircher: HGK, s.v. "Kircher, Athanasius S.J.", pp. 367–369; on Kircher's conclusions by analogy see also Sobiech, *Herz, Gott, Kreuz*, pp. 291–292.

analogy, he was struck by the thought that those persons sinned against God's dignity who did not observe the works of nature themselves but instead, content with the reading of books, arranged and constructed for themselves various illusions causing them to not only forfeit the pleasure of the joyous viewing of the wonders of God, but also to lose the time they should be investing into the benefit of their neighbors and to firmly believe many things depreciatory of God.[305] He counted among these people the scholastics, most philosophers, and those who spent their entire life on the study of logic, wherefore one was not to waste much time concerning oneself with them nor interpret one's own observations prematurely and over-hastily in their direction, but instead rely as much as possible on assumptions, experiments and their analysis – albeit under consideration of the observations by the "Ancients" (veteres).[306] Stensen's early complaint about the neglect of experience in medicine shows striking similarity to a lament by Dr. med. Giacomo Sinibaldi (1641–1720), lector at the Roman *La Sapienza* university and, among other engagements, physician for the entire Apostolic Palace. In context with the controversial corpuscular theory and English chemistry, he likewise deplored the neglect of experience and rational penetration of the object of observation: "For like in unfortunate manner the scholastics, who, drawing solely upon speculation, do not recognize experience as a teacher, so do the physicians and chemists sometimes consternate even worse, who, neglecting the basic materials of things and nearly becoming stuck in experimental mechanics, declaim differing and often contradictory aphorisms stemming from various observations relating to the senses."[307]

Stensen's complaint as a student had referred to the obstruction of medical thinking by complicated syllogisms that obfuscated what had been gained by

---

[305] Ziggelaar, *Chaos-Manuscript*, pp. 159–160 (col. 57, fol. 44ʳ, N.59) and Kircher, *Magnes*, pp. 657–662 (Liber III, Pars VI, Caput I). – Stensen also voiced this opinion he held in his letter to Athanasius Kircher from Florence on 04/141676 (E 112, p. 314) by way of a 'brief review' of the title (pointed out to him by Kircher) of the work Baldewein, *Phosphorus hermeticus, sive magnes luminaris* (Frankfurt 1675) by the jurist and alchemist Christian Adolph Baldewein (1632–1682), presumably not least due to his unpleasant memories of Kircher's problematic *Magnes*. He did not know, Stensen wrote, whether books with such grand promises deserved the time one spent reading them, and added in Latin: as much time as they took from us, that much of eternity they secretly deprived us of, cf. Sobiech, *Herz, Gott, Kreuz*, pp. 291–292. On Baldewein, who was searching for the so-called philosophers' stone: Alphons Oppenheim, "Baldewein, Christian Adolf (Balduinus)"; in: ADB 2 (1875), pp. 3–4.

[306] Ziggelaar, *Chaos-Manuscript*, p. 160 (col. 58, fol. 44ʳ, N.59). Cf. ibid., p. 100 (col. 32, fol. 37ᵛ, N.32); ibid., p. 169 (col. 63, fol. 45ᵛ, N.67).

[307] Sinibaldi, *Dissertationes* (Dissertationes habitae in Congressu Medico Romano [pp. 1–132]. Acidi contrariae virtutes [pp. 1–18]), here 5–6: "sicut enim infeliciter errant Scolastici, qui solis speculationibus innixi magistram experientiam ignorant, ita deterius interdum coecutiunt [*corr.*: concutiunt] medici, & chymici quam plurimi, qui [6] rerum principia negligentes, & mecanicae experimentali tantummodo inhaerentes iuxta varias ad sensum observationes varios, ac sepe inter se discrepantes aphorismos pronunciant." Cf. Donato, "Onere", p. 79. On Sinibaldi: Adelmann, *Correspondence*, vol. 3, no. 436, p. 928n19; Cavarzere, *Censura*, pp. 146–147.

experiment.[308] Already the astronomer Ismaël Boulliau (1605–1694),[309] a convert who had been ordained a priest in 1630 and from whose work *De natura lucis* (On the Nature of Light; Paris 1638) Stensen had excerpted on June 15, 1659 st.v.,[310] two and a half months after reading *Magnes*, emphasized that there was no doubt that Aristotle, had he been able to avail himself of the knowledge of later times, would have philosophized differently.[311] From this point of view, Stensen benefitted from the anti-Aristotelian culture of natural science prevalent in Tuscany when he came to Italy as a young researcher in 1666.[312] It allowed him to gain a higher level of impartiality in comparison with the thinking of most contemporary theologians, which was heavily influenced by Aristotle's outdated natural philosophy. For with few exceptions, the new medical theories were finding next to no reflection in the theological publications of his time.[313]

This resulted in friction between the natural sciences and theology: the physician and naturalist Giovanni Battista Capucci (* early seventeenth century–1680s)[314] responded to Marcello Malpighi on February 14, 1669 with regard to certain friars who, according to Malpighi, had raised some – undefined – objections during the printing by the *Royal Society* of his work *Dissertatio epistolica de bombyce* (Treatise by Letter on the Silkmoth; London 1669), that their heads were always full of scholastic sophistry which they sometimes applied to the handling of other affairs.[315] There is no evidence, however, that Malpighi had any altercations with the ecclesiastical authorities concerning the printing of his opus. In 1693 and by then personal physician to the Pope, Malpighi did complain with regard to the corpuscular theory, which was eyed with suspicion by the Holy Office, about the "lacking wisdom of some overly hot Neapolitan brains".[316] One may assume that Stensen would have been able to mitigate the dispute between medical and theological scholasticism on the one hand and modern philosophy and natural science on the other. In regard to atomism, Stensen had – presumably due to his Cartesian imprint – taken an

---

[308] Cf. Stroppiana, "Lancisi", p. 13.

[309] On Boulliau: Robert Alan Hatch, "Boulliau, Ismaël"; in: DFPh. 1 (2008), pp. 201–205.

[310] According to Boulliau, much in Aristotle's writings had become obsolete as a result of newer findings: Boulliau, *De natura lucis*, foreword "Ad lectorem" (unpaginated), here third and fourth page, and excerpt in Ziggelaar, *Chaos-Manuscript*, p. 317 (col. [130], fol. 62ʳ).

[311] Boulliau, *De natura lucis*, foreword "Ad lectorem" (unpaginated), here fourth page. On the same argumentation with reference to theological scholasticism see Hack, *Beseelung*, p. 431.

[312] Cf. Findlen, *Possessing Nature*, p. 219.

[313] Lanza, "Momento", p. 91 = Lanza, *Momento*, p. 159; Louis Cognet, "Cor et cordis affectus. 4. Le coeur chez les spirituels du XVIIe siècle"; in: DSp. 2 (1953), cols. 2300–2307, here 2300; with an example Sobiech, *Herz, Gott, Kreuz*, pp. 239–240; Sobiech, "Experience", pp. 182–183.

[314] On Capucci: Augusto De Ferrari, "Capucci, Giovanni Battista"; in: DBI 19 (1976), pp. 268–270, here 269 on his correspondence with Malpighi.

[315] According to the English paraphrasing of the letter manuscript in Adelmann, *Embryology*, p. 347. The letter by Malpighi is not preserved. On Malpighi, *De bombyce* cf. Cesare Preti, "Malpighi, Marcello"; in: DBI 68 (2007), pp. 271–276, here 274.

[316] Quoted according to Donato, "Onere", p. 84: "la poca prudenza d'alcuni napoletani cervelli troppo caldi".

ambiguous stance that could be described as cautious.[317] Furthermore, several of his anatomical and geological works had been reprinted in the Calvinist Republic of the Seven United Netherlands after his conversion and also after his episcopal ordination.[318]

Stensen's great awareness for philosophy of science is demonstrated by the fact that he regularly discusses sources of error with regard to research in his writings.[319] In the inaugural address of his Copenhagen public section of a female corpse in 1673, for instance, he explains that these sources of error included the hitherto published book learning, which had often proved incorrect during his century, as well as prejudices that obstructed the researcher in his search for the truth. Nothing was more difficult to discard than prejudice; not even the medical writings of his time could be composed, even if one exercised the utmost care, so free of error as to not include traces of preconceived notions, and if he, Stensen, were to except himself from this statement, he would deserve reproach for his unabashed haughtiness.[320]

Instead of blending his research with metaphysical speculations, as was typical for the time, Stensen – e.g. during his *Ova viviparorum spectantes observationes* (Observations Regarding the Eggs of the Vivipara), conducted in Tuscany in 1667 but only published in 1675 in the *Acta Hafniensia*, in a paragraph towards the end of the treatise presumably added during his time as "Anatomicus regius" in Copenhagen – expounds on how much more "proof of divine wisdom and benevolence" one could have discovered in individual sections performed up until then, had one used the time wasted on regarding the natural attractiveness of the dissection object and one's own discoverer's pride entirely to direct one's gaze towards the creator of such beautiful and aesthetic objects.[321] By this Stensen means the third pillar of his motto from the Copenhagen inaugural address: "[…] most beautiful by far that which is not known". It thus becomes possible to speak of a "unity of knowledge and faith"[322] (Adolf Faller) in a completely new and comprehensive sense which respects the methodical individuality[323] of both sciences. In this regard, the image of the scientist embodied by Stensen remains ageless to this day.

---

[317] Ziggelaar, *Chaos-Manuscript*, pp. 472–473; Miniati, *Steno's Challenge*, pp. 45–46; Sobiech, *Herz, Gott, Kreuz*, p. 132n35.

[318] Cf. the compilation of editions in Maar, *Opera philosophica*, vol. 2, pp. 351–359.

[319] See note 274, in Chap. 2 above.

[320] Prooemium, p. 256, lines 3–13.

[321] OvaViv. II, p. 178, lines 28–31, here 28: "divinae sapientiae & bonitatis argumenta". Cf. ibid. (Notes) preliminary remarks, p. 330 r. col.

[322] This phrasing can only be found beginning with the 2nd edition edited by Faller in Bierbaum/Faller, *Stensen*, p. 167 = Bierbaum/Faller/Traeger, *Stensen*, p. 135.

[323] On this, see also Miniati, *Steno's Challenge*, p. 286.

2.3.1.2   The Copenhagen Public Dissection of a Female Corpse in Comparison

From July 13, 1672 until mid-July 1674, the self-appointed end of his career as an anatomist and scientist, Stensen held an uncertain position in Copenhagen due to the claim to exclusivity of the Danish Lutheran state religion in all public and academic areas. Unofficially he was referred to as "Anatomicus regius" (Royal Anatomist)[324] by his former private preceptor Thomas Bartholin, rector and university librarian of the University of Copenhagen since 1671,[325] in the second volume of the *Acta Hafniensia* for the year 1673. Studies by Stensen from this time appeared in this scientific medical journal edited by Bartholin, one of the first in the world with primarily medical contents, the *Acta medica & philosophica Hafniensia* (5 vols., Copenhagen 1673–1680) which report on the years 1671–1679.[326] As dean of the medical faculty at the time, Bartholin published an elaborate invitation, dated January 28, 1673 st.v. and extolling the virtues of the anatomist and academic teacher Stensen, to the latter's great public dissection of a female corpse (Fig. 2.5).

Stensen devoted these years, which were dedicated to more didactics and the conveying of his extensive knowledge and experience as an anatomist rather than to scientific research, entirely to the education of students. Hence in his invitation, Bartholin writes that Stensen "eases the hopes of the students [for a successful conclusion of their studies]", thus contributing to the common good.[327] In the second volume of the *Acta Hafniensia*, Stensen is introduced as "our great prosector Mr. Niels Stensen", who, "in order to apply great care to being anxious for the good of the students", had developed a special method for making the dissection objects less perishable and demonstrated it to those interested most recently around the turn of the year 1673/1674.[328] Already in the preamble of that volume, addressing the Danish Lord Chancellor Peter Schumacher Griffenfeld (1635–1699) who had appointed Stensen, Bartholin identifies the latter as "that great prosector, who traverses the bowels of animals and humans with a fine hand and unpretentious language".[329] In the above-mentioned invitation, "our Stensen, second to no other prosector in speed of dissecting and ease of discovering" is publicized.[330] The notes

---

[324] HGK, s.v. "Bartholin, Thomas", pp. 70–71, here 71.

[325] E.g. in AMPH 2 (1673), pp. 81–92 (XXXIV. In ovo & pullo observationes d[omini] Nicolai Stenonis Anatom[ici] Reg[ij])=OvPul.

[326] Kragh, "New Science", pp. 60–61.

[327] Bartholin, *Decanus Anatomes s*[alutem], line 25: "studiosorum spem solatur".

[328] AMPH 3 (1674/1675), pp. 9–10 (V. De mumia Bilsiana), here 9: "magnus prosector noster d[ominus] Nicolaus Stenonius, studiosorum commodis invigilaturus".

[329] AMPH 2 (1673), p. (*) 3: "magno illo prosectore, [...] subtili manu & modesto ore per viscera animalium hominumq[ue] grassante". On Griffenfeld: Scherz, *Epistolae*, vol. 1, p. 78.

[330] Bartholin, *Decanus Anatomes s*[alutem], lines 41–42: "Stenonius noster, prosectorum nulli secundus secandi promtitudine [sic!] & inveniendi facilitate".

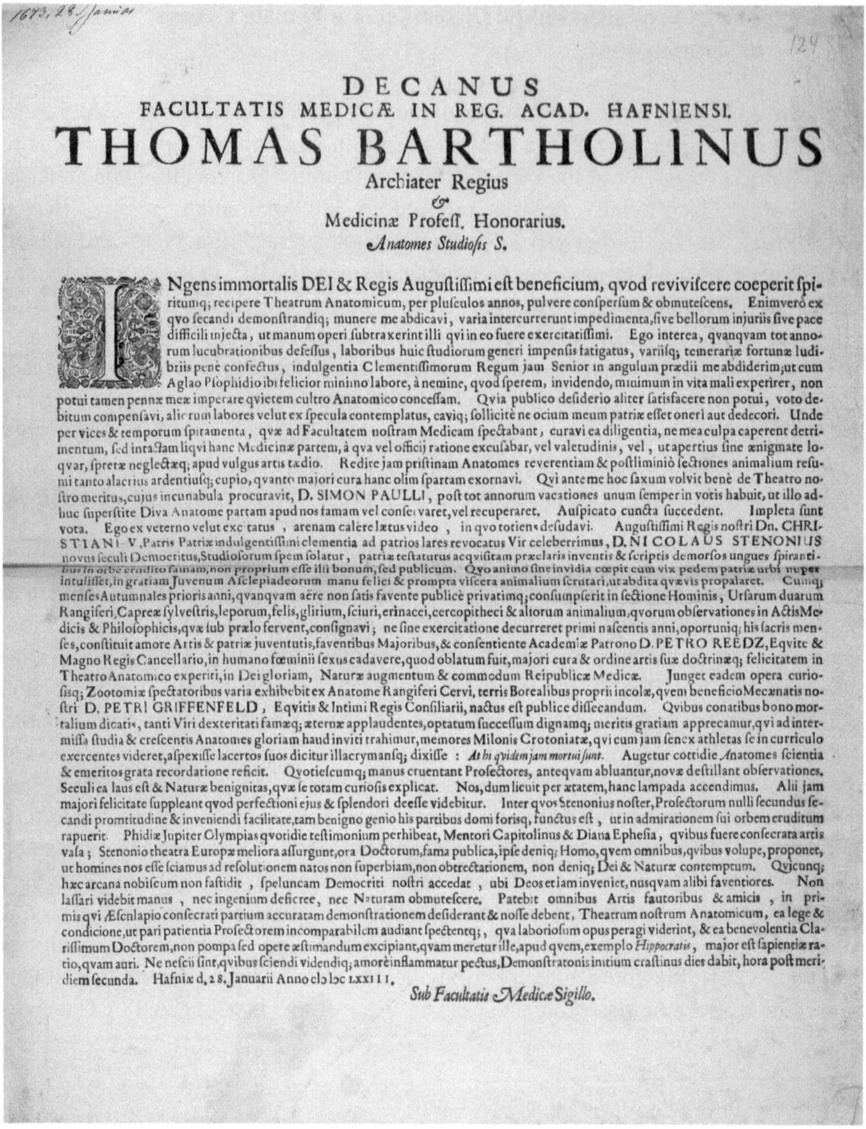

1673, 28. Januar

124

DECANUS
FACULTATIS MEDICÆ IN REG. ACAD. HAFNIENSI.

THOMAS BARTHOLINUS

Archiater Regius
&
Medicinæ Profess. Honorarius.
Anatomes Studiosis S.

Ngens immortalis DEI & Regis Augustissimi est beneficium, qvod revivisцere coeperit spiritumq; recipere Theatrum Anatomicum, per plusculos annos, pulvere conspersum & obmutescens. Enimvero ex qvo secandi demonstrandiq; munere me abdicavi, varia intercurrerunt impedimenta, sive bellorum injuriis sive pace difficili injecta, ut manum operi subtraxerint illi qvi in eo fuere exercitatissimi. Ego interea, qvanqvam tot annorum lucubrationibus defessus, laboribus huic studiorum generi impensis fatigatus, variisq; temerariæ fortunæ ludibriis pene confectus, indulgentia Clementissimorum Regum jam Senior in angulum prædii me abdiderim, ut cum Aglao Psophidio ibi felicior minimo labore, à nemine, qvod sperem, invidendo, minimum in vita mali experirer, non potui tamen pennæ meæ imperare qvietem cultro Anatomico concessam. Qvia publico desiderio aliter satisfacere non potui, voto debitum compensavi, aliorum labores velut ex specula contemplatus, caviq; sollicite ne ocium meum patriæ esset oneri aut dedecori. Unde per vices & temporum spiramenta, qvæ ad Facultatem nostram Medicam spectabant, curavi ea diligentia, ne mea culpa caperent detrimentum, sed intra?am liqvi hanc Medicinæ partem, à qva vel officij ratione excusabar, vel valetudinis, vel, ut apertius vitæ ænigmate loqvar, spretæ neglectæq; apud vulgus artis tædio. Redire jam pristinam Anatomes reverentiam & postliminiò sectiones animalium resumi tanto alacrius ardentiusq; cupio, qvanto majori cura hanc olim spartam exornavi. Qvi ante me hoc saxum volvit bené de Theatro nostro meritus, cujus incunabula procuravit, D. SIMON PAULLI, post tot annorum vacationes unum semper in votis habuit, ut illo adhuc superstite Diva Anatome partam apud nos famam vel conservaret, vel recuperaret. Auspicato cuncta succedent. Impleta sunt vota. Ego ex veterno velut excitatus, arenam calere lætus video, in qvo totiens desudavi. Augustissimi Regis nostri Dn. CHRISTIANI V. Patris Patriæ indulgentissimi clementia ad patrios lares revocatus Vir celeberrimus, D. NICOLAUS STENONIUS novus seculi Democritus, Studiosorum spem solatur, patriæ testaturus acqvisitam præclaris inventis & scriptis demorsos ungues spiranti-bus in orbe erudito famam, non proprium esse illi bonum, sed publicum. Qvo animo sine invidia coepit cum vix pedem patriæ urbi nuper intulisset, in gratiam Juvenum Asclepiadeorum manu felici & prompta viscera animalium scrutari, ut abdita qvævis propalaret. Cumq; menses Autumnales prioris anni, qvanqvam aëre non satis favente publice privatimq; consumpserit in sectione Hominis, Ursarum duarum Rangiferi, Capreæ sylvestris, leporum, felis, glirium, sciuri, erinacei, cercopitheci & aliorum animalium, qvorum observationes in ActisMedicis & Philosophicis, qvæ sub prælo fervent, consignavi; ne sine exercitatione decurreret primi nascentis anni, oportuniq; his sacris menses, constituit amore Artis & patriæ juventutis, faventibus Majoribus, & consentiente Academiæ Patrono D. PETRO REEDZ, Eqvite & Magno Regis Cancellario, in humano fœminii sexus cadavere, qvod oblatum fuit, majori cura & ordine artis suæ doctrinæq; felicitatem in Theatro Anatomico experiri, in Dei gloriam, Naturæ augmentum & commodum Reipublicæ Medicæ. Junget eadem opera curiosisq; Zootomiæ spectatoribus varia exhibebit ex Anatome Rangiferi Cervi, terris Borealibus proprii incolæ, qvem beneficio Mecænatis nostri D. PETRI GRIFFENFELD, Eqvitis & Intimi Regis Consiliarii, nactus est publice dissecandum. Qvibus conatibus bono mortalium dicati, tanti Viri dexteritati famæq; æternæ applaudentes, optatum successum dignamq; meritis gratiam apprecamur, qvi ad intermissa studia & crescentis Anatomes gloriam haud inviti trahimur, memores Milonis Crotoniatæ, qvi cum jam senex athletas se in curriculo exercentes videret, aspexisse lacertos suos dicitur illacrymansq; dixisse : *At hi qvidem jam mortui sunt.* Augetur cottidie Anatomes scientia & emeritos grata recordatione reficit. Qvotiescumq; manus cruentant Prosectores, anteqvam abluantur, novæ destillant observationes, Seculi ea laus est & Naturæ benignitas, qvæ se totam curiosis explicat. Nos, dum licuit per ætatem, hanc lampada accendimus. Alii jam majori felicitate suppleant qvod perfectioni ejus & splendori deesse videbitur. Inter qvos Stenonius noster, Prosectorum nulli secundus secandi promtitudine & inveniendi facilitate, tam benigno genio his partibus domi forisq; functus est, ut in admirationem sui orbem eruditum rapuerit. Phidiæ Jupiter Olympias qvoridie testimonium perhibeat, Mentori Capitolinus & Diana Ephesia, qvibus fuere consecrata artis vasa ; Stenonio theatra Europæ meliora assurgunt, ora Doctorum, fama publica, ipse deniq; Homo, qvem omnibus, qvibus volupe, proponet, ut homines nos esse sciamus ad resolutionem natos non superbiam, non obtrectationem, non deniq; Dei & Naturæ contemptum. Qvicunq; hæc arcana nobiscum non fastidit, speluncam Democriti nostri accedat, ubi Deos etiam inveniet, nusqvam alibi faventiores. Non lassari videbit manus, nec ingenium deficere, nec Naturam obmutescere. Patebit omnibus Artis fautoribus & amicis, in primis qvi Æsculapio confecrati partium accuratam demonstrationem desiderant & nosse debent, Theatrum nostrum Anatomicum, ea lege & condicione, ut pari patientia Prosectorem incomparabilem audiant spectentq; qva laboriosum opus peragi viderint, & ea benevolentia Clarissimum Doctorem, non pompa sed opere æstimandum excipiant, qvam meretur ille, apud qvem, exemplo *Hippocratis*, major est sapientiæ ratio, qvam auri. Ne nescii sint, qvibus sciendi videndiq; amore inflammatur pectus, Demonstrationis initium crastinus dies dabit, hora post meridiem secunda. Hafniæ d, 28. Januarii Anno cIɔ Iɔc LXXIII.

Sub Facultatis Medicæ Sigillo.

Fig. 2.5 Program for Niels Stensen's public dissection on January 29, 1673 st.v. in Copenhagen, created by Thomas Bartholin
KB, 34:3,–79 2°

of his student Holger Jakobsen, the *Exercitia academica*, depict Stensen as a teacher who set great story by methodical thinking.[331] He had previously already demonstrated his didactic abilities in Paris with his *Discours*.

---

[331] Scherz, *Vom Wege*, p. 121. Kardel, Steno, p. 88 reckons that Stensen had been "not a very good speaker", and that his strengths had lain in anatomical demonstrations in small groups and in written presentations.

In the January 28, 1673 st.v. invitation, Bartholin announced the "beginning of the presentation" by Stensen for the following day[332]; on January 29 st.v., Stensen officially reopened the so-called *Domus Anatomica*,[333] the Copenhagen Anatomical Theater on today's Frue Plads which had been unused for many years prior to his arrival in the city and made available for his use, with his inaugural address *Prooemium demonstrationum anatomicarum in Theatro Hafniensi anni 1673* (Preamble to the Anatomical Demonstrations in the Copenhagen [Anatomical] Theater in the Year 1673).[334] According to Jakobsen's count, this presentation taking part over the course of several days was Stensen's "16th public dissection"[335]; it also was to remain the only one announced by way of a printed leaflet. Hence Stensen's inaugural address formed the introduction to his major anatomical demonstration from January 29 to February 8, 1673 st.v. (see Figs. 2.5, 2.6, and 1.2).

Bartholin writes of a "human corpse of female gender which has been made available [for dissection] for the honor of God, the advancement of nature and the benefit of the republic of physicians" in the leaflet.[336] Stensen had expected the dissection to commence as early as November 19, 1672 after it had already been postponed for a few weeks, but it was subsequently delayed even further, presumably due to the difficulty of obtaining a corpse.[337] That the female corpse was eventually provided by the executioner, as Gustav Scherz claims, is not documented[338]; the example of Anne Mickelsdatter, who had been executed for the murder of a child and then dissected by Thomas Bartholin on February 12, 1653, shows that this pos-

---

[332] Bartholin, *Decanus Anatomes s*[alutem], line 51: "demonstrationis initium". – Reference to the program is made in the compilation in AMPH 2 (1673), pp. 375–376 (Programmata Academiae publica), here 37, no. 5: "Th[omae] Bartholini decani programma ad sectiones anatomicas publicas d[omini] Nicolai Stenonii. Hafn[iae] 1673. in fol[io]".

[333] Bartholin, *Anatomihuset*, p. 253, Fig. 87 (original location in what today is the main building of the University of Copenhagen at Frue Plads); ibid., p. 123, Fig. 11 (longitudinal section); ibid., p. 254, Fig. 89 (layout); ibid., p. 11, Fig. 2 & p. 116, Fig. 8 & p. 252, Fig. 88 (exterior views); ibid., p. 11, Fig. 2 & p. 119, Fig. 9 & p. 256, Fig. 90 (interior views of the anatomical theater); ibid., p. 122, Fig. 10 (autopsy table and instruments). See supplementarily Bartholin, *Cista Medica Hafniensis*, p. 189, Fig. 4 & p. 191, Figs. 8, 9, 10, 11. – The *Domus Anatomica* burned down together with the other university buildings during the Copenhagen Fire of 1728, cf. Maar, "Domus Anatomica", p. 17.

[334] Prooemium. On Stensen's activity in the *Domus Anatomica* see most recently Bartholin, *Anatomihuset*, pp. 250–251.

[335] Kardel, *Steno*, p. 142 (b/w photograph of KB, Ny kgl. Samling 309aa 4°, fol. 145ᵛ): "Anatome XVI publica". In ExAc., p. 300, line 21,"feminei" is transcribed erroneously instead of "faeminini". The text of Holger Jakobsen's notes taken at the inaugural address and omitted in ExAc., p. 300, lines 21–22 can be found in Kardel, *Steno*, pp. 129–130 (English trans.), 142 (b/w photograph of KB, Ny kgl. Samling 309aa 4°, fol. 145ᵛ–147ʳ, without transcription).

[336] Bartholin, *Decanus Anatomes s*[alutem], lines 32–33: "humano foeminii sexus cadavere, quod oblatum fuit […] in Dei gloriam, Naturae augmentum & commodum reipublicae medicae."

[337] E 87, p. 278, lines 21–22.

[338] Without source in Scherz, *Vom Wege*, p. 117. – It was only on 09/021685 st.v. that a decree was passed by the Danish king Christian V ordering the corpses of executed persons to be delivered upon request to the Copenhagen *Domus Anatomica* by convicts from the prison on Bremerholm (Melchior, *Fakultet*, pp. 43–44).

Fig. 2.6 Niels Stensen, presumably as "Royal Anatomist" (Anatomicus regius) in Copenhagen
(1672–1674)
Contemporary oil painting by an unknown artist, located in the Copenhagen *Medicinsk Museion*,
cf. Scherz, *Vom Wege*, plate between pp. 112 & 113; Sacra Congregatio pro Causis Sanctorum,
Positio, pp. 1051–1051, no. 3
KU-MM, Billede no. I-000263

sibility cannot be excluded, however.[339] More likely is that the body was the result
of a death among the poorer population of Copenhagen, possibly during lying-in
after childbirth; such deaths were to be reported to Stensen to aid him in his instruction of midwives,[340] and in this case the corpse would probably have been that of a
comparatively young woman. This is also supported by the fact that Stensen repeatedly emphasizes the beauty of the human body and broaches the issue of sexuality
in his preamble.

---

[339] Bartholin, *Anatomihuset*, p. 140, ad p. 71, line 7.
[340] See note 538, in Chap. 2 below.

In his invitation, Bartholin addresses the targeted audience with the statement
that the Anatomical Theater would be open to "all patrons of the craft and friends,
especially to those consecrated to Aesculapius [= students of medicine] who desire
and are required to know a thorough demonstration".[341] It is unknown whether – as
in the first years of the *Domus Anatomica* – an entry badge was handed out against
payment of a fee by the anatomist himself, namely until 1648 by the extraordinary
professor of anatomy, surgery and botany and royal personal physician Dr. med.
Simon Paulli Jr.,[342] Thomas Bartholin's predecessor, and subsequently by the latter
in order to refuse access to "children and ordinary folk",[343] or whether the presentation featured no restriction on admission. Access control for public sections was
common from the end of the fifteenth century throughout the entire early modern
period as a way of controlling the attendees due to repeatedly occurring conflicts
between the anatomist and the public.[344] The existence of an entry badge as mentioned above at the University of Greifswald is documented for the time of Stensen's
public dissection (Fig. 2.3).

The beginning of Stensen's inaugural address had been scheduled by Bartholin
for "the second hour after midday", ie 2 p.m.[345] It can be assumed that Stensen
began punctually, as Holger Jakobsen, who was among the spectators, put down at
the beginning of his hand-written notes "January 29, 2 o'clock in the afternoon".[346]
After the end of his lecture, Stensen did not start the dissection of the corpse, which
he only commenced on January 30, but instead – according to Jakobsen – proceeded
to discuss the structure of the human nervous system with the help of a preparation
and an explanatory panel.[347] Thus the inaugural address served as a prelude to the
dissections during the following days, announced by Stensen[348] at the end of his talk
and demonstrating to the audience the current state of research.

Dissections of female corpses – particularly if conducted in public, as was popular for inaugurations of anatomical theaters, – were often problematic in the seventeenth century as they had an air of scandalousness about them.[349] Therefore, a
glance may be cast at this point at lesser known dissections, dissections which have
hitherto not been subjected to analysis, and contemporary opinions in the Holy

---

[341] Bartholin, *Decanus Anatomes* s[alutem], lines 47–48: "omnibus artis fautoribus & amicis, in
primis qui Aesculapio consecrati partium accuratam demonstrationem desiderant & nosse debent".

[342] Egill Snorrason/Anne Fox Maule, "Paulli, Simon"; in: DBL³ 11 (1982), pp. 181–183, here 181.

[343] For the time before 1648: Bartholin, *Anatomihuset*, p. 56, line 11–p. 57, line 3, here p. 56, lines
16–17: "pueri ac profanum vulgus"; on this, see ibid., pp. 134–135; for the time after 1648 cf. the
similar phrasing ibid., p. 68, lines 10–16, here 12: "promiscuum vulgus"; on this, see ibid., p. 140.
Cf. Faller, "Eintrittskarten"; Wolf-Heidegger/Cetto, *Sektion*, p. 54.

[344] Bergmann, *Patient*, pp. 182–183.

[345] Bartholin, *Decanus Anatomes* s[alutem], lines 51–52: "hora post meridiem secunda".

[346] Kardel, *Steno*, p. 142 (b/w photograph of KB, Ny kgl. Samling 309aa 4°, fol. 146ʳ): "D[ies] 29
Januar[ii], h[ora] 2 pomeridiana".

[347] Details ibid. (b/w photograph of KB, Ny kgl. Samling 309aa 4°, fol. 147ʳ).

[348] Prooemium, p. 255, line 37–p. 256, line 2.

[349] Cf. with further examples Bergmann, *Patient*, pp. 183–184 & p. 371n627.

Roman Empire, the Netherlands and Denmark after the onset of the seventeenth century.[350]

One academic lending himself to comparison in this regard is Dr. med. Johannes Jessenius von Jessen (1566–1621),[351] professor of anatomy and surgery at the University of Wittenberg, who had received his doctorate in Padua and wanted to introduce the practice of public dissections he had witnessed there in his home environment. He was forced, however, to answer back to the Wittenberg Lutheran professor primarius and city superintendent Dr. theol. Ägidius Hunnius (1550–1603),[352] who blamed him for scandalous events during the dissection on November 15, 1599 st.v. of a woman executed for adultery. Jessen defended the dissection of female corpses in his invitation, dated January 5, 1600, to the next such event on the following Friday, January 9 st.v.,[353] with the following words: "In the meantime, what then actually forbids dissection of the corpses of women? Dignity perhaps? But it is greater in men due to their priority in gender. Or because through the sight of the hidden parts all sense of shame is betrayed and the door is opened to the impetuousness of the sexual drive? Pretty words. I let all honest and virtuous men among the spectators pass their judgment on this accusation. I am quite certain they will say what they also say of their own volition, that Jessenius undertook the description of those areas so augustly and chastely that anyone could not interpret a psalm by David any more piously: also that water extinguishes fire no better than this spectacle subdues sensual fervors."[354] In contrast to Jessenius' appraisal of the difference in dignity between the sexes, such uprating of men and likewise any depreciation of women were not a part of Stensen's views; in fact, there is evidence that he held a completely unbiased stance towards women.

---

[350] On the public dissection of the corpse of an unknown black African woman, who had presumably died of pneumonia, on 12/25/1675 st.v. in Kiel by Nicolaus Pechlin see AMPH 4 (1676), pp. 50–51 (XVII. Anatome Aethiopis faeminae. Ex literis d[omini] Joh[annis] Lud[ovici] Hannemanni. Kiloni 25. Decembr[is] 1675); Mazzolini, "Kiel", pp. 372, 379–380. On the program: Cimbria lit. 2 (1744), s.v. "Johannes Nicolaus Pechlinus", pp. 633–639, here 636: Pechlin, Programma; retaining the early modern notation: *Programma Anatomæ cadaveris foeminæ Æthiopicæ præmissum. Kil[onii] 1675. in f[olio] pat[enti]*; Dietrich Korth, "Pechlin, Johann Nicolaus"; in: SHBL 7 (1985), pp. 164–166, here 165. As noted likewise in Mazzolini, "Kiel", p. 379, no copy could be located.

[351] On Jessenius von Jessen: Heinz Rührich, "Jessen(ius), Johannes (Jan Jessenský)"; in: NDB 10 (1974), pp. 425–426; Koch, "Wittenberger medizinische Fakultät", s.v. "Jessenius, Johannes von", p. 314.

[352] On Hunnius: Theodor Mahlmann, "Hunnius, Ägidius"; in: TRE 15 (1986), pp. 703–707.

[353] Jessenius von Jessen, *Invitatio*, pp. A 6–7; Friedensburg, *Universität Wittenberg*, pp. 456–457; Ruttkay, "Jessenius", pp. 37–38.

[354] Jessenius von Jessen, *Invitatio*, p. A 4: "Verum quid tande[m] foeminaru[m] incidi ita vetat corpora? dignitasne? at haec in viris ob sexus praestantiam maior. An quod ex abditarum partium intuitione pudor omnis detectus sit, apertaq[ue] ad libidinum aestus fenestra? Bona verba. Sino spectatorum quosq[ue] candidos atq[ue] bonos huius ferre judicium exprobrationis. Certius sum, dicturos, id quod & ultro dicunt, Iessenium locorum istorum explicationem tam sancte tamq[ue] caste pertractavisse, ut ne quidem religiosius Davidicum hymnum quispiam interpretaturus fuerit: nec ignem tam aquam fortiter extinguere, quam venereos furores hoc spectaculum compescere."

Pronounced reservation in regard to publicity can be observed in Dr. med. Zacharias Rosenbach (1595–1638), who among other places visited Padua, Rome and Naples in 1616, continued his anatomical studies in Montpellier in 1617 and became professor of medicine and oriental languages at the Reformed Herborn Academy in 1623. For the dissection of the genitals of a woman executed as a criminal, undertaken in the week of Christmas 1623, he excluded his students and conducted the dissection only in front of an invited circle of professors.[355]

The Vicar Apostolic of the Dutch Mission, Titular Archbishop Licent. theol. Philippus Rovenius ([* 1573] 1614–1651),[356] a 'counterpart' of Stensen's two generations earlier who had never studied medicine but was active as a writer, produced the following warning in his *Reipublicae Christianae libri duo [...]* (Two Books on the Christian Republic [...]; Antwerp 1648), a treatise on the administration of missions: "With regard to anatomy, the physicians and surgeons should take care during the cutting and examining of the bodies of the deceased not to be easily susceptible to that which has any air whatsoever of crudity about it. Therefore they should use anatomy extremely rarely, and only in case of some great urgent necessity or usefulness, in order to find the roots of the diseases, so that they may help others, but not to satisfy their curiosity, never mind their desire while examining the bodies of women or virgins."[357] These lines by Rovenius, who set great store by medical assistance and personal hygiene for the preservation of health, must be understood in the context of his rightful condemnation of contemporary charlatanry by apothecaries and surgeons without education; Rovenius in fact wanted to see the position of the physician strengthened: he also warned, for example, that surgeons "should not casually without approval or the mandate or knowledge of the physician open the veins of the sick, and also not resort to cauterizing or removing the limbs without a physician being present and ordering it so."[358] Rovenius apparently was of the opinion that anatomy was a sometimes similarly barbaric craft. Fluctuating between leniency and rigorousness in his other opinions[359] – presumably due to his unsystematic way of working –, he apparently intended to curb an immoral attitude which

---

[355] Grün, "Herborn", pp. 69–70, 104–109. On Rosenbach: NassB, s.v. "Rosenbach, Zacharias", p. 658.

[356] On Rovenius: Paul Berbée, "Rovenius (van Rouveen), Philipp"; in: BHRR II, pp. 598–600, here 599 on his work.

[357] Rovenius, *Reipublicae Christianae* [on the year of printing cf. Visser, *Rovenius*, p. 61] (Liber secundus, Caput XIII, [Nr.] V), p. 363, line 32–p. 364, line 2: "Circa anathomiam attendant medici & chirurgi ne faciles sint ad ea quae crudelitatis aliquam speciem habent, in secandis & scrutandis defunctorum vel occisorum corporibus. Itaque raro admodum anathomiam instituant, & non nisi urgente aliqua magna necessitate, aut utilitate, ad investigandas morboru[m] radices, ut alios [364] possint juvare, non autem ut suae curiositati, ne dicam libidini in scrutandis mulierum vel virginum corporibus, satisfaciant."

[358] Rovenius, *Reipublicae Christianae* (Liber secundus, Caput XIII, [Nr.] V), pp. 362–363, here 363: "sine consensu vel praescripto aut prudentia medici non facile aegrotis venam aperiant: nec ad ustionem vel praecisionem membrorum accedant, nisi medico praesente & jubente." Cf. Visser, *Rovenius*, p. 71.

[359] For details see Visser, *Rovenius*, pp. 151–157.

he considered a possibility, most likely because he had adopted without reflection prejudices abounding among the population, and hence delivered his well-meaning but nevertheless strangely unrealistic advice.[360] Stensen, by contrast, dissected the ovary of a young woman's corpse in Florence[361] and was more concerned with restricting vivisections of animals to a minimum. With his basic knowledge about the usefulness of anatomy, Rovenius should in fact have recognized the sense in frequent rather than infrequent scientific dissections.[362] For on the whole, the position of the Church was that of promoting anatomy: Andrew Cunningham emphasizes that "the Catholic Church has never been opposed to the practice of anatomy, whether for post-mortem, demonstration, teaching or research purposes".[363] Taking into consideration the papal archiaters like Malpighi she was – perhaps even more so than Protestantism – "actually virtually a proponent and patron of anatomising", not least because dissection benefitted the Christian position and mission.[364]

The common prejudices mentioned above are apparent in a report by the physician and polyhistor Dr. med. Johann Joachim Becher (1653–1682),[365] a contemporary of Stensen who complained about the adversity he faced from the population of Würzburg in 1661 in his *Methodus didactica [...]* (Didactic methodology [...]; Frankfurt 1668): "For the people here in Germany are quite strange in this regard [= in their attitude towards anatomy]. Some time ago in Würzburg I dissected an executed woman with the permission of the superiors; the entire city became hostile to me, and did not stop before they had chased me off."[366] No details on the mentioned public dissection are known. The overreaction of the Würzburg population can be

---

[360] Cunningham, *Anatomist*, p. 14n15 similarly refers to the "outrage of local clerics" concerning the exhibition "Körperwelten" (since 1995) by the anatomist and artist Gunther von Hagens, which was intentionally provocative.

[361] See note 278, in Chap. 2 above.

[362] Visser, "Relation", p. 50 similarly states the following about Rovenius' ignorance of the Roman condemnations regarding questions of the theological doctrine of grace: "He was not informed about the decrees by the Holy See, but he was willing to submit to its judgment" (Il n'était pas informé des décrets du Saint-Siège, mais il était prêt à se soumettre à son jugement). For details cf. Visser, *Rovenius*, pp. 26–27. – On the scientific need for anatomy cf. Boudewijns, *Ventilabrum*, pp. 227–234 (Pars prima, Quaestio XXXVII), here 228: "anatomy is allowed as well as necessary for physicians" (Anatomia medicis tam licica [*corr.*: licita] est, quam necessaria). Ibid., p. 231 Boudewijns highlights in a notably 'ecumenical' fashion the "Christian anatomists" (Christiani Phylotomi [*corr.*: Phytotomi; *presumably a printing error instead of* "Anatomi"]) like Thomas Bartholin, Jean Pecquet, Johannes van Horne, Anton Deusing and others.

[363] Cunningham, *Anatomist*, p. 14. Cunningham, who ibid. describes himself as "a life-long evangelical atheist" who "certainly hold[s] no brief for the Catholic Church", ie is by no means a lawyer for the Church, emphasizes that "all of this anti-Catholic nonsense", which is "nothing but a lie", had been propagated so widely that even some historians of science still accepted it. The same is emphasized by Michelangelo Peláez, "Medicina"; in: DISF 1 (2002), pp. 901–919, here 909.

[364] Cunningham, *Anatomist*, pp. 14–15, here 14.

[365] On Becher: HGK, s.v. "Becher, Johann Joachim", pp. 80–81.

[366] Becher, *Methodus didactica*, p. 51 r. col.: "dann die Leut bey uns in Teutschland hierinnen gar seltzam seynd/ich habe zu Würtzburg einsmals permissu Superiorum ein Justificirtes Weibsbild anatomirt, darüber ist mir die gantze Stadt feind worde[n]/wie sie dann auch nit nachgelassen/biß sie mich von dan[n]en getribe[n]". On this, see Kölliker, *Universität Würzburg*, p. 11 with n. 1.

explained by the fact that, as emphasized by the Bonn moral theologian Werner Schöllgen (1893–1983), medical ethos "will never offer a measurable schema" with regard to cultural sociology, in particular concerning notions of intimacy.[367] It was a long time, Schöllgen states, before physicians were allowed to conduct the indispensable intimate examinations of the female body necessary for treatment.

Similarly to Becher, Thomas Bartholin in Copenhagen also spoke of the public dissection as an "art disdained and neglected due to the revulsion of the people" in 1673.[368] Bartholin conducted several public dissections of female corpses, and he highlighted in one of his announcements printed in his booklet *Domus Anatomica Hafniensis brevissime descripta* (The Copenhagen Anatomical Theater Briefly Explained; Copenhagen 1662) the bashfulness of the anatomist's hands touching the corpse, and of his and the audience's eyes, which he considered to be a matter of course during such dissections.[369] Hence revulsion does not represent a sustainable foundation for appropriate behavior during public dissections.

Another scientist calling for respect for the corpse was Heinrich Meibom Jr., who at the end of his invitation to the dissection of a female corpse scheduled for Friday, February 28, 1673 st.v. in Helmstedt, which will be further discussed in the following, admonished his potential spectators: "Yet while I present for examination this female body (a denuded Diana without robes) with all its parts, at the same time I request of the observers a chaste mind and chaste eyes. For this exposure of that which nature herself wanted hidden occurs not without an important and plausible reason for the use expected from it, and therefore it is appropriate to view and examine it with a certain sacred shiver and silence."[370] The phrasing "Diana without robes"[371] alludes – typically for the period – to the myth of Actaeon sung by the Roman poet Ovid (43 B.C.–c. 17 A.D.), in which the protagonist inadvertently surprises the naked goddess Diana during her bath. The previously mentioned Christoph Helwig Sr. appealed particularly elaborately to his audience's sense of shame in his invitation to the dissection of the genitalia of a female corpse resulting from an execution scheduled for Tuesday, April 3, 1677. After introducing the "denuded Diana without robes" in similar fashion to Meibom by way of several quotes from Ovid, he admonishes thus: "That above all you bring a clean mind, bashful eyes and chaste ears to *these holies* [haec sacra], that I demand."[372] Meibom's and Helwig's

---

[367] Schöllgen, *Soziologische Grundlagen*, pp. 316–317 with n. 2. On Schöllgen: Gerhard Mertens, "Schöllgen, Werner"; in: LThK³ 9 (2000), col. 204.

[368] Bartholin, *Decanus Anatomes* s[alutem], line 19: "spretae neglectaeq[ue] apud vulgus artis taedio."

[369] Bartholin, *Anatomihuset*, p. 70, lines 7–19, and see ibid., p. 140; cf. Bergmann, *Patient*, p. 184.

[370] Meibom, *Programma*, fol. A 4ᵛ: "Dum tamen hoc muliebre corpus (nudam sine veste Dianam) cum omnibus suis partibus spectandam propono, castam pariter mentem & oculos ab intuentibus posco. Nec enim haec eorum, quae Natura ipsa occulta voluit, propalatio sine gravi caussa & evidenti, quae inde speratur, utilitate fit, ideoque quasi sacro quodam horrore & silentio inspicere illa & tractare par est."

[371] Holzberg, *Tristia*, p. 70 (Tristium II,105): "sine veste Dianam"; on this, see ibid., p. 555.

[372] Helwig, *B[enevolo] l[ectori] s[alutem]*, fol. 2ʳ: "nudam & sine veste Dianam. Puram ante omnia mentem, verecundos oculos, castas aures in haec sacra ut inferatis postulo."

exhortations, however, are characterized by a certain abashed stiltedness – perhaps itself due in part to their dissection objects – when compared to the matter-of-fact language devoid of baroque mannerism used by Stensen in his Copenhagen inaugural address, which will be presented in the following.

2.3.1.3   Physicality and Sexuality of Humans

In his 1664 accountability report on his studies abroad – which was reprinted in Leiden in the Republic of the Seven United Netherlands in 1683 while Stensen was suffragan bishop in Münster –, Stensen points out how the arrangement of vessels and nerves augmented the "gracefulness" (gratia) of the human face.[373] With a view to this statement, August Ziggelaar asks: "Was Niels Stensen's delight with the beauty in nature purely platonic? I mean: could human beauty not awaken erotic feelings or desire within him? I know nothing of eroticism in the life of Niels Stensen, but he possessed a sense for and understanding of it, and he was wary of its dangers."[374] What, then, constituted Stensen's personal, medical and theological attitude towards human sexuality? In the same way he did not explicitly develop a system of medical ethics, however, a cohesive system of sexual ethics cannot be expected from his statements either.

That Stensen faced the opposite sex unbiasedly is evidenced by a psychologically telling note from his examination of conscience, composed in 1683/1684 in Hamburg in the style of a general confession, in which he claims to have interacted too carelessly with women – mentioning relatives, penitents, his female cook and other more general occasions –, thereby giving others "harmful opportunity for suspicion and gossip".[375] Stensen was not worried that he or the mentioned women had done or said anything wrong, but about the permanent danger of malicious defamations a priest was faced with; he had previously broached this subject in a lecture to other priests in Münster.[376] Already during the time before his episcopal ordination he had reported to Grand Duke Cosimo III from Rome how the more informal interaction with women in Denmark, which had already proved ruinous to others, had also been to the disadvantage of a Catholic missionary who otherwise led a saintly life, as a Lutheran preacher had accused him thusly during a conversation with Stensen.[377] For himself, however, Stensen seemed by his unsuspecting nature

---

[373] MuscGland., p. 184, lines 14–17.

[374] Ziggelaar, "Udvikling", p. 6: "Var Niels Stensens glæde ved skønhed i naturen rent platonisk? Jeg mener: Kunne menneskelegemets skønhed ikke vække erotiske følelser eller længsler i ham? Jeg kender ikke til erotik i Niels Stensens liv, men han havde sans og forståelse for den, og han var på vagt over for dens farer."

[375] Op. 4, p. 442, lines 5–7, here 6–7: "occasio mala suspicandi et loquendi".

[376] In Sermo 44, p. 367, lines 21–22, Stensen speaks of dwelling on "ordinary conversations with the weaker sex" (conversationes ordinariae cum segniori sexu).

[377] E 128, p. 343, lines 26–36. – Presumably for their protection, Stensen mentions neither person by name.

to underestimate this danger. Already in Florence he, whose conversion was prepared through spiritual conversations with women – in Florence, Sr. Maria Flavia
del Nero OSCl.[378] of the pharmacy of the convent of San Vincenzio and the wife of
the ambassador of Lucca, Lavinia Arnolfini, *née* Cenami (1631–1710) –, had been
forced to be cautious in this regard: Mrs. Arnolfini, who was only a few years older
than Stensen, had a husband who was about 27 years her senior and presumably for
this reason had admitted Stensen to religious dialog only "against my style".[379]

Stensen had found opportunity to dissect the female genitalia already as a
23-year-old student of medicine: on January 27/28, 1661, the "corpse [of a woman
named] Helena", who had died of a disease, was made available for the students of
the University of Leiden to dissect under the guidance of Sylvius and Van Horne in
the autopsy room of the *St. Cecilia Gasthuis*.[380] As "Anatomicus regius", Stensen
then for the first time publicly dissected a female corpse starting on January 29/30,
1673 st.v. He had accepted his appointment to Copenhagen – which held no pros-
pect in regard to his career, as he was paid a lowly wage and was only to be afforded
a "teaching lectureship" of sorts for an unknown period of time without becoming a
member of the medical faculty of the University of Copenhagen – "only" under the
"motive of faith", ie he saw it from a missionary perspective.[381]

Stensen's stance on sexuality, communicated in his Copenhagen inaugural
address over the female corpse laid out on the dissection table – in anticipation of
the public dissection of the woman's genitals and reproductive organs which he
would undertake on February 6, 1673 –, is enlightening. His straightforward
anatomical-physiological approach to the topic is extraordinary: he begins by stating that anyone who remembered ever having observed "the charms of any [human]
beautiful form" with a mind insufficiently prepared against allurement would
concede what power and effect the human appearance had on the human mind.[382]
And yet this view onto a human occurred only as if from afar, when one used a
microscope and recognized that we see of the skin only the less developed tips,
similar to the spikes of grain on a distant wheat field.[383] If one were allowed to penetrate deeper, one felt like in labyrinths of which one barely knew anything, and

---

[378] On Del Nero: Manni, *Stenone* (Libro I, Cap. XVIII), pp. 62–68, here 62, 64, according to which
she had been in the convent since 1631 and was "di età provetta" (of advanced age); E 83, p. 271n1.

[379] Bambacari, *Signora Arnolfini*, pp. 37–38, here 37 = Sacra Congregatio pro Causis Sanctorum,
Positio, pp. 102–103, here 102: "contro il mio stile". On Arnolfini: Gaspare De Caro, "Cenami,
Lavinia Felice"; in: DBI 23 (1979), pp. 498–499, here 499 on Stensen. Stensen had taken a vow of
chastity following his conversion, presumably in private before his father confessor; see Add. 15,
p. 933, lines 29–30; Sobiech, *Herz, Gott, Kreuz*, p. 78. On possible advances by the daughter of his
first landlord in Florence see Bernardi, *Paggio*, pp. 111–113.

[380] Details in Schepelern, *Olai Borrichii Itinerarium*, p. 69–71, here 69: "cadavere Helenae". On
Van Horne, cf. ibid., p. 70, no. 4 & p. 70–71, no. 5 & p. 71, no. 9; on Sylvius, cf. ibid., p. 70, nos.
1, 3, 4 & p. 71, no. 9.

[381] As mentioned in a letter to Cosimo III from Florence on 12/28/1671: E 70, p. 254, lines 27–32,
here 27: "il solo motivo della fede".

[382] Prooemium, p. 253, lines 2–5, here 3–4: "ullius formae veneres".

[383] Ibid., lines 11–15.

even that only through conjecture as it eluded mere sensory perception. From this Stensen infers in the guise of a rhetorical question: "Who then [under these circumstances] would still rely on sensory perception of the exterior surface and render judgment about the remainder based on the pleasantness or arduousness of that perception?"[384]

He continues by making a distinction between himself and those "ethicists" (Ethici) and moral philosophers who would endeavor to find something reprehensible or ugly in the object of human love in order to prevent desire: "So that the ethicists may turn the spirits away from harmful liaisons, they find in the beloved object nothing but that which is worthy of reproach".[385] By the term "harmful liaisons" (noxii amores) – they might also be described as "disordered liaisons" (amores inordinati) –, Stensen presumably means, without explaining or even wanting to explain in detail, disordered sexual inclinations. In his mention of the "ethicists", he may have been thinking eg of the Stoic philosopher Lucius Annaeus Seneca Jr. (c. 1 B.C./1 A.D.–65 A.D.), who likens the "love for beauty" to insanity:

In this vein, in his book *Amaltheum medico-historicum [...]* (Medico-Historical Amaltheum [...]; Padua 1668), influenced among others by philosophers like Seneca Jr., Dr. med. Francesco Boselli (1620–1680),[386] professor of surgery at the University of Padua from 1662, mentions briefly the legitimate, reason-born use of sensual desire on the one hand,[387] but on the other offers up a broad and verbose description of various dangers originating from human sexuality.[388] Taking a view common at the time, Boselli considers conjugal sexuality chiefly under the aspect of sinfulness.[389] While he initially explains – similarly to Stensen – the effects of beautiful bodies as the highest stimuli provided by nature,[390] he then goes on to repeatedly cite the aphorism by the Stoic Seneca condemning sensual-affective love in marriage: "*The love for beauty* [amor formae] is the forgetting of reason and a neighbor to insanity".[391] Philippus Rovenius also makes reference to this aphorism with identical intent in his *Reipublicae Christianae libri duo*.[392] Seneca's statement is founded in the philosophical lessons of Stoicism, which place the ἔρως within the realm of objectionable affects and require the drives to be subordinate to human

---

[384] Ibid., lines 20–26, here 24–26: "quis amplius in solius externae superficiei perceptione sensibili haereret, & ex illius perceptionis suavitate, vel molestia de reliquo judicaret?"

[385] Ibid., lines 31–32: "Ut avertant animos a noxiis amoribus Ethici, in objecto amato reprehendenda omnia investigant."

[386] On Boselli: ScrittIt. 2,3 (1762), s.v. "Boselli (Francesco)", p. 1830; Cigogna, *Inscrizione*, pp. 12–13, no. 14. Boselli's works have hitherto not been analyzed.

[387] Boselli, *Amaltheum* (Apparatus II, Caput 8, Syntagmata 2, 3), p. 144.

[388] Ibid., passim.

[389] Cf. Venard, "Fragen der Ethik", pp. 1196–1197.

[390] Boselli, *Amaltheum* (Apparatus II, Caput 1, Syntagma 2), p. 76.

[391] Full version ibid. (Apparatus II, Caput 8, Syntagma 6), p. 146, line 20 & ibid. (Apparatus III, Caput 29, Syntagma 17, no. 138), p. 552, lines 20–21: "Amor formae rationis oblivio est, & insaniae proximus". This is in fact a fragment of Seneca's unpreserved *De matrimonio* (Trillitzsch, *Seneca*, p. 374; cf. Stelzenberger, *Sittenlehre*, pp. 429–438, here 435n127).

[392] Rovenius, *Reipublicae Christianae* (Liber secundus, Caput XVII, [no.] 5), p. 393, line 20.

reason (ratio).[393] The aphorism also appears in a slightly altered version in *Ethica ordine geometrica demonstrata [...]* (Ethics, Demonstrated in Geometrical Order [...]; published as part of the *Opera posthuma [...]*, [Amsterdam] 1677 [corr. 1678]) by the philosopher Baruch de Spinoza (1632–1677), who acknowledged the Stoics and Descartes.[394]

A further example for what Stensen was referring to can be found in Ovid's *Remedia amoris* (Remedies against Love), where the author suggests the following with reference to the reader's internal attitude towards a woman in order to rid himself of "amor" for her: "Wherever you can, reverse the charms of the woman into their opposite and deceive your judgment with the fine distinction", with the latter meaning the mental expansion of actually existing bodily shortcomings.[395]

Due to the insights into physiological processes within the human body gleaned from his autopsies, Stensen was of the opinion that intentional and demonstrative dispraise of the object of the "noxii amores" was likely not conducive to a successful treatment due to its inherent element of untruth, and suggested the following: "The anatomist, however, asked for a remedy for such liaisons, would not stoop to inculpating [the beloved beautiful objects] but would instead lift up the loving mind to *nobler objects of love*", should the subject not be completely "incapable" (ineptus) of elevating himself at least marginally over the sensual.[396] His use of the word "incapable" may indicate that Stensen was referring to neurotic aberrations. For him, physical beauty was also displayed in human sexuality.

If, however, as Stensen adds in his inaugural address, "the desire for impermissible amenities" or "the love for forbidden things" should cause one to take a 'skeptical' stance consisting of accusing the human senses and claiming that they did not allow things to be recognized for what they truly are, instead leaving everything incorrect or at least uncertain, then the response to this would have to be that it was precisely *not* the task of the senses to render things the way they are, or to judge them, but rather to transmit to the mind for examination that nature of things sufficient for one to acquire knowledge of these things commensurate with the (supernatural) goal of man.[397] The only true skeptic was he who used the inveteracy of his doubt as an excuse for his vice.[398] Humans possessed their "reason" (ratio) in order to use it as a judge and appraising instance for the things perceivable with the senses, and thus to rise up from the sensually perceivable to the sensually not

---

[393] Mauch, *Frauenbild*, pp. 51–54, here 53.

[394] Spruit/Totaro, *Ethica*, p. 286 (Pars quarta, Appendix, nos. 19, 20); on this, see Dilthey, "Autonomie", p. 988. On Spinoza: HGK, s.v. "Spinoza, Baruch de", pp. 628–630.

[395] Holzberg, *Ars amatoria* (Remedia 291–356), pp. 194, 196, 198, here 196 (Remedia 325–326): "Qua potes, in peius dotes deflecte puellae/Iudiciumque brevi limite falle tuum." On the emphasizing of physical deficiencies cf. also ibid. (Remedia 417–418), p. 202.

[396] Prooemium, p. 253, lines 32–35: "Anatomicus autem talium amorum remedia rogatus non ad culpanda se dimitteret, sed ad amoris argumenta nobiliora animum amantem elevaret [...] ineptum."

[397] Ibid., p. 253, lines 35–37 & p. 254, lines 1–8, here p. 253, lines 35–36: "suavitatum illicitarum desiderium" and 38: "rerum illicitarum amor".

[398] Ibid., p. 253, line 37–p. 254, line 1.

perceivable; far be it from us, Stensen says, to discard our humanity and place ourselves – by not using our singular human reason – on a level below the animals.[399]
Stensen follows Seneca in his use of reason (ratio), hence attaching his experience
as an anatomist to the theological assertion of the final goal of man.

The difference between the two anatomists Boselli and Stensen lies in the contrast between Boselli's rather fearful approach to sexuality and Stensen's knowledge – acquired through his own research – about the functionality of the human
body and human reproduction, his therefore gentler, therapeutic approach and his
insight, untainted by sexual phobia, that particularly in the area of sexuality as a
very powerful human drive, disregard is preferable from a psychological point of
view to strained effort which may result in neurotic tension. From his experience, eg
with the reflexes caused by unpleasant smells emanating from dissection objects,
Stensen must have known about not entirely willful bodily automatisms comparable
to sexual stimulation.[400] Furthermore, he was indifferent towards conventional
thinking and intent on reconciliation as an anatomist, geologist, and equally as a
theologian.[401] His views on sexuality are expressed clearly and in linguistically
appropriate fashion; in his biography of Stensen written in the 1960s, Gustav Scherz
pointed out the "wise calmness and balanced naturalness" of Stensen's elaborations;
the inaugural address, Scherz emphasizes, does not condone a "distorted sexual
life".[402]

What exactly did Stensen mean by the incapability to "elevate oneself at least
marginally over the sensual"? From January 1675 to 1677, he had instructed the
Florentine crown prince Ferdinando III de' Medici (1663–1713) in "natural" and
"Christian philosophy".[403] In the course of this teaching, he told the boy that, in just
like many people applied their talents too late, only temporarily or never at all, so it
was with those faculties of man which served to lead life according to the rules of
God in theoretical as well as in practical terms. A considerable number of people
remained on the level of animals their whole lives by using the entirety of their
mental gifts to learn, teach and practice vices. Stensen continued: "But many also

---

[399] Ibid., p. 254, lines 9–11.

[400] Ibid., p. 24–28.

[401] In his report to the Propaganda Congregation from Rome, c. July 1677, Stensen argues that a
missionary should remain indifferent in "scholastic disputes" (dispute scholastiche), and in the
"studies of the mind" (studio dello spirito) should know different paths if not from experience, then
at least in theory in order to be able to do justice to the individual case (E 128, p. 343, lines 10–25,
here 20, 22). In disputes in the field of philosophy of science, Stensen favored Lucius Annaeus
Seneca Jr. (Prooemium, p. 256, lines 18–20; Prodromus, p. 187, line 28–p. 188, line 9=Scherz,
Geological Papers, p. 142, lines 31–36 & p. 144, lines 1–11); cf. Iofrida, "Stensen", p. 895. – On
the "virtus in medio" in Thomas Aquinas, see ThLex., also at http://www.corpusthomisticum.org/
tl.html, s.v. "virtus", sec. e.

[402] Scherz, *Biographie*, vol. 1, pp. 312–313, here 313 (cf. Kardel/Maquet, *Biography and Original
Papers*, p. 301).

[403] E 95, p. 293, lines 30–33, here 32: "philosophie naturelle"; E 120, p. 322, lines 32–38, here 35:
"philosophia Christiana"; ibid., p. 323, lines 6–10, here 9: "philosophia Christiana"; ibid., lines
14–16, here 14: "filosofia Christiana". On Ferdinando III: Francesco Martelli, "Medici, Ferdinando
de'"; in: DBI 73 (2009), p. 43–47, here 43 on his teachers (Stensen is not mentioned).

spend their whole lifespan in the realm of the sensual alone, as is the way of children, and thereby never break through the crust to reach the unperceivable [ie that which transcends the senses]".[404] Although sexuality is not explicitly mentioned here, it can be presumed to be implicit. Stensen saw the danger of grave sin in case of sexual aberrancies clearly: it was not the corpse, but the sins that stank, he proclaimed during his inaugural address.[405]

When Stensen talks about "sexual love", food and drink as "animal pleasures" in a sermon,[406] he is referring to those pleasures which are enjoyed "with God's concurrence, according to the will of God, [but] without any reference to God".[407] He does not intend to devalue the sexual, precisely because he emphasizes the necessity of its connection with God as a criterion of being human. Hence in another sermon he includes matrimony in a request, formulated in keeping with Col 3:17, to "do everything in the name of Jesus".[408] In the spirituality of the 'circulation of love' as a process between God and humans, which he fully developed after his conversion, the human body and its sensory organs as "interpreters of divine love" play a fundamental role.[409] Stensen views the body as a God-given instrument for achieving holiness.[410] The body as an appendage of Christ (1 Cor 6:15) must therefore not be abused: while the members of the various worldly occupations knew how to handle their tools professionally, "the soul alone abuses the instrument closest to it in the most inadequate way".[411] Thus the human body contributes significantly to reaching the goal of life.

2.3.1.4 Conjugal Sexuality

As a student in Copenhagen on Ascension Day, May 12, 1659 st.v., Stensen had resolved – possibly as a result of the sermon he had heard in the Lutheran divine service he had attended – that if God ever made him father of a family, he intended to accustom himself, his wife and his children to the very healthiest of lifestyles and to raise his children, as soon as they had learned to speak, in such a way that no time

---

[404] Sermo 40, p. 348, lines 8–21, here 19–21 (corresponds to Scherz, *Geological Papers*, p. 260, lines 9–19, here 18–19): "Sed et multi puerorum more in solis sensibilibus toto vitae tempore morantur, numquam perrumpentes corticem, quo ad insensibilia penetrent."

[405] Prooemium, p. 254, line 29.

[406] Sermo 1, p. 180, lines 5–6: "Gaudia bestialia [...] veneres".

[407] Op. 9, p. 504, lines 22–23: "cum concursu Dei [see above, note 234, in Chap. 2], secundum voluntatem Dei, sine ulla reflexione ad Deum", here metaphorically referring to humans living the "life of an animal" (ibid., lines 20, 22: "vita [...] bestiae").

[408] Sermo 8, p. 208, lines 28–30, here 28–29: "Omnia in nomine Jesu [...] facite".

[409] Sermo 12, p. 229, lines 7–13, here 8: "divini amoris interpretes". On the 'circulation of love', which is central to Stensen's spirituality, see more detailed Sobiech, *Herz, Gott, Kreuz*, pp. 157–174.

[410] Sermo 12, p. 229, lines 20–23.

[411] Ibid., lines 16–18, here 18: "Sola anima proximo instrumento abutitur pessime."

would be lost.[412] His own scientific nature is very much apparent in the latter resolution. It is possible, however, that Stensen – as surmised by Ole Peter Grell – never seriously intended to marry at all.[413] The reasons for this were his travels and the study of nature to which he was, as he said himself, entirely devoted.[414] Added to this was the time spent on pastoral activities following his conversion. It is therefore of particular interest in view of this personal background to examine the stance adopted by Stensen with regard to conjugal sexuality.

The student Holger Jakobsen noted down the following elucidation by Stensen during the latter's dissection of the genitals of the female corpse in the *Domus Anatomica* on February 6, 1673: "Precisely that act which, seen for itself, is of the highest value among those actions aimed at preserving [the human race] also holds as its accompaniment the greatest lust, offered as reward by the Creator to those who practice it legitimately."[415] The trigger for this activity is the allure of the human appearance.[416] As an anatomist, Stensen attempts during his inaugural address to approach the topic of human physicality and sexuality more from the point of view of natural law, starting from the physiological structure of marital consummation. Astonishingly, his chosen phrasing – particularly regarding the adjective "of dignity" (dignus) – represents a nearly verbatim anticipation of the corresponding passage of the Pastoral Constitution *Gaudium et spes* (Joy and Hope), promulgated at the Second Vatican Council (1962–1965) on December 7, 1965, which states that "the acts by which the spouses unite intimately and chastely are honorable and of dignity, and, conducted in truly humane fashion, denote and foster mutual self-giving, with which they enrich themselves in a cheerful and thankful manner."[417] The contemporaneousness to the three anonymized theological reports created in the years 1962–1966 in the course of Stensen's beatification process is striking.[418] During the time after his conversion in Florence, Stensen had formulated a prayer of grace reflecting his research on reproduction which he continued to recite even as a bishop in the household shared with his servants: "O God, who shows by preserving man as an individual and as humankind through this corruptibility the comfort

---

[412] Ziggelaar, *Chaos-Manuscript*, p. 277 (col. 112, fol. 57ᵛ, N.145).

[413] Grell, "Conversions", p. 214. On the Lutheran family ideal of Caspar Bartholin Sr., within which Stensen's early statement is to be contextualized, see ibid., pp. 209–210.

[414] DeConv., p. 126, lines 22–23 & p. 127, lines 4–5 = E 73, p. 257, lines 28–29 & p. 258, lines 3–4.

[415] ExAc., p. 306, lines 34–36: "Ipsa per se considerata actio omnium conservationis actionum dignissima, etiam comitem sibi habet suavitatem maximam in praemium ab Authore propositam rite illam exercentibus."

[416] See note 382, in Chap. 2 above.

[417] AAS 58 (1966), p. 1070, no. 49: "Actus […], quibus coniuges intime et caste inter se uniuntur, honesti ac digni sunt et, modo vere humano exerciti, donationem mutuam significant et fovent, qua sese invicem laeto gratoque animo locupletant." See supplementarily ibid., p. 1072 no. 51 (lines 23–25).

[418] Sacra Congregatio Rituum, *Vota*; the reports are dated 03/12/1962 (ibid., pp. 10–11, here 11), 01/06/1965 (ibid., pp. 12–15, here 15) and 05/29/1966 (ibid., pp. 16–63, here 63). Concerning the role of the general postulator of Stensen's beatification process Paolo Molinari SJ (1924–2014) and his potential contribution see Sobiech, "Science, Ethos, and Transcendence", p. 118.

to be received: that you are our nourishment, our giver of food, our husband, our eternal preservation [...]."[419] This too shows the lack of any sexual phobia in Stensen as a human being and as a priest.

In his inaugural address, Stensen follows his aforementioned words with this: "So the same is proffered as punishment in the shape of exceedingly painful diseases upon those who abuse [this act], so that finally they seek recourse in quicksilver preparations. Nature speaks to them: If the comfort does not suffice for love [for the Creator], the pain shall induce fear [of the Creator]."[420] Quicksilver, better known as mercury today, was used to treat syphilis.[421] Stensen may also have been among the students who had dissected the corpse of one Janicke Jansen, who was assumed to have died of "syphilis" (lues venerea), under instruction by Van der Linden in the autopsy room of the *St. Caecilia Gasthuis* on February 7, 1661.[422] With his words about personified "Nature", Stensen once again takes up the natural philosophy of his early anatomical writings, this time however with the pastoral intent of making his audience aware of the beauty of Creation (cf. Rom 1:20).[423] As a bishop he describes nature as the "living voice of the Creator" in a similar vein.[424]

With regard to Stensen's appraisal of sexual pleasure during marital consummation, the question of the scope of original sin arises. Richard Toellner speaks of the "tension between Christian doctrine of original sin and the new image of nature as a perfect order" as a research desideratum.[425] Just like St. Albertus Magnus OP (c. 1200–1280)[426] had gained a greater freedom and impartiality regarding the question of the naturalness of sexual pleasure through his Aristotelian psychological as well as natural scientific knowledge,[427] Stensen's research on reproduction – to an even greater extent – gave him an advantage over purely speculative theology which does not take anatomy into consideration. As opposed to his inaugural address, in which he characterizes lust – without mentioning original sin – as something mandated by the will of the Creator, Stensen describes "conception with sensual pleasure" as a consequence of original sin in one of his sermons.[428] This is another parallel to Albertus Magnus in that both he and Stensen create a certain tension by describing

---

[419] Spir. 1, p. 74, lines 17–22, here 17–20: "O Dio, che per la conservazione dell'uomo nell'individuo e nella specie in questa corruttibilità mostri la suavità che si riceverà: Che Tu sei nostro cibo, nostro donatore de' cibi, nostro sposo, nostra conservazione eterna [...]." The Latin version, composed from c. 1680, can be found in Op. 9, p. 482, lines 15–22.

[420] ExAc., p. 306, lines 36–39: "sic abutentibus eadem in poenam acerbissima symptomata proponuntur, ut ad mercuralia tandem confugiant. Ipsos Natura alloquitur, si suavitas non sufficit ad amorem, valeat dolor ad timorem."

[421] Sylvius, *Opera medica* (Praxeos medicae appendix, Tractatus III, no. CXLVII), p. 673.

[422] Schepelern, *Olai Borrichii Itinerarium*, pp. 71–74, here 71, 72, no. 6 (Van der Linden).

[423] Sobiech, *Herz, Gott, Kreuz*, p. 58.

[424] Op. 9, p. 499, lines 34–35: "vivam authoris vocem"; cf. also ibid., p. 500, line 12.

[425] Toellner, "Autorität", p. 170n21.

[426] Declared a Doctor of the Church in 1931 and patron saint of natural scientists in 1941, cf. AAS 24 (1932), pp. 5–17; AAS 34 (1942), pp. 89–91.

[427] Brandl, *Sexualethik*, pp. 37, 42.

[428] Sermo 30, p. 300, lines 8–11, here 9: "concipere cum voluptate sensuali".

lust as a result of original sin in their theological writings, but not in their ethical writings.[429]

Around March 1681, Stensen wrote to the still very young Francelina Odilia Theodora Freiin (Baroness) von Galen,[430] who was in the process of deciding whether to enter the chapter of secular canonesses for unmarried noble women in Nottuln in Westphalia: "Those who seek marriage in order to amuse themselves with lust are compared by the angel Raphael with animals [Tob 6:17 (Vul.)] and by Tobias with heathens who do not know God [Tob 8:5 (Vul.)]."[431] He continues: "A true Christian woman, in order to enter into matrimony with the will of God, must be able to say as did Sarah: God, you know that with this [matrimony] I have not coveted any man nor ever stained my soul with lust [Tob 3:16 (Vul.)], and I must do it only so that she [= my soul] may have heirs who praise God eternally in heaven [cf. Tob 8:9 (Vul.)]."[432] The clause "and I must do it only so that" is an addition by Stensen to Tobias' prayer. This may correspond to the directive on marital consummation to be found in one of his Christian religious instructions, namely that it was to occur "without the sensation of even a very small voluptuous act".[433] Could Stensen's addition of "and I must do it only so that" have been intended as an "instruction stipulating the elimination of lust if possible or its mere toleration"?[434]

The second censor of Stensen's writings during the beatification process thought this was the case and attested "excessive rigorism in conjugal matters" to this statement "which was not even at the time commensurate with Catholic teaching".[435] Stensen does not speak explicitly about sexual pleasure in matrimony, however, but of the fact that Sarah and Tobias married in order to conceive and raise successors in the true faith. Great value was attached to this purpose of marriage in the

---

[429] For details see Brandl, *Sexualethik*, pp. 42–48.

[430] Von Galen married in 1719; on her: Kohl, *Nottuln*, pp. 308–309; supplementarily and on her correspondence with Stensen (unmentioned ibid.) see Sobiech, *Herz, Gott, Kreuz*, pp. 98–99.

[431] E 222, p. 477, lines 19–21: "Die die Ehe begehren, um die Lust zu vergnugen, werden von dem Engel Raphael verglichen mit den Thieren, und von Tobie mit den Heyden, die Gott nicht kennen."

[432] Ibid., lines 21–25: "Eine wahre Christinne, umb nach Gottes Willen in dem Stand der Ehe zu tretten, mus mit Sara konnen sagen: Gott, du weist, das ich hie keinen Man begehret, noch meine Seele mit Lust jemahlen habe beflecket, und mus es darumb allein thuen, damit sie Erben habe, die Gott ewig loben in dem Himmel."

[433] Op. 5, p. 450, lines 19–21, here 20–21: "sine vel subtilissimae voluptuariae actionis motu". The passage is part of an (undeclared ibid.) excerpt by Stensen from the *Collationes patrum* by Johannes Cassianus (360/65–432/35). Ibid., lines 12–26 Stensen rephrases with a view to matrimony the seven "grades of chastity" (castimoniae gradus) stipulated for monks by Cassianus, cf. Pichery, *Cassien* (XII [Conlatio abbatis Chaeremonis secunda. De castitate],7), pp. 131–133, here 132.

[434] As in Jansenist morals, see Klomps, *Ehemoral und Jansenismus*, p. 204. – According to Albertus Magnus and Thomas Aquinas, sexual pleasure can in fact be approved of in the context of legitimate matrimonial intercourse, cf. Brandl, *Sexualethik*, p. 41.

[435] Sacra Congregatio Rituum, *Vota*, pp. 12–15, here 14: "rigorismum excessivum in rebus matrimonialibus, qui ne illo quidem tempore correspondebat doctrinae catholicae." The expertise by an anonymized censor is dated Rome, 01/06/1965 (ibid., p. 15).

Catechismus Romanus (1566) promulgated by Pope Pius V ([*1504] 1566–1572) following the Council of Trent (1545–1563) as well.[436] It seems that this is what Stensen wanted to emphasize to the Baroness of Galen, who as a young unmarried woman was contemplating a devoted life as canoness in the chapter and in whom he apparently saw a true spiritual vocation. That Stensen's understanding of marriage was indeed in keeping with the Church's teaching is shown by his mention of the "purpose of matrimony" in a sermon, namely "holy offspring, a remedy for concupiscence and a sign of love and unity between Jesus and his Church [cf. Eph 5:25.32]".[437]

From his perspective as anatomist and theologian, Stensen asked with regard to "every act necessary for the body" about the "will of the Creator for this act and all its accompanying circumstances".[438] Since he parallelizes the preservation "of man as an individual and as humankind", the consummation of marriage, one of whose accompanying circumstances – described as "accompaniment"[439] in the inaugural address – is the sensation of pleasure, must be counted among the acts "necessary" for the preservation of humankind. Stensen wishes to call attention to the fact that due to the predisposition of man resulting from original sin, this sensation of pleasure can turn into "stirrings of immoral desires".[440]

When further passages from sermons in which Stensen quotes from the Book of Tobit with reference to marriage are taken into consideration, his concept is clarified even more: here Stensen criticizes people who consciously exclude God from their conjugal sexuality and – in the manner of animals – only satisfy their desire (Tob 6:17 [Vul.]), which according to him leads to lust taking priority, with no thought being given to children initially and their raising being neglected later. The married couple would thus be imposing "obstacles" to God's grace, the consequence being many marriages unhappy in regard to conjugal unity and offspring.[441] During the Last Judgment, children would therefore accuse their parents of leading their marriage without fasting, prayers or alms, and in the same animalistic spirit in which they had concluded it.[442] These statements show the responsibility towards children that Stensen called for, as expressed already in his study diary. In one sermon, he evokes an image of the Last Judgment in which the people who were close to one in life (cf. 1 Cor 6:2), along with inanimate nature, act as plaintiffs

---

[436] Pius V, *Catechismus* and Kochuthara, *Sexual Pleasure*, pp. 276–278, here 277.

[437] Sermo 39, p. 339, lines 22–23: "Finis matrimonii proles sancta, remedium concupiscentiae, signum amoris et unionis Jesu cum sua ecclesia."

[438] Op. 9, p. 477, lines 28–29, 33–36, here 28, 33–34: "in omni actione necessaria corpori [...] voluntas auctoris cum hac actione et omnibus ejus circumstantiis".

[439] See note 415, in Chap. 2 above.

[440] Sermo 5, p. 197, lines 24–27; Op. 9, p. 478, lines 4–11, here 4–5: "desideriorum malorum agitationibus"; Sobiech, *Herz, Gott, Kreuz*, p. 167n143.

[441] Sermo 8, p. 208, lines 10–18, 21–25; Sermo 39, p. 339, lines 14–26, here 16: "obstacula"; ibid., lines 32–33, 35–36 & p. 340, lines 4–6. On the "obstacula" see Sobiech, *Herz, Gott, Kreuz*, pp. 222–225.

[442] Sermo 24, p. 275, lines 20–23.

against the sins one had committed, like "the persons who gave the life; father and mother, who contributed the means for life; the sky, whence the sun, the moon and the stars shone, and they will say: those persons have abused our rays, we have provided them with light."[443]

It should be taken into consideration in terms of the historical background that a somewhat rough interaction between married couples was common during this time, especially in rural areas; they and their children often shared beds with their guests, maidservants and farmhands – a practice which Stensen experienced during his visitations in the Münster area, and which he attempted to prohibit with his *Parochorum hoc age*.[444] One must also bear in mind that Stensen increasingly felt the burden of responsibility associated with his priestly and episcopal functions weighing down on him, for as he admitted as dean and parish priest of St. Ludgeri in Münster to his assembled congregation on November 1, 1680, he felt personally responsible for every sin of every one of the faithful entrusted to his pastorship.[445] He had, for example, ordered a brothel in the district of the parish St. Ludgeri to be shut down.[446] These facts explain the only seemingly divergent passages about sexuality in Stensen's sermons, when he eg denotes the "sight of a [sensually] beloved person" as "food of unchastity" on the one hand,[447] while on the other hand surprisingly citing the "love for beauty" – the "amor formae" condemned by Seneca Jr., Boselli, Rovenius, and by Spinoza in the manuscript which had found its way into Stensen's hands in 1677! –, body movements and the desire for physical closeness between man and woman, besides their spiritual bond, as benchmarks for the relationship of man to God.[448] This statement by Stensen in its atypicality for his time demonstrates clearly that he maintained the views expressed in his Copenhagen inaugural address.

Appreciation of Stensen's attitude and statements likewise cannot be considered complete without knowledge of his experience with immoral conduct by individual priests in the Münster diocese. As proved by his illustration of the beauty of God's creation to his audience in Copenhagen by way of the anatomy of a female corpse, Stensen possessed a keen sense for the fact that it was not the naked human body, but only the body language as a mirror for the thoughts of humans that could be indecent (Mark 7:20–23 par). In an address of Münster priests from the years 1680–1683 as suffragan bishop, he complained that priests were "killing souls" entrusted

---

[443] Sermo 29, p. 295, lines 12–16, 19–23, here 19–23: "homines, […] qui contulerunt vitam; pater et mater, qui contulerunt vivendi media; caelum, unde luxerunt sol et luna et stellae, et dicent: Nostris illi radiis abusi, subministravimus illis lucem."

[444] ParHAg., p. 50, line 37–p. 51, line 2; cf. Dülmen, *Haus*, p. 191.

[445] E 200, p. 442, lines 33–36; see Sobiech, *Herz, Gott, Kreuz*, pp. 277–278, 284–286, 300.

[446] Sobiech, *Herz, Gott, Kreuz*, p. 285; cf. Dülmen, *Haus*, p. 193. The topic of contraception appears in Stensen's writings as vicar apostolic in Op. 3, p. 430, line 23 among the questions of the confessor to the penitent.

[447] Sermo 27, p. 284, lines 30–31: "cibus impuritatis […] personae amatae intuitus".

[448] Sermo 1, p. 177, lines 2–12, here 5: "formam […] amas" (*literally*: "you love beauty"). In the context of Stensen's spirituality see in more detail Sobiech, *Herz, Gott, Kreuz*, pp. 177–178.

to them as pastors with the eyes of a basilisk[449] by inviting them "to unchastity with seductive movements of their eyes".[450] Augustine of Hippo also speaks of the "concupiscence of the eyes".[451] In his address, Stensen phrases his thoughts drastically: "But what if the tongue of a priest developed into the tongue of the snake in Paradise [Gen 3:1.4–5] that misleads innocent souls?"[452] Christ, he continues, left limbs which had separated from his body to Lucifer as a head, for they resembled the latter.[453] In his report to Pope Innocent XI ([* 1611] 1676–1689) on the state of affairs in the bishopric of Münster, Stensen writes that several cases of "Sollicitatio ad turpia" (incitement to unchaste acts or words) by priests of the diocese had been brought to his attention outside the seal of confession.[454] This was countered by the Münster vicar general Johannes Alpen ([* 1632; † 1698] 1661–1683), who stated – with a condescending undertone – in an expertise sent to the Propaganda Congregation that "the good Suffragan Bishop" believed all to quickly what he heard.[455] The so-called incitement is the abuse of position and of trust by a priest who tempts a penitent into sinning against the Sixth Commandment.[456] That Stensen was not gullible in regard to such accusations – though the incidents he mentions likely cannot be clarified due to a lack of source material – is shown by the fact that he was conditioned by his anatomical research to place emphasis on autopsy for the securing of evidence. As suffragan bishop of Münster, he would register in detail the guests and priests invited to receptions and their behavior, for instance.[457] At any rate, it was precisely the attitude of persons like the mentioned vicar general of Münster which inflicted immeasurable harm upon the Church, while Stensen's position of clarification represents an attempt of protection against this harm.

---

[449] Mythical creature (composite of a snake, a dragon and a rooster) with a deadly gaze, considered an attribute of the personification of unchastity well into the seventeenth century, cf. Franz Niehoff, "Basilisk"; in: LThK³ 2 (1994), col. 65.

[450] Sermo 44, p. 370, lines 9–10, here 10: "occidimus animas, […] per lubricos […] [oculorum] gyros ad impuritatem".

[451] Martin, *De doctrina christiana* (De vera religione XX,40,109; XXXVIII,70; LV,107,293), pp. 212, 233, 256.

[452] Sermo 44, p. 370, lines 19–21: "At quid, si lingua sacerdotis evadit lingua serpentis in paradiso pervertens innocentes animas?"

[453] Ibid., lines 21–22. On this, see Sobiech, *Herz, Gott, Kreuz*, pp. 284–286.

[454] E 304, p. 597, lines 7–8.

[455] E 323, p. 641, lines 23–30, here 31–32: "bonum dominum suffraganeum". On Alpen: Kohl, *Diözese*, s.v. "Johannes (von) Alpen 1661–1683", pp. 159–161.

[456] On the history of this offense, considered a crime in canon law, cf. Wietse de Boer, "Sollicitazione in confessionale"; in: DSI 3 (2010), pp. 1451–1455.

[457] Corresponding observations by Stensen on the choice of food and communication behavior of priests can be found eg in Sermo 44, p. 367, line 16–p. 368, line 28.

2.3.1.5 Simultaneous Ensoulment Versus Successive Ensoulment of the Embryo

Regarding the question of ensoulment of the embryo created by the act of conception, scholasticism for many centuries followed – due to the Greek translation Exod 21:22–23 LXX – the Aristotelian view of successive ensoulment.[458] Karl Lehmann argues that the adoption of this view, which apart from an intermezzo due to the Apostolic Constitution *Effraenatam* (1588)[459] by Pope Sixtus V ([* 1521] 1585–1590) was the official one until the promulgation of the Constitution *Apostolicae Sedis* (1869) by Pope Pius IX ([* 1792] 1846–1878) and implied a penal differentiation between ensouled and unensouled embryos,[460] "put a strain on the Church's teaching from the point of view of tradition".[461] The theological and doctrinal[462] developments concerning ensoulment theory must be viewed in a more differentiated manner, however, whereby a source study for the early modern period taking medico-historical as well as theological aspects into consideration has yet to be conducted. What was Stensen's stance on this important issue?

During Stensen's time, there existed theological voices opposing the prevailing doctrine of successive ensoulment: in his *Disputatio de ministrando baptismo humanis foetibus abortivorum [...]* (Treatise on the Administering of Baptism of Human Fetuses from Premature Deliveries [...]; Lyon 1658 and Lucca 1665, 2nd ed. 1666), Girolamo Fiorentini MIC (1620–1678)[463] set forth from a theological perspective – based on the traditional medical notion of the mixing of male and female sperm – that the moment of conception should be assumed as the moment of ensoulment of the unborn child, and that therefore all delivered fruits of the womb, no matter how small, had to be baptized "conditionally" (sub condicione), with neglect to do so constituting a mortal sin.[464] Fiorentini was also familiar, he tells his reader, with observing "small animals" (bestiolae) with the help of a microscope.[465]

---

[458] For details see Hack, *Beseelung*, pp. 25, 29, 47–49.

[459] Gasparri, *Codicis iuris canonici fontes*, vol. 3, N. 165, pp. 308–311 (*Effraenatam*); it was abrogated by the Constitution *Sedes Apostolica* (1591) by Pope Gregory XIV ([* 1535] 1590–1591), ibid., N. 173, pp. 330–331).

[460] ASS 5 (1869), pp. 287–312 (*Apostolicae Sedis*), here 298 (Excommunicationes latae Sententiae Episcopis sive Ordinariis reservatae, II)=Gasparri, *Codicis iuris canonici fontes*, vol. 1, N. 552, pp. 24–31, here 28: "Procurantes abortum", without specifying a criterion for ensoulment; for details see Hack, *Beseelung*, pp. 23–24.

[461] Lehmann, *Mensch*, p. 18.

[462] On 11/18/1974 the Congregation for the Doctrine of the Faith stated that, although there had been different opinions on the moment of ensoulment throughout the centuries, "the Holy Fathers of the Church, her Pastors and Teachers" (Sancti Ecclesiae Patres eiusque Pastores ac Doctores) had left no doubt about the illegitimacy of induced abortion even before that moment (AAS 66 [1974], pp. 730–747 [Declaratio De abortu procurato], here 734–736, no. 7).

[463] On Fiorentini: Sarteschi, *De scriptoribus* (Cap. I, Art. IV. De Hieronymo Florentinio), p. 154.

[464] Fiorentini, *Disputatio*, p. 93, and, discussing Fiorentini, Cangiamila, *Embriologia sacra* (Libro primo, capo VIII), pp. 52–56, here pp. 53–54, no. 2–4.

[465] Fiorentini, *Disputatio*, p. 20.

According to Fiorentini's biographer, this opus was praised highly by the Roman Congregation of the Index, but was nevertheless indexed 'donec corrigatur' by decree on June 22, 1665 pending Fiorentini defining his theories as merely "probable" and "his opinion", without a final conclusion and without intent of introducing a new rite into the Church.[466] The altered edition (Lucca 1665 and 2nd ed. 1666), revised "by command" of the Congregation of the Index and furnished with the required declaration by Fiorentini was the only one allowed by the Congregation in its second decree on April 1, 1666 with reference to the 1666 edition 'donec corrigatur'.[467]

Several years earlier, however, Dr. med. Paolo Zacchia (1584–1659),[468] who had studied medicine and law at *La Sapienza* in Rome, was an authorized expert at the Rota Romana and had been appointed personal physician to the Pope and head medical officer of the Papal States in 1644, had expressed a similar opinion: in the 9th book *De foetus humani animatione* (on the Ensoulment of the Human Fetus) of the posthumously published edition Lyon 1661 (2 vols.) of his *Quaestiones medico-legales [...]* (Medico-legal Investigations [...]; Rome 1621–1635 [books 1–7] and Leipzig 1630 [books 1–4], Amsterdam 1651 [book 8], Lyon 1661 [book 9[469]]; Frankfurt 1666 [books 1–9 identical to the 1661 edition] and others), Zacchia, who is often referred to as the Father of Legal Medicine, changed his mind with regard to the previous editions of his book from successive ensoulment to simultaneous ensoulment, claiming that the ensoulment of the male as well as the female embryo occurred "at the first moment of conception".[470]

---

[466] Sarteschi, *De scriptoribus* (Cap. I, Art. IV. De Hieronymo Florentinio, § 1, no. IV), pp. 155–156, here 156: "rem probabilem, sententiam suam"; date given according to ILP, s.v. "Fiorentini, Girolamo", pp. 347–348.

[467] Fiorentini, *Disputatio*, title page: "iussu"; ibid., p. 8 (from title page): "Auctoris de ordine Superiorum declaratio" (Declaration by the author at the Superiors' orders). On this and on the editions Rome 1672, Lyon 1674 (and 2nd ed. 1675), see Sarteschi, *De scriptoribus* (Cap. I, Art. IV. De Hieronymo Florentinio, § 1, no. IV), p. 156. Cf. ILP, s.v. "Fiorentini, Girolamo", pp. 347–348 (*divergent date*: 04/05/1666); on the further developments Prosperi, "Baptism", pp. 224–226, here p. 226n57 (*divergent date*: 04/07/1666); Reusch, *Index* 2,1, pp. 427–428.

[468] On Zacchia: Claudio Pogliano, "Zacchia, Paolo"; in: DBSM 4 (1989), pp. 238–239; Duffin, "Zacchia", pp. 150–151; HGK, s.v. "Zacchia, Paolo", p. 718 (rudimentary).

[469] To be differentiated from the Book 9 added to the edition Amsterdam 1651, which contains consilia and decisions by the Rota Romana and becomes Book 10 in the edition Lyon 1661; this as a correction to Duffin, "Zacchia", pp. 151, 157 (erroneously edition "Amsterdam 1661" & "Tome 3" instead of Book 10), Hack, *Beseelung*, pp. 29, 63 ("1660" to be corrected to "1661") and supplementary confirmation of Spitzer, *Zacchia*, p. 31n9.

[470] Spitzer, *Zacchia*, pp. 40, 70, here 70: "in primo conceptionis momento". Cf. Cangiamila, *Embriologia sacra* (Libro primo, capo VIII), p. 53, no. 2; Lanza, "Momento", pp. 87–91 = Lanza, *Momento*, pp. 155–159. – In regard to simultaneous ensoulment, there is a tendency in literature to draw erroneous conclusions from the editions of Zacchia's *Quaestiones medico-legales* published after 1661 to the contents of the first edition, as seen eg in Jones, *Soul*, p. 164. – Hack, *Beseelung*, pp. 63–64, here 63n8 follows Spitzer, *Zacchia*, p. 25, by stating that Zacchia appears in his "formulaic final sentence" (ibid.) to "shy away from the consequences following from his work". This final sentence, however, is simply the common so-called "Protestatio auctoris" (author's caveat) by

Michael Boudewijns, who appears not to have been aware of Fiorentini's work, concludes in his *Ventilabrum medico-theologicum* (1666), after lecturing on different views on the moment of ensoulment, that according to the "general opinion of almost all theologians" it was forbidden for physicians and midwives as well as for pregnant women themselves to terminate a pregnancy – regardless of whether the embryo was ensouled or not, since it was something "intrinsically evil" to knowingly and intentionally kill an innocent person.[471] A converse opinion temporarily argued as probable by a theologian can be found among the 65 Laxist propositions condemned by Pope Innocent XI in 1679.[472]

With regard to medical science, the second half of the 1670s saw the term "ovary" becoming generally accepted, with research by several of Stensen's colleagues providing further evidence in support.[473] As a result, the theory of simultaneous ensoulment of the embryo established itself more and more, especially in the field of medicine.[474] As far as contemporary theology was concerned, Cangiamila observed in the eighteenth century that Fiorentini had already been aware of the discovery of the ovary in 1675.[475]

Since the 1970s, Stensen's discovery of the ovary has increasingly been mentioned in connection with the end of the doctrine of successive ensoulment: (1) John R. Connery referred to the research by Stensen and De Graaf which preceded the theory of the so-called ovulists.[476] (2) In his study *Das Lebensrecht des ungeborenen Kindes als Verfassungsproblem* (The Right to Life of the Unborn Child As a Constitutional Problem; Tübingen 1984), the jurist Hans Reis[477] notes that the doctrine of successive ensoulment was only given up entirely in the time after Stensen's discovery of the ovarian follicles.[478] (3) In his opening address at the autumn plenary meeting of the German Bishops' Conference held on September 23–26, 1991, Bishop Karl Lehmann emphasized that Stensen had deprived the traditional doctrine of successive ensoulment of its foundation through his discoveries in the field of human embryology. Lehmann also pointed out the existing academic void and

---

which the author of a book sought to protect himself against the Inquisition, cf. eg Farinacci, *Tractatus de haeresi*, pp. 27–28 (Quaestio CLXXVIII, § VII).

[471] Boudewijns, *Ventilabrum* (Pars prima, Quaestio XXVI. An medico aliquando liceat procurare abortum?), pp. 170–179, here 172: "omnium fere theologorum communis sententia [...] malum intrinsecum".

[472] DH 2135; the theologian in fact based his words on a physician's, see details in *2135¹.

[473] Cobb, *Egg & Sperm Race*, pp. 101, 157.

[474] Bruch, "Schutz", p. 113.

[475] According to Cangiamila, *Embriologia sacra* (Libro primo, capo IX), pp. 67–68, no. 28, Fiorentini disavowed his hitherto shared view of the existence of female sperm in the edition Lyon 1675.

[476] Connery, *Abortion*, p. 208 (both incorrectly placed "in the early eighteenth century"), 331n26; more elaborately in Lanza, "Momento", pp. 65–67 = Lanza, *Momento*, pp. 132–134.

[477] Scientific assistant at the German Federal Constitutional Court from 1958 to 1965, Deputy Head of the Commissariat of the Catholic Bishops in Lower Saxony (Germany) from 1970 to 1991; on Reis: LFam., p. XXIX.

[478] Reis, *Lebensrecht*, p. 146n670.

added that one might ask oneself "why the Magisterium and particularly also theology have not already argued more resourcefully with the use of this knowledge".[479] (4) In his lecture on the history of embryology at the third plenary assembly (Vatican City, February 14–16, 1997) of the *Pontificia Academia pro vita* (Pontifical Academy for Life; est. 1994), Salvino Leone described Stensen as a representative of the Scientific Revolution of the seventeenth century which provided objective proof as a base for the forming of ethical opinions, and as discoverer of the fact that all vivipara including man originated from an "egg".[480] (5) David Albert Jones likewise mentions Stensen's discovery of the female ovaries in his overview work *The Soul of the Embryo: An Enquiry into the Status of the Human Embryo in the Christian Tradition* (London 2004).[481]

Explicit statements regarding the moment of the beginning of human life or the ensoulment of the embryo are not to be found in Stensen's anatomical writings; this is a result of his caution.[482] In a letter to the Florentine court botanist Paolo Boccone (1633–1704),[483] written in October 1671 in Florence, Stensen agreed with his addressee that while the human senses often were aware of the growth of natural things, they almost entirely lacked any knowledge about their beginnings to describe them initially.[484] This insight has lost none of its topicality pertaining to the field of embryology even to this day.

Stensen's excerpt[485] about the soul in his *Chaos* manuscript is not relevant to the process of ensoulment, not least because he neglected to furnish it with any personal comments.[486] Thomas Bartholin provides some information on the views Stensen may have been taught as a student in Copenhagen: in his *Cista medica Hafniensis*

---

[479] Lehmann, *Lebensrecht*, p. 16 = Lehmann, "Lebensrecht [I]", p. 47 = Lehmann, "Lebensrecht [II]", pp. 113–114. Lehmann has been a cardinal since 2001, cf. Deckers, *Lehmann*, pp. 285–379; on the opening address cf. ibid., p. 322. Mieth, "Moralenzyklika", p. 11 thrusts in the same direction.

[480] Leone, "Ancient Roots", pp. 42–43, here 43; Leone, *Bioetica*, pp. 335–337, here 336.

[481] Jones, *Soul*, pp. 164–165. Discussed in Hack, *Beseelung*, p. 66. Stensen's discovery is also mentioned in Dupont, "Embryon", p. 261 in the context of ensoulment doctrines.

[482] As a bishop Stensen excludes considerations on possible collisions of Descartes' philosophy and methodology with "Jesuit free will" (Jesuitica arbitrii libertas) and "Roman transubstantiation" (transsubstantiatio Romana) in DefConv., p. 390, lines 16–20. By way of this philosophical eclecticism, Stensen avoids the scholasticism-breaking power of Cartesian thinking, cf. Sobiech, *Herz, Gott, Kreuz*, p. 151n46. On the question of the effects of Cartesian thinking on the doctrine of transsubstantiation, negotiated at the Holy Office from 1671 to 1673, cf. Armogathe, "Cartesian Physics", p. 149; Donato, "Onere", pp. 71–72. These negotiations were inititated by Honoré Fabri (Lefèvre) SJ (c. 1608–1688), who was in contact with the *Cimento* academy, thought highly of Stensen's Myology, and had engaged in conversations on the Faith with Stensen in Rome in 1666, cf. Armogathe, "Cartesian Physics", p. 149; Miniati, *Steno's Challenge*, pp. 138–139; Scherz, *Epistolae*, vol. 1, p. 109; Kardel/Maquet, *Biography and Original Papers*, p. 155. On Fabri: Carla Rita Palmerino, "Fabri, Honoré"; in: DFPh. 1 (2008), pp. 453–460.

[483] On Boccone: Isabella Sermonti Spada, "Boccone, Paolo"; in: DBI 11 (1969), pp. 98–99.

[484] E 64, p. 247, lines 5–8.

[485] See note 189, in Chap. 2 above.

[486] Ziggelaar, *Chaos-Manuscript*, pp. 431–435 (col. 177, fol. 74^r–col. 178, fol. 74^r) & p. 387.

[...] (Copenhagen Medical Box [...]; Copenhagen 1662), while discussing a case of induced abortion heard at a Copenhagen court in 1632/1633, he ponders the "generally" (vulgo) held notions regarding the different moments of ensoulment – after the 40th day for boys, after the 90th day for girls calculated from the day of conception – on the one hand, and the caution apparent in the words of Augustine of Hippo regarding the question of the moment of the shaping of the body by the soul.[487] Nothing else is known.

More enlightening is a sermon by Stensen during his time as bishop in which he states that man, who in so many centuries until today had been nowhere near able to investigate the corporeal part of humanity, could never explain without God's help the spiritual part, eg the nature of the soul.[488] Furthermore, Stensen had refuted Aristotle's theory of the cardiocentric[489] seat of the soul and likewise opposed those who believed it to be located in the blood.[490] He spoke in general terms of "my difficulty" which consisted "entirely" in explaining the collaboration of body and soul, with the latter being of immaterial nature.[491] Aristotelean natural philosophy to him represented – not only in medicine, but also in questions of creation theology – a falsifiable opinion which one could, and in most cases should, take into consideration so as not to overlook something others might already have discovered previously.

A passage from another of Stensen's sermons allows conclusions as to what was probably his preferred concept: he personally addresses each and every one of his listeners with the statement that God "since the first moment of your conception in every instant is observing, protecting, guiding and setting in motion your body as well as your soul".[492] It is striking that Stensen mentions the moment of conception and that he speaks only of one – the rational – soul. Already during his time as "Anatomicus regius" in Copenhagen, he had rejected the speculations by William Harvey concerning the "vegetative soul" (anima vegetativa), the first of the three successive Aristotelean levels of the soul and the only one to be infused at the moment of conception according to the Aristotelean and Thomist-scholastic

---

[487] Bartholin, *Cista Medica Hafniensis* (Loculus XXXIII. Judicium medicorum Hafn[iensium]. De abortu Margaretae Johanneae), pp. 151–156 with citation on p. 156; cf. Spitzer, *Zacchia*, p. 180n17.

[488] Sermo 35, p. 322, lines 28–30, 33–35; ibid., p. 323, lines 18–20.

[489] MuscGland., p. 168, line 37–p. 169, line 2 & p. 181, lines 28–33. The same applies to the cephalocentric theory of Descartes: There was so much presumptuousness, Stensen notes with dry humor, possibly alluding to complaints by citizens about disrespectful behavior by students roaming the city of Leiden, in a letter to Thomas Bartholin from Leiden on 03/05/1663 (VesPul., p. 136, lines 3–7 = E 11, p. 171, line 35–p. 172, line 1), that the black bile he had noticed during his examination of the pineal gland in a horse's head did not even spare the seat of the soul. On Descartes' theory cf. Irmgard Müller, "Seelensitz"; in: HWPh. 9 (1995), cols. 105–110, here 105–107.

[490] GlandOr., p. 51, lines 7–8.

[491] E 65, p. 249, lines 1–4, here 1 = Adelmann, *Correspondence*, vol. 2, no. 271, p. 597: "La mia difficultà tutta".

[492] Sermo 32, p. 307, lines 8–11, here 9–10: "a primo conceptionis tuae momento et corpus et animam omni momento intuetur, conservat, dirigit, actuat."

view; he would "not elaborate" questions in connection with this topic, as he tersely informed his students.[493]

As Stensen also refers to the dignity of the body and soul of Mary in the above-mentioned sermon,[494] it seems possible that he follows the same train of thought as Paolo Zacchia: starting with the 1661 edition of his *Quaestiones medico-legales*, the latter argues that, since the Church was not celebrating an embryo without a rational spiritual soul on the Feast of the Immaculate Conception, it applied not only to Mary but to all humans that the rational spiritual soul was received at the first moment of conception.[495] In his records as suffragan bishop, Stensen also notes that "as long as one was in the mother's womb, as long as one was an infant" nothing was known of the presence of God in the physiological processes occurring without human knowledge or assistance in creatures, eg during the formation of the fetus.[496] It is apparent that Stensen views humans in the womb as humans in the true sense of the word without chronological restrictions. Hence under consideration of all available evidence it can be presumed as nearly certain that Stensen located the embryo's ensoulment in the moment of conception.

In this context, and due to the co-occurrence of several events in the summer of 1677, it seems relevant to ponder whether Stensen's research into human reproduction and his discovery of the ovary had any influence on the condemnation of one of the 65 propositions of the probabilist system of moral theology listed in the decree by Pope Innocent XI on March 2, 1679, namely: "It is allowed to execute an abortion prior to the ensoulment of the embryo, so that the girl, if she is caught pregnant, will not be killed or suffer infamy."[497] With reference to Stensen's sojourn in Rome in the summer of 1677, Hans Reis assumes that "the temporal coincidence of both events" may not have been "mere happenstance".[498] John R. Connery emphasized that while the decree did not mean that the hitherto accepted doctrine of successive ensoulment was overturned, although the corresponding notions did eventually lose

---

[493] ExAc., p. 291, lines 28–30, here 30: "non determinavero". Cf. Lesky, "Säugetierovar", pp. 247–248.

[494] In more detail in Sermo 32, p. 304, lines 6–9, 12–13; ibid., p. 306, lines 15–16, 25–26.

[495] Spitzer, *Zacchia*, p. 160. Zacchia is referring to the passive conception (conceptio passiva completa) by the mother of Mary through the infusion of a rational spiritual soul, cf. De Fiores, "Maria", p. 157n241, which Zacchia implicitly assumes to have been concurrent with the active conception (conceptio activa) resulting from the act of procreation (Hack, *Beseelung*, pp. 293–294). In his sermon on the Feast of the Immaculate Conception, Stensen speaks of the "at no moment ever" (nullo unquam momento) stained soul of Mary (Sermo 30, p. 298, lines 25–26, here 25).

[496] Op. 9, p. 498, lines 31–35 & p. 500, lines 31–33, here 32–33: "dum [...] in utero, dum infans esses".

[497] Innocent XI, *Decret*, p. 12 (in a different edition of the time p. 15), no. 34: "Licet procurare abortum ante animationem foetus, ne puella deprehensa gravida occidatur, aut infametur." Likewise in DH 2101–2167, here 2134, with sources for the censored sentence and the local censorial notes up to the year 1659 in *2134¹. Cf. Antonio Menniti Ippolito, "Innocenzo XI, beato"; in: EPapi, pp. 368–389, here 384 & 386 on the condemned sentences.

[498] Reis, *Lebensrecht*, p. 146n670.

their proponents in its aftermath,[499] it was possible that the condemnation may have contributed to a certain degree to subsequent theologians being less apt to approve of direct abortions of the so-called unensouled fetus in the context addressed in the condemned statement.[500]

Around the occasion of his episcopal ordination, Stensen spent about 4 months in Rome from May 22 until at least September 22, 1677[501]; during this his longest sojourn in Rome, he found many opportunities to speak about his research. Among others, he met with two cardinals[502]: (1) the jurist and father of a family Francesco degli Albizzi (1593–1684),[503] who had been ordained a priest after being widowed early in life, eventually serving as Assessor of the Holy Office from 1635 to 1654 and as Cardinal Inquisitor from 1654 until his death, and (2) Federico Baldeschi Colonna Ubaldi (1625–1691),[504] who had been made Assessor of the Holy Office on March 22, 1673. Stensen appreciated Albizzi as "exceptionally devoted to the sincerity of faith",[505] as he informed Cosimo III in 1684. Shortly before his departure, Stensen had spoken with Albizzi about Jansenism, which had its source in the indexed monumental work *Augustinus* [...] on the doctrine of grace (Leuven 1640; Paris 1641; Rouen 1643 and 1652) by the Leuven professor of theology and bishop of Ypern Dr. theol. Cornelius Jansenius Jr. (1585–1638).[506] The reason for this conversation was probably the presence of a delegation from the Leuven theological faculty in Rome since May 27, 1677 with a list of 116 probabilistic propositions, some of which were also concerned with the doctrine of grace, in order to have them condemned.[507] This delegation's audience with Innocent XI in August 1677 resulted in the eventual issue of the decree concerning 65 propositions – which did not, however, include those on the doctrine of grace.

[499] Connery, *Abortion*, pp. 189–190. By contrast Lehmann, *Lebensrecht*, p. 16=Lehmann, "Lebensrecht [I]", p. 47="Lehmann, Lebensrecht [II]", p. 113, according to which in DH 2134 "the differentiation [between "ensouled" and "not yet ensouled" embryo]" had been "revoked in terms of its validity" (likewise Mieth, "Moralenzyklika", p. 11; cf. Lehmann, *Recht*, p. 18). It remains unclear how this would relate to the two Constitutions *Effraenatam* and *Apostolicae Sedis* (see notes 459 & 460, in Chap. 2 above). Cf. also Hack, *Beseelung*, p. 28 with n. 38.

[500] Connery, *Abortion*, pp. 197–198; likewise Jones, *Soul*, p. 184.

[501] Scherz, *Biographie*, vol. 2, pp. 12, 24.

[502] E 123, p. 329, lines 3–4.

[503] On Albizzi: Adelisa Malena, "Albizzi, Francesco"; in: DSI 1 (2010), pp. 29–31, here 23, 29–30.

[504] On Baldeschi Colonna: Alberto Merola, "Baldeschi Colonna, Federico"; in: DBI 5 (1963), pp. 456–457; Metzler, "Kongregation", pp. 261–262.

[505] E 359, p. 703, lines 6–11; E 369, p. 715, line 38–p. 716, line 3, here p. 716, line 1: "zelantissimo per la sincerità della fede".

[506] On Jansen: Antony McKenna, "Jansenius, Cornelis Janssen (known as Cornelius)"; in: DFPh. 1 (2008), pp. 619–624. – The Jansenists were pleased with Stensen's piousness and his disapproval of the simoniacal election of the Elector of Cologne Maximilian Heinrich of Bavaria (1621–1688) as bishop of Münster, cf. Régine Pouzet, "Sténon, Jean-Nicolas"; in: DPR, p. 950.

[507] See details in Bouuaert, "Députation", pp. 1135, 1137; Ceyssens, *Albizzi*, pp. 178–179. On Albizzi's vehement disapproval of the case see ibid., pp. 179–181.

The insertion of a brief character study of Stensen for the summer of 1677 is appropriate at this point: that Stensen had sought contact with the Holy Office in these months is demonstrated by the fact that on September 4, 1677, he turned in to the Office a report on the philosophy of his friend during his studies in Leiden, Baruch de Spinoza, and a few weeks later on September 23, the manuscript of the latter's *Ethica ordine geometrica demonstrata [...].*[508] For at the end of July or beginning of August 1677, Stensen had received several of Spinoza's works as manuscripts from a Lutheran foreigner, most likely the physicist Ehrenfried Walther von Tschirnhaus (1651–1708), a good friend of the Dutch philosopher who was visiting Rome from 1677 to 1678.[509] As mentioned by Giuseppina Totaro, Stensen used medical terminology in his report in keeping with the language of the Church's censors of the sixteenth and seventeenth centuries.[510] Stensen, whose own Myology was based on geometry, saw the use of the geometric method restricted to the natural sciences. In a letter from Rome in August 1677, Tschirnhaus reported to Leibniz the impression he had gained of Stensen during his stay. The passage referring to Stensen is preserved in a copy of an excerpt by Leibniz and reads as follows: "From the letter by Tschirnhaus, dated 1677, in the month of August. On Stensen:[511] The Jesuit Fathers introduced me to Mr. Stensen against my will. He has recently left the Florentine court tacitly; his enemies[512] claimed to know he had taken something

---

[508] See details in Spruit/Totaro, *Ethica*, pp. 6–17, 25–26, 55–56. The Calvinist States General of the United Netherlands banned Spinoza's *Opera posthuma* as well as his philosophy as such, cf. ibid., p. 57. On Spinoza's 'geometrical' method cf. Moritz Epple, "More geometrico"; in: ENZ 8 (2008), cols. 791–793.

[509] Spruit/Totaro, *Ethica*, pp. 17–20, 54. Tschirnhaus (on him: HGK, s.v. "Tschirnhaus, Ehrenfried Walther von", pp. 664–665) had received the completed manuscript from Spinoza in the summer of 1675, see ibid., p. 52.

[510] Spruit/Totaro, *Ethica*, p. 13. – The term "denouncement" (ibid., pp. 13, 55; Giuseppina Totaro, "Spinoza, Baruch"; in: DSI 3 [2010], pp. 1472–1473: "denuncia") does not take into consideration Stensen's pastoral intent: he informed the Holy Office after Spinoza's death in February 1677, having contacted Spinoza with the same intention in a private letter – since he knew him in person from his study days in Leiden – in 1671 and published that letter in Florence in 1675 when Spinoza failed to respond (DePhil. = E 61, pp. 231–238), albeit without naming Spinoza as the recipient, cf. Sobiech, *Herz, Gott, Kreuz*, p. 34n36 & p. 297. It is interesting to note in this context and in regard to Stensen's stance what he reports in his *Defensio et plenior elucidatio scrutinii reformatorum* (Hanover 1679) "from personal experience" (propria experientia), namely how he had witnessed in Rome, Florence, Pisa and Livorno, where the Inquisition ruled, complete freedom afforded to non-Catholics so long as they did nothing scandalous (DefScrut., p. 286, lines 23–26). Stensen advocated improvement, not punishment; for example, in the mentioned Defensio he also reports a case he witnessed in Rome where a nobleman incarcerated by the Inquisition for a grave public scandal had received no other punishment except being put under oath to read "the entire Bellarmin" (Bellarminum totum) at home (ibid., p. 288, lines 24–31), ie all books by Cardinal Robert Bellarmine SJ (1542–1621), and that the man had converted to Catholicism following his reading. Thus "the truth of the doctrine" (doctrinae veritas) had achieved in the complete freedom of his home country what "the threats of the Inquisition in prison" (minae inquisitionis in carcere) had been unable to. On Bellarmino: Gustavo Galeota, "Belarmino, Roberto"; in: DHCJ 2 (2001), pp. 387–390.

[511] Leibniz' introductory words presumably end here.

[512] It is unknown which enemies are being referred to.

with him: later it was found out, however, that he travelled on foot to the Holy House in Loreto, collecting alms because he had nothing with him. The duke [John Frederick[513]] ordered that he be abundantly feasted upon his arrival in Rome: The Catholics themselves speak varyingly of him, and many cannot get it into their head that he intends to institute so rigorously the life of the disciples of Christ, about which and in defense of Catholic religion he has recently had a treatise[[514]] published. No one has ever, in order to bring me to such a religion, used as great a force as this man, and never have I heard anyone so adroit at persuasion and so well-reasoned: etc." (Fig. 2.7)[515] In this characterization by Tschirnhaus, which pays Stensen obvious respect, the latter appears as an eloquent candidate for the episcopate who actively promoted the Catholic Faith.

There are various indications that Stensen held conversations about his research in Roman circles: his Myology, whose second appendix contains the description of the ovaries as the female gonads which revolutionized research, had been sent by Prince Leopoldo de' Medici (1617–1675), supporter of the first edition of Galilei's works and of the Florentine *Accademia del Cimento*,[516] to Cardinal Barbarigo in the conclave in the Vatican (May 22 to June 20, 1667) immediately after its publication with a letter dated May 24, 1667. After the end of the conclave, Barbarigo thanked Leopoldo in a letter from Rome on June 25, 1667, praising the "clearness of talents" possessed by Stensen of which his Myology was proof.[517] One example of the book along with its two appendices was located in the private library of Pope Clement IX ([* 1600] 1667–1669), the former cardinal Giulio Rospigliosi who had been appointed in the conclave. This is evidenced by the leather binding, whose front and rear cover each bear Clement IX's golden coat of arms (with tiara and decussate

---

[513] Duke of Brunswick-Lüneburg ([* 1625] 1665–1679); on him: Klaus Mlynek, "Johann Friedrich, Herzog zu Braunschweig und Lüneburg"; in: HBL, p. 189.

[514] Presumably Scrut. (Florence 1677; German trans. Hanover 1678).

[515] The "etc." is Leibniz'; LBH, LBr. 943, fol. 168[r] (section of Leibniz' transcript): "Ex literis Tschirnhausii Romae a[nn]o 1677 datis, mense Augusto. De Stenone: Mit dem H[err]n Stenone haben mich die Patres Jesuit wieder meinen willen bekandt gemacht. Er hatt unlengst den Hoff zu Florenz stillschweigend verlaßen; seine feinde haben vermeint, er würde etwas mitgenom[m]en haben: man hatt aber darnach erfahren, daß er zu fuß nach dem Heiligen Hause zu Loretto gangen stipem colligendo, weil er nichts beÿ sich gehabt. Der Hertzog hatt befohlen daß er beÿ seiner ankunft nach Rom wohl soll tractirt werden: die Catholici reden selbst divers von ihm, und will vielen nicht ein, daß er das leben der Jünger X[ris]ti so rigorose will introduciren, wovon und zu defendirung der catholischen religion er neulich ein Scriptu[m] ausgehen laßen. Niemand hat iemals mich zu solche religion zu bringen so grose macht angewand alß dieser mann, und habe niemalen dergleichen starcken persuasorem und so artlich gegründet, gehöret: etc." Cf. the partially inaccurate transcription in Leibniz-Archiv der Niedersächsischen Landesbibliothek Hannover, *Briefwechsel*, N. 93, p. 254.

[516] Leopoldo was a cardinal from 12/12/1667, thereafter member of, among others, the Congregation of the Index; on him: Alfonso Mirto, "Medici, Leopoldo de'"; in: DBI 73 (2009), pp. 106–112, here 109 r. col.

[517] ASF, Mediceo del principato 5510, fol. 156[r]: "chiarezza degli ingegni". Also in Scherz, *Epistolae*, vol. 1, p. 48.

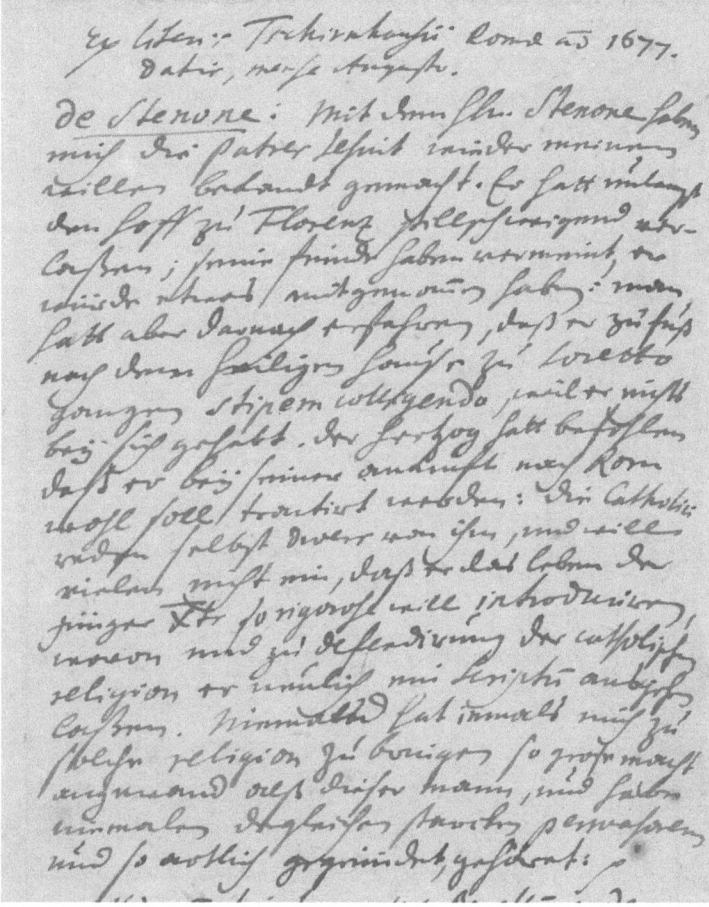

Fig. 2.7 Gottfried Wilhelm Leibniz' transcript of a letter addressed to him by Ehrenfried Walther von Tschirnhaus (Rome, August 1677) about Niels Stensen in Rome
Transcription: see note 515 above
LBH, LBr. 943, fol. 168ʳ, detail

keys), consisting of four rhombuses in the four subdivisions of the shield (Fig. 2.8).[518] The question of whether it might be Barbarigo's copy and how it found its way into the papal library has hitherto not been asked and would likely be difficult to answer.[519] Today this copy is located in the *History of Science Collections* of the

---

[518] See details in Martin, *Heraldry*, pp. 153–156.

[519] In the first half of the 1930s, many of the titles of Clement IX's private library went through the Roman *Libreria "Il Bibliofilo"* (Piazza Barberini, 46), as the latter's catalogs – not all of which are available in public libraries – document.

Fig. 2.8 Front cover of Pope Clement IX's copy of Stensen's *Elementorum myologiae specimen* *[...]* (Florence 1667) with two appendices, displaying the pontifical coat of arms
UOL, (no shelfmark); Image courtesy History of Science Collections, University of Oklahoma Libraries

University of Oklahoma Libraries (Norman, Oklahoma).[520] On December 20, 1668, Clement IX granted Stensen – following an inquiry by the latter, presumably due to his being notified in the summer of 1668 that he would once again be offered a position in Denmark[521] – "permission to own and read forbidden books about medicine, anatomy and surgery, everywhere in Denmark, that in actual fact deal with religion, for a period of 5 years".[522] As a cardinal, Rospigliosi had been a member of the Congregation of the Index and the Holy Office, and under his pontificate the sciences gained heavily in importance in the Curia.[523] Pope Innocent XI continued this policy of support for the sciences.[524] Stensen's conversations in the summer of 1677 presumably benefitted from this development with regard to his upcoming episcopal ordination.

Stensen was received in audience by Pope Innocent XI at least twice; once on Friday, August 6 and once on Monday, September 13, 1677, with the latter audience held at the palace of *Santa Maria Maggiore*.[525] Like in his daily routines, Stensen proved to be an accurate observer in Rome as well, for example when he overheard conversations between prelates and Protestants during one of his dates in the "papal palace".[526] The cardinals of the Holy Office as well as Pope Innocent XI himself presumably were keen to get to know Stensen as a physician and geologist, not least because he was recommended as a candidate for the episcopate in the associated informative process as a result of his knowledge of secular, ie natural scientific issues,[527] and also with a view to his future assignment. A letter written by the sec-

---

[520] It possesses neither a bookplate nor a handwritten dedication, cf. Scherz, "Stensenverbindungen", pp. 60–61; Gustav Scherz was able to view it on 02/12/1960. A b/w image of the front outer cover of the book with a pontifical coat of arms can be found on the last page of University of Oklahoma Libraries, *Exhibition* (unpaginated). The b/w image of the title page of the Myology in ibid., no. 7 is from the other of the two copies of the Myology located in the *History of Science Collections* at the *University of Oklahoma Libraries*. A reconstruction of the book's journey to the USA is difficult since, according to an e-mail (07/16/2012) by curator Dr. Kerry V. Magruder, "no provenance information in any accessible form" is currently extant.

[521] In the biographical context see details in Sobiech, *Herz, Gott, Kreuz*, p. 69.

[522] Spruit/Totaro, *Ethica*, p. 50: "licentiam […] tenendi, et legendi libros prohibitos de re medica, anatomica, et chirurgica ubique in Dania vero tractantes de religione ad quinquennium." The Index promulgated in 1664 by Pope Alexander VII ([* 1599] 1655–1667), an admirer of Galilei (Gross, *Rome*, p. 248), specifies the following in Alexander VII, *Index*, pp. 4–9 (Regulae indicis sacrosanctae synodi Tridentinae iussu editae), here 5 (Regula IX) in delimitation against necromancy, astrology etc.: "Permittuntur autem iudicia, & naturales observationes, quae […] medicae artis juvandae gratia, conscripta sunt." (But reports and natural observations are allowed which were written to […] promote the art of medicine.). This is the 9th rule of the Index of 1564 (with trans. in Wolf/Schmidt, *Buchzensurverfahren*, pp. 204–213, here 208).

[523] Schmidt, *Büchersäle*, pp. 94, 199, 446.

[524] See details in Donato, "Onere", pp. 73–74.

[525] Scherz, *Epistolae*, vol. 1, pp. 47–48 and Stensen's letters from Rome during this time: E 129, p. 345, lines 24–25; E 130, p. 346, lines 39–40; E 137, p. 358, lines 18–19; E 138, p. 360, lines 9–10.

[526] See details in DefScrut., p. 286, lines 26–28, here 26: "aula pontificia".

[527] Add. 17, p. 936, lines 12–13; ibid., p. 937, lines 11–12.

retary of the Congregation for the Propagation of the Faith, Odoardo Cibo (1619–1705) on May 6, 1684 shows that Pope Innocent XI appreciated Stensen greatly as a bishop.[528] All of the pieces of evidence cited above thus point to the very probable, though hitherto not proven, assumption that Stensen spoke about his research, and among it that on human reproduction, during his 1677 stay in Rome – be it with members of the Holy Office or with Pope Innocent XI himself.

Further substantiation for such conversations may be provided by the fact that Grand Duke Cosimo III was intent on preventing induced abortions by way of legislation: around the turn of the seventeenth to the eighteenth century, he inquired with the Holy Office whether it would be possible to baptize all aborted fetuses without consideration of the commonly held notions on the moment of ensoulment. The answer he received stated that adequate advice in this question was currently not possible without deliberation.[529] It should be noted with regard to the motive for Cosimo's query that Stensen had still been actively advising the Grand Duke in secular criminal cases after assuming his position as bishop in Hanover.[530] Although there is no concrete evidence that Stensen's research on human reproduction had any influence on the condemnation of the above-mentioned proposition on March 2, 1679, his demonstrable contribution to the gradual acceptance of the doctrine of simultaneous ensoulment nevertheless awaits further investigation.

2.3.1.6   Midwife Education and Dedication to Pregnant Women

Stensen not only proved his scientific intuition in the question of female anatomy and embryology, but also showed empathy concerning complications during pregnancy and childbirth. Already during his years of study in Leiden he was confronted with one of the many cases of death of a mother in childbed, and an unusual case at that: Thomas Bartholin wrote to him on August 5, 1662 st.v. with regard to the miscarriage of his sister-in-law Karen Bartholomaeusdatter Haagensen (†1662),[531] who had given birth to a fetus wrapped inside its membranes that had lived for 4 h after being liberated, and the following day had died herself: "Permit me the sorrow, forgive that I present you with this page full of grief. I know you will be seized by our mourning, which therefore I could not conceal from you as a most lovingly partaking friend."[532] Bartholin was certain of Stensen's compassion.

---

[528] E 347, p. 674, lines 29–31. On Cibo: Enrico Stumpo, "Cibo, Odoardo"; in: DBI 25 (1981), pp. 258–260.

[529] See details in Prosperi, "Baptism", pp. 228–230, here 229n67 ("document [...] undated"). On the decree by the Holy Office on 04/05/1713 see note 583, in Chap. 2 below.

[530] E 177, p. 412, lines 3–16; on this, see Sobiech, *Herz, Gott, Kreuz*, p. 140n83.

[531] E 8, p. 158n7.

[532] Ibid., p. 157, line 19–p. 158, line 8, here p. 158, lines 6–8: "Da veniam luctui, ignosce, quod paginam hanc moerore plenam tibi offeram. Scio te nostro luctu affici, quem igitur reticere non potui tibi amantissimo."

During his time as "Anatomicus regius" in Copenhagen, where he arrived on July 3, 1672 st.v., Stensen found the opportunity to care for pregnant women by contributing to the 'academization', ie the professionalization, of midwives. The pharmacy listing *Catalogus & valor medicamentorum simplicium & compositorum in officinis Hafniensibus prostantium [...]* (Catalog and value of the simple and composite medications which are found available for purchase at the [four] privileged apothecaries in Copenhagen [...]),[533] issued in Copenhagen in the summer of 1672 in Latin, Danish and German with a foreword by Thomas Bartholin from August 1, 1672 st.v., cites "Mr. Niels Stensen" among the 15, as stated in the listing, "names of the physicians held in high esteem in Copenhagen in 1672, listed in alphabetical order".[534] Bartholin, personal physician to the Danish king Christian V ([* 1646] 1670–1699),[535] had with his influence also managed to achieve the enactment (likewise in three languages) on December 4, 1672 st.v. of the Medical Ordinance *[...] Constitutiones regiae de medicis & pharmacopoeis &c. [...]* ([...] Royal ordinance concerning the physicians and apothecaries [et]c. [...]) for Denmark and Norway, according to whose guidelines the midwives in Copenhagen were to be trained and examined by the medical faculty in Copenhagen and by licensed local physicians in other cities before being allowed to practice.[536]

In an article for the *Acta Hafniensia* published shortly thereafter, Bartholin explained the need for this ordinance by stating that it was relatively certain that most of the mistakes made daily with the parturients were to be attributed to the ignorance of the midwives and the "weak women" (mulierculae) assisting them or to the disobedience of the parturients (towards the midwives).[537] The exam on September 30, 1673 st.v. at the Copenhagen anatomical theater, for which Bartholin and Stensen were among the five examiners and two older midwives, Christiana and Elisabeth, served as observers, was sat by 15 midwives from Copenhagen. As Bartholin emphasizes in his report, all of the testees possessed ample birthing experience: one of the women had assisted in nine births, one in 17, and the remaining 13 in a dozen each. Most of them answered the questions they were asked adequately, so that the examiners could have no doubt of their competence; as they were found to be not appropriately proficient in anatomy, however, they were told to take part in Stensen's dissections the following winter. They were also urged to inform Stensen immediately when a pregnant or birthing woman from the poorer population had died, so that he could use the opportunity to demonstrate to them – free of charge –

---

[533] Bartholin, *Catalogus* (date of publication according to frontispiece 1671, according to title 1672).

[534] Ibid. (unpaginated), here the first page following the foreword "Decanus Facultatis Medicae in Regia Acad[emia] Hafniensi Thomas Bartholinus lectori sal[utem]", dated 08/01/1672 st.v.: "Nomina medicorum, qui Hafniae 1672 floruerunt, ordine alphabetico posita. [...] d[ominus] Nicolaus Stenonius." The frontispiece (recto) shows, among other things, medicinal plants with a sheaf of sunlight falling onto them.

[535] HGK, s.v. "Bartholin, Thomas", pp. 70–71, here 71.

[536] Christian V, *Forordning* (unpaginated), no. 6 (in Danish, Latin and German versions).

[537] AMPH 1 (1671/1672), pp. 286–292 (CXXXVIII. De obstetricis absentia querelae), here 288, no. I & pp. 290–291.

what he deemed important with respect to the case.[538] Whether such training by Stensen took place in the *Domus Anatomica* or in the field is unknown. As the anatomical exercises during the winter of 1673/1674 were almost all cancelled, Stensen's additional anatomical education of the midwives most probably never took place.[539] That there was continued contact between the midwives and Stensen in the time following the exam and the women valued Stensen's work – and that through it they may have been among the first women to be aware of the existence of their ovaries – is demonstrated by the fact that the midwife Müller gifted Stensen's student Holger Jakobsen with a parrot for dissection, which Jakobsen promptly conducted in his private quarters under Stensen's instruction on December 29, 1673 st.v.[540]

Although Stensen set great store by the autonomy of science in the sense of basic research independent of any application, he nevertheless would rise to speak if he thought that possibilities for treatment arose from it. In his treatise *Uterus leporis proprium foetum resolventis* (Uterus of a Female Hare that Rejects its Own Fetus), published 1673 in the first volume of Thomas Bartholin's *Acta Hafniensia* and going back to notes from Stensen's time in Paris and Florence,[541] one can find – as remarked by Stefano Miniati, "a quite rare example of a concrete clinical employment" of Stensen's research.[542] Here again, Stensen's empathy so cherished by Bartholin becomes evident: "Indeed one may in any case find a proof of hope in this observation for the birthing [women], who after all manner of unsuccessful attempts retain a dead fetus, if they are not for some other reason overfull of putrid fluids; and it is no slight consolation in a desperate case to awaken the hope for rescue at least by means of a single secured example, where one fears bodily harm [to the pregnant woman] from the pain of desperation."[543] Stensen, who saw little value in merely treating the symptoms of disease with a view to the necessary advancement of medical science, was by no means blind to the suffering of the sick.

As suffragan bishop of the Münster diocese from 1680 to 1683, at the latest, Stensen encountered induced abortion more closely. During this time, on a major visitation tour through the so-called "Niederstift" (Lower Diocese) of the Diocese of Münster during which he also visited the deanery of Vechta (Emsland), he concerned himself with the appropriate training of midwives by written order in August 1682 to the parish priest of Dinklage, Bernhard Ribbers (†1715).[544] In it he emphasized that the imperiled bodily well-being of the mother as well as the endangered

---

[538] AMPH 2 (1673), pp. 53–55 (XXII. Obstetricum examen), here 53.

[539] Kardel/Maquet, *Biography and Original Papers*, p. 304.

[540] Add. 12, p. 928.

[541] UtLep. (Notes) preliminary remarks, p. 318 r. col.

[542] Miniati, *Steno's Challenge*, p. 106n39. A further example can be found in GlandOc., p. 90, lines 23–29.

[543] UtLep., p. 58, lines 25–30: "Sane vel spei saltem argumentum ex hac observatione sumere licet pro parturientibus post conatus omnes irritos foetum mortuum retinentibus, si alias putridis humoribus alia de causa non abundaverint; nec leve solamen est in re desperata posse salutis spem vel uno certo exemplo excitare, ubi a desperationis dolore virium detrimentum timetur."

[544] On Ribbers: BAM, Coll. Kleruskartei, s.v. "Ribbers, Bernhard".

spiritual well-being of the children had to be considered.[545] He thus obviously did not lose his concern for the health of the mother due to a preoccupation with the ulterior well-being of unborn children. Only 1 month later, Stensen investigated the urgent suspicion, based on circumstantial evidence, of concubinage with induced abortion by the housekeeper Katharina Gertrud Averbeck,[546] 23-year-old niece of the Emstek parish priest Johannes Lübbermann (1636–earliest 1696),[547] whom Stensen had suspended on September 11 due to multiple grievances. As part of the correspondence dealing with this case, which continued beyond his visitation, Stensen unmistakably identifies induced abortion as a "crime"[548] in a letter to chaplain Caspar Strübbe (†1714)[549] on December 5, 1682, ordering the housekeeper to be examined[550] by circumspect and properly trained and licensed midwives for signs of having given birth, and the midwife present during the presumed time of the abortion to be questioned.[551] Stensen did not hesitate to use his official authority as bishop for the protection of the helpless unborn. After concluding his visitation, he wrote a letter to Odoardo Cibo in Rome on November 15, 1682 concerning financial aid for – among others – a young girl abandoned by her mother on the street shortly after birth in Hanover during his time in that city, whom he had placed with a Catholic woman at the time because the mother could not be found.[552]

Stensen's caring commitment to pregnant women and mothers finds a parallel in Florence, where the priest and Oratorian Filippo Franci (1625–1694),[553] a tireless organizer of aid initiatives for women and the youth, had established a ward for women who wished to conceal the shame of extramarital offspring could give birth anonymously and receive spiritual support. This probably occurred after the promulgation of Pope Innocent XI's decree on March 2, 1679, ie during Stensen's time as vicar apostolic in Hanover.[554] Hence the latter was apparently not the only one dedicated to this cause.

---

[545] More details in E 271, p. 542, lines 34–37. Likewise in August 1682, parish priest Johannes Stodtbrock (†1706) of Cappeln (on Stodtbrock: BAM, Coll. Kleruskartei, s.v. "Stodtbrockh, Johannes") approached Stensen with a request to ensure that the midwives were sworn in (Willoh, *Pfarreien*, vol. 4, p. 166). On two further midwives named in the course of Stensen's visitation, Grete Koldehof and Grete Tölking in Vechta, see Willoh, *Pfarreien*, vol. 3, p. 121.

[546] E 277, p. 550, esp. lines 8–14. On the sworn-in midwives of the city of Münster in the seventeenth century see Gördes, Heilkundige, pp. 62–64, here 64. Regarding the midwives in the visitation records of the "Niederstift" (Lower Diocese) of Münster in the first half of the seventeenth century see Lackmann, *Niederstift Münster*, pp. 52–53, 316 & passim; on visitations in the "Oberstift" (Upper Diocese) of Münster see Lackmann/Schrörs, *Fürstbistum Münster*, pp. 79, 82, 295, 380.

[547] On Lübbermann: BAM, Coll. Kleruskartei, s.v. "Lübberman, Johann"; E 275, p. 548n1 & 2.

[548] E 288, p. 571, lines 36–37, here 37: "crimine".

[549] On Strübbe: BAM, Coll. Kleruskartei, s.v. "Strübbe, Kaspar"; E 287, p. 571n3.

[550] E 288, p. 572, lines 5–6.

[551] Ibid., lines 32–33. – The midwife is not named.

[552] E 286, p. 569, lines 17–20; more details ibid., lines 20–32.

[553] On Franci: Daniela Lombardi, "Franci, Filippo"; in: DBI 50 (1998), pp. 133–134.

[554] Lombardi, *Povertà*, pp. 199–200.

2.3.1.7   'Demythicization' of Deformities

Already in his study days, Stensen had been confronted with the topic of deformities in connection with a concrete case of bereavement in Thomas Bartholin's family.[555] During his stay in Paris in 1665, he dissected a dysplastic female human embryo[556] and published his results in the article *Embryo monstro affinis Parisiis dissectus* (An Embryo Akin to a Monster, Dissected in Paris), based on notes made in Paris and later in Florence and appearing in the first volume 1673 of the *Acta Hafniensia*.[557] This first ever description of the heart malformation known since 1888 as *Tetralogy of Fallot*, is still present in medical essays today.[558] It can be assumed that, as is usual with this condition, death occurred shortly after birth and the fetus had been given to Stensen for dissection.

At the end of the examination, Stensen betokens what he had observed as "intelligent nature" (intelligens natura) whose effect in the organism is always achieved "in ingenious fashion" (modo semper ingenioso).[559] Hans-Michael Bonse interprets this statement to mean that Stensen ascribed a rationality to nature which distinguished it as God's creation, and "that only in Faith is a comprehensive insight into nature possible and not alone on the level of empirical examinations".[560] While this may apply to the revised article published in 1673, contemplation of Stensen's statement against the background of the history of his conversion allows a more in-depth appreciation: in the first appendix to his Myology, Stensen mentions the "elegant work of art that is intelligent nature", adding that this did not occur "without a certain law" – meaning the natural laws.[561] No one could likely have described the teeth of the shark he had dissected, he states, as elegantly[562] as nature – whom he

---

[555] See note 532, in Chap. 2 above.

[556] There is no information on the age of the fetus. While 'embryo' and 'fetus' were used synonymously in the early modern era (cf. Faller, "Tetralogie", pp. 324–325), the term 'embryo' is used today to refer to an unborn human being during the first eight weeks after insemination, cf. Rager, *Person*, p. 197.

[557] Embryo. Cf. ibid. (Notes) preliminary remarks, p. 318 r. col.

[558] Named after the physician Dr. med. Étienne-Louis Arthur Fallot (1850–1911), who – not only with a view to Stensen – cannot be regarded as the actual discoverer; on this and on Fallot: Acierno, "Fallot", here 321 (163) on Stensen's priority of discovery. Acierno advocates retaining the eponym, see ibid., p. 322 (164). Stensen's priority of discovery is also mentioned in: Tubbs/Gianaris/Shoja/Loukas/Cohen[–]Gadol, "Tetralogy of Fallot", pp. 313–314; Gaither/Ardite/Dhuper, "Agenesis", p. 2 r. col.

[559] Embryo, p. 53, lines 11–13.

[560] Bonse, "Tetralogie", p. 36.

[561] CanCap., p. 121, lines 8–9: "Elegans [...] intelligentis naturae artificium [...] non [...] sine certa lege".

[562] Here Stensen follows he common language use of his time, cf. eg Schepelern, *Olai Borrichii Itinerarium*, p. 64, no. 5. The "elegant distribution of the nerve" (elegans nervi distributio) was demonstrated to the students in the autopsy room of the *St. Caecilia Gasthuis* in Leiden using the parotid glands of the corpse of Janicke Jansen, presumably by Van der Linden (ibid., pp. 71–74, here 72, no. 6 [Van der Linden], 7).

refers to in personifying fashion – had created and arranged them.[563] In his 1664 accountability report on his studies abroad, Stensen claims that it is nature who "enlivens" the dazzling colors of the pupil.[564] This metaphorical use of the term "nature" in his early anatomical publications, which Stensen intentionally composed with ambivalence regarding the religious terminology as was common fashion for the time, may also be an expression of the deist viewpoint 'sui generis'[565] which he adopted in the Netherlands as a result of his reading and initial impressions of the differing religious currents and confessions.

Later, after his conversion and as a bishop, Stensen regarded deformities from a theological perspective – which does not appear in his anatomical writings, methodically independent from theology without exception –, namely in connection with original sin. Here again one may pose questions regarding the "tension between the Christian doctrine of original sin and the new concept of nature as a perfect order" (Richard Toellner).[566] In the *Chaos* manuscript, Stensen had excerpted on March 10, 1659 st.v. from Pierre Borel's *Historiarum, et observationum medicophysicarum centuriae IV* that the "seeds of diseases" had been created by God after the Fall of Man and were disseminated to the bodies of men through food, drink, contagion, exhalation and wind.[567] Borel refers to the Paracelsian Danish royal physician Peder Sørensen (1540/1542–1602), who differentiated between the first, perfectly created seed and the post-lapsarian death-bringing seed, the mixing of which had resulted in the current state of creation marked by disease and death.[568] Stensen made a note of the latter's work *Idea medicinae philosophicae [...]* (Notion of Philosophical Medicine [...]; Basel 1571) for later reading.[569]

In one of his Christian religious instructions as suffragan bishop, Stensen states that prenatal defects of the "[five] outer senses" manifesting themselves in children born blind, deaf, mute or paralyzed were consequences of "original sin" (peccatum

---

[563] PiscCan., p. 150, lines 8–9. Personifications of nature can also be found in CanCap., p. 138, lines 6–7, 20–21, in the "industriousness of intelligent Nature" (naturae intelligentis industria) in UtLep., p. 58, line 30 and in the "industriousness of ingenious Nature" (naturae solertis industria) in PiscCan., p. 150, line 3. Nature was the same everywhere (ibid., p. 151, lines 8–9); a statement used by Stensen referring to the natural laws.

[564] AnRaj., p. 206, lines 1–5, here 4: "animat".

[565] Developed in Sobiech, *Herz, Gott, Kreuz*, pp. 51–68, here 57, 62, 65 (theory by the author and differentiation into two phases: (1) c. 1660–1662/1663 and (2) 1662/1663–1667 with a break in 1664). On the personification of nature cf. Numbers, "Science without God", p. 268.

[566] See note 425, in Chap. 2 above.

[567] Ziggelaar, *Chaos-Manuscript*, p. 36 (col. 6, fol. 29ʳ) with the subheading "Semina morborum" (Seeds of the Diseases); on this, see Borel, *Historiarum* (Centuria I, De canum rabidorum morsu. Observatio LXXIV), pp. 72–75, here 74–75.

[568] Shackelford, *Paracelsian Medicine*, p. 188; Toellner, "Autorität", pp. 169–170; Toellner, "Körper", pp. 138–139.

[569] Ziggelaar, *Chaos-Manuscript*, p. 83 (fol. 35ᵛ). On Sørensen: Gordon Norrie, "Sørensen, Peder"; in: DBL³ 14 (1983), pp. 329–331, here 329–330 on Sørensen, *Idea medicinae philosophicae*.

originale).[570] When consulting anatomy in order to learn more about the workings of the human organism equipped with its many functions, one could observe that the body's self-maintaining harmonious processes sometimes became damaged during the development of the fetus or due to later causes, disturbing the orderliness of these processes and resulting in pain and various inabilities.[571] The positive mention of anatomy in these notes is proof that Stensen's appreciation of the natural sciences as a bishop remained the same as before; of great significance considering the time in which these statements were made, however, is the fact that Stensen 'demythicized' the contemporary speculation on deformities by use of the theological term of original sin.

The impulse for this 'demythicization' was Stensen's dissection of a malformed calf's head, a so-called hydrocephalus, in Innsbruck in 1669. In the introduction to his treatise De vitulo hydrocephalo (On the Hydrocephalic Calf) published in the Acta Hafniensia for the years 1671/1672, he writes that he had not begun to believe the stories of humans without brains – whose deformity is now referred to as anencephaly – before his dissection of this calf's head, for which Sylvius' frequent anatomical exercises dealing with the brain had been of decisive benefit.[572] Already as a student in Leiden Stensen had emphasized with regard to alleged ducts connecting the brain and mouth that while he did not want to deny that such passages could be found, he would not believe they existed before he had seen them himself.[573] As a result of his examination of the calf's head, he explained that it was not the mother's imagination that was the cause of the fetus' deformity, but the defective anlage of the fetus itself[574]; he also created a detailed drawing of the misshapen head.[575] In context with Stensen's treatise De vitulo hydrocephalo and the relationship between anatomy, theology and the Magisterium, there has hitherto been no study into the fact that at the end of April 1624, a drawing of a two-headed calf was presented to Pope Urban VIII ([* 1568] 1623–1644),[576] his Secretary of Briefs and his personal theologian in the Vatican Gardens by the pontifical botanist Dr. med. Johannes Faber (1574–1629), member of the Roman Accademia dei Lincei (est. 1603) and

---

[570] Op. 5, p. 454, lines 1–4, 7, 17–18, here 1: "Peccati originalis" and 17: "sensus externos". This had already been asserted by Augustine of Hippo, cf. Gélis, Naissance, pp. 362–363.

[571] Op. 9, p. 498, line 31–p. 499, line 2.

[572] VitHyd., p. 230, lines 9–16.

[573] HepRed., p. 70, lines 13–14.

[574] VitHyd., p. 237, lines 28–29; pursued in Kardel, "Hydrocephalus", p. 172, no. 1. Stensen had excerpted from Kircher, Magnes, pp. 725–733 (Liber III, pars VII, caput VII), here 728–733 on 04/20/1659 st.v. the notion that due to processes in the internal organs, women were more than men, and pregnant women more than not pregnant women, imperiled by fantasy, wherefrom deformities of the fetus could result, cf. Ziggelaar, Chaos-Manuscript, pp. 256–257 (col. 100, fol. 54ᵛ– col. 101, fol. 55ʳ with N'.94).

[575] VitHyd., here the fold-out fascimile between pp. 237 and 238, a drawing in red chalk; cf. ibid. (Notes) preliminary remarks, p. 341l. col.

[576] De Renzi, "Rome", p. 79.

dissector of the calf, and the Pope's personal physician Dr. med. Giulio Mancini (1559–1630).[577]

In regard to the question of deformities, a letter sent by Stensen from Dresden, while en route from Florence to Copenhagen to assume the position of "Anatomicus regius", to Cosimo III on June 21, 1672 warrants further examination. In this letter, which once again emphasizes Stensen's propensity for precise observation and perception of events going on around him, he expounds: "Here they are speaking of nothing but signs and horrors. One says that two children have been born, one of them crying, the other sweating blood. The particular letters from Regensburg report that a heretic woman, giving birth assisted by a Catholic midwife, when the difficulty of the birth was growing, was asked to call for the aid of the Mother of God, and at the moment when she furiously answered: *The sow-mother*, gave birth to three piglets. *I pass on what was reported* [Relata refero], which some do not believe and for which others make reference to nature and happenstance. It could of course be true."[578] Further information on this occurrence, which may pertain to children born with severe deformities, cannot be found. The "particular letters" may refer to mail from envoys or, more likely (due to the identifier "from Regensburg"), a journal by mail in the periphery of the Perpetual Diet of Regensburg (1663–1806).[579]

Stensen's words "Relata refero" and his mention of other – unspecified – appraisals which either disbelieved the story or, if they did believe it, attributed it quite typically for the time to "nature and happenstance", ie to a "joke of nature" (lusus naturae), demonstrate his skepticism and intent to distance himself from these notions. Nevertheless, the second censor appointed to Stensen's writings argued in 1965 that his assessment of the consequences of this blasphemy "reveals a certain tendency towards pseudo-mystical interpretation of natural events; that a man of such education in the medical sciences believed that a woman who uttered the blasphemy 'the sow-mother' against the Mother of God could give birth to three pigs is extraordinary even for the time in question."[580] It therefore seems appropriate to investigate the question of deformities in more detail.

---

[577] On Faber: Gabriella Belloni Speciale, "Faber (Fabri, Fabro), Giovanni"; in: DBI 43 (1993), pp. 686–689, here 687; on Mancini: Silvia De Renzi/Donatella L. Sparti, "Mancini, Giulio"; in: DBI 68 (2007), pp. 500–509, here 503–504.

[578] E 79, p. 266, lines 2–17, here 12–17: "Quì non si parla di altro che di prodigii e di spaventi. [...] Dicono [...] essere nati [...] duo bambini, l'uno lacrimando, l'altro sudando sangue. [...] Le lettere particolari di Ratisbona referiscono come una parturiente heretica servita da una levatrice catholica, e crescendo la difficolta del parto pregata a invocare l'assistenza della madre di Dio, mentre infuriata rispose: *Die Saw Mutter* partorisce tre porcelli. *Relata refero*, le quali cose da alcuni non si credono, da altri si riferiscono alla natura e al caso. Possono ben sì esser vere".

[579] On the "particular correspondenz" cf. Friedrich, *Regensburg*, pp. 199–205; on the journals by mail cf. ibid., pp. 428–448. No corresponding report can be found in the Hauptstaatsarchiv Dresden, at least not in holding 10024 Geheimer Rat (Geheimes Archiv), 201 (Wunderzeichen und Curiosa; Loc. 10690/45 missing).

[580] Sacra Congregatio Rituum, *Vota*, pp. 12–15, here 14: "revelat certam inclinationem ad pseudo-mysticam interpretationem factorum naturalium; quod homo tantae eruditionis in arte medica cre-

There had been a similar case in 1609 in which a woman in Geneva, who had been in labor for 11 days, told a neighbor in the presence of the midwife and a Catholic chambermaid that she would prefer to die or give birth to a calf than have a prayer be said to Margaret the Virgin[581] – the ancient martyr is one of the Fourteen Holy Helpers –, whereupon she gave birth to an animal body in the shape of a calf which was subsequently thrown, presumably still alive, into the Rhône by ruling of a committee convened to deliberate the case.[582] One may assume that the interpretation of a child born with severe deformities as an animal was a topical element, disseminated in the fashion of rambling legends, which was used frequently in the context of miscarriages.

The *Rituale Romanum* (1614), promulgated by Pope Paul V ([* 1552] 1605–1621) with its last Editio typica published in 1952, dictated "great caution" (magna cautio) in the case of miscarriages, and that the "Ordinarius loci", meaning for example the bishop, or other experts were to be consulted if the life of the miscarried child was not threatened.[583] With regard to this caution, Girolamo Fiorentini had made use of the fact that the Rituale specified no temporal boundary concerning the development of the fetus before a baptism was forbidden.[584] Contemporary textbooks seem to give preference to conditional baptism of a "monstrum" over its not being baptized at all; Hermann Busenbaum SJ (1600–1668) for instance, in his oft reprinted and widely used handbook *Medulla theologiae moralis [...]* (The Core of Moral Theology [...]; Münster 1650), endorsed a "conditional" baptism for a "monstrum" with only an animal-like head if its life was in danger.[585]

Thomas Bartholin warned as follows in 1665 concerning miscarriages in Denmark: "One must however take great care not to be misdirected by designating as monstra that which deviates from the normal path of nature only in people's judg-

---

debat mulierem contra Dei Matrem blasphemantem, verbis 'die Saumutter', parere posse tres porcos, etiam pro illo tempore res sat extraordinaria est."

[581] On Margaret the Virgin: Thomas Berger, "Margareta in Antiochien"; in: LThK³ 6 (1997), col. 1311–1312.

[582] Anonymous, *Miracle* (the title page [recto] cites as a motto pertaining to the miscarriages 4 Ezr 5:8 translated into French), pp. 7–9, condemning the event. Cf. Gélis, *Naissance*, pp. 364–365.

[583] Sodi/Flores Arcas, *Rituale Romanum*, p. 16 (= p. 8 of the Rituale [De baptizandis parvulis]). The Editio typica of 1614 ruled that in the case of "a deformed child, if it does not exhibit a human form" (Monstrum, quod humanam speciem non praeseferat), the aforementioned child did not have to be baptized, but was to be conditionally baptized in case of doubt (ibid.). On the consultation of "physicians and theologians" (medici et theologi) in case of "premature birth" (foetus abortivi) see the decree by the Holy Office on 04/05/1713 (Gasparri, *Codicis iuris canonici fontes*, vol. 4, N. 777, p. 60). The text of the 1614 Editio typica remained unaltered until the Editio typica of 1913 (on it, Pius X, *Rituale Romanum*, p. 7, nos. 18–19; Sodi/Flores Arcas, *Rituale Romanum*, pp. LIX–LX).

[584] According to Cangiamila, *Embryologia sacra* (Libro primo, capo VIII), p. 56, no. 7; cf. Prosperi, "Baptism", pp. 224–225.

[585] Busenbaum, *Medulla theologiae moralis* (Liber VI, Tractatus II, Caput I, Dubium IV, Responsio 2, no. 7), pp. 434–435, here 434; cf. Weber, *Spezielle Moraltheologie*, p. 216n381. On Busenbaum: Philip Schmitz, "Busembaum [sic!], Hermann"; in: DHCJ 2 (2001), p. 578.

ment and ignorance, but not in the eyes of the knowledgeable."[586] On the other hand, however, he did to some degree believe the stories going around.[587] Fanciful contemporary illustrations can be found in the book *De monstrorum caussis, natura, & differentiis [...]* (On the Causes, the Nature and the Differences of Deformed Babies [...]; Padua 1616) by Dr. med. Fortunio Liceti (1577–1657), who served as professor of medicine at the University of Padua from 1645.[588] Gerard Leonard Blasius published the fourth edition of this treatise in Padua in 1668, adding an appendix discussing miscarriages described by authors like Thomas Bartholin and Lodewijk de Bils.[589] That Liceti's treatise, whose descriptions often lapse into the fanciful, was printed in a French translation as late as 1708 goes to show that it was still considered credible in the medical science of the early eighteenth century – as were various older notions in the field of embryology.[590]

During Stensen's time, detailed reports on miscarriages were commonplace: the *Philosophical Transactions*, for example, quoted in their issue on Monday, June 3, 1667 st.v. from a letter sent from Paris according to which a dead miscarried child was on display in the house of the physician and lay abbot Dr. theol. Pierre Michon Bourdelot (1610–1685),[591] founder of a private academy which existed until 1684. The letter stated that the mother claimed to have seen a monkey wearing clothes on a theater stage in the fifth month of her pregnancy, and had eventually given birth to a monstrum in the shape of a monkey with fourfolds of skin hanging down from the back of the head over the shoulders nearly to the middle of its torso, a small cape of sorts akin to the theater monkey's cloak. This posed various questions, the letter continued, pertaining eg to the power of imagination, to whether this creature possessed a human soul and if not, what had happened to the soul of the 5-month-old embryo.[592] Stensen had incidentally taken part in Bourdelot's private sessions several times while in Paris, but no lasting contact had resulted.[593] After reporting a

---

[586] Bartholin, *De cometa*, p. 126: "Diligenter tamen cavendum, ne in his decipiamur, ea pro monstris traducentes, quae hominum tantum judicio & vulgi ignorantia, neutiquam rerum peritis, a solita naturae via recedunt."

[587] Gélis, *Naissance*, p. 357.

[588] Liceti, *De monstris*. On Liceti: Giuseppe Ongaro, "Liceti, Fortunio"; in: DBI 65 (2005), pp. 69–73.

[589] Liceti, *De monstris*, pp. 263–316 (with printing errors in the page numbers): "Appendix monstra quaedam nova, & rariora, cum satyro indico, & muliere cornuta, proponens."

[590] Roger, *Life Sciences*, p. 32.

[591] On Bourdelot: Katia Béguin, "Bourdelot, Pierre"; in: DFPh. 1 (2008), pp. 205–208.

[592] PhTRS 2 (1667), p. 473: "Numb. 26. Munday, June 3. 1667"; ibid., pp. 479–480 (Extract of a Letter, Written from Paris, containing an Account of some Effects of the Transfusion of Bloud; and of two Monstrous Births, &c.), at http://rstl.royalsocietypublishing.org/content/2/23-32/479.full. pdf+html. Between 1665 and 1780, 96 case studies on monstrosities appeared in the *Philosophical Transactions* (Enke, *Soemmering*, p. 32n174). According to the physician Soranus of Ephesus (c. 78–138), a woman who looked at monkeys in the moment of conception would give birth to a simious child; it is unclear whether this is an aesthetical or pathological resp. anatomical categorization, however (ibid., p. 25).

[593] Scherz, *Epistolae*, vol. 1, p. 11; Andrault, *Stensen*, p. 17n2.

further miscarriage which disagreed with the Cartesian theory of the pineal gland, the article ends with a reference to a letter by Stensen arrived from Florence and reporting about an experiment on a tortoise which had likewise addled the Cartesians and dealt a heavy blow to their views.[594] At any rate, the leading English-speaking scientific journal of its time printed, in stark contrast to the reserved caution of the *Rituale Romanum*, a rather imaginative-seeming speculation.

A more realistic report can be found in the *Acta Hafniensia*; it is written by Dr. med. Matthias Jakobsen (1637–1688),[595] who had been given a professorship at the University of Copenhagen on August 19, 1664 st.v. in Stensen's place. According to Jakobsen, a "misshapen and hideous monstrosity of female gender had been born to Fredericus Halla of Norway on May 30, 1673", and it was unclear whether the child had been conceived in or out of matrimony. The girl lived for 3 days, expelling the mother's milk given to her out through her nose. The dejected mother had her house sprinkled with holy water and called upon St. Catherine[596] – an ancient martyr and one of the Fourteen Holy Helpers – for assistance.[597] This report, which sheds light on the still largely Catholic popular belief in Norway, which at the time belonged politically to Denmark and hence to the Lutheran state church, is followed in the *Acta Hafniensia* by a treatise by Stensen entitled *In ovo & pullo observationes* (Observations on Egg and Chicken),[598] thereby succinctly demonstrating the embedding of scientific research into the circumstances of the times.

Stensen's caution[599] in his evaluation of the events accounted from Dresden in 1672 on the one hand is a sign that he did not wish to condemn contemporary conceptions a priori, but on the other hand maintained a distance to them as a natural scientist with his use of "Relata refero" until he was able to verify or falsify them through autopsy. Every man is shaped by his time, and Stensen retained a high degree of autonomy by discarding nothing prematurely while at the same time applying the criterion of critical examination to everything. In this he coincided with the caution stipulated by the *Rituale Romanum*. Upon overall evaluation of Stensen's statement in the context of the seventeenth century, the censor's misgivings are refuted. A sermon by Stensen as bishop includes the criticism that "bodily

---

[594] PhTRS 2 (1667), p. 480. On placement of this letter see Rome, "Royal Society", p. 249.

[595] On Jakobsen: FL 4 (1927), s.v. "Jacobæus, Matthias", pp. 255–256; Kardel, *Steno*, pp. 25–26.

[596] There are no historical documents on St. Catherine; on her: Hans Reinhard Seeliger, "Katharina v. Alexandrien"; in: LThK³ 5 (1996), cols. 1330–1331.

[597] AMPH 2 (1673), pp. 80–81 (XXXIII. De monstro Norwegico. D[omini] Matthiae Jacobaei Prof[essoris] Regij) with three plates with captions (unpaginated) between pp. 80 and 81, here 80: "monstrum foeminini sexus informe & aspectu horrendum".

[598] Ibid., pp. 81–92 (XXXIV. In ovo & pullo observationes d[omini] Nicolai Stenonis Anatom[ici] Reg[ij]) = OvPul.

[599] Miniati, *Steno's Challenge*, p. 238. Like the censor (see note 580, in Chap. 2 above), Miniati does not do justice to the wording of Stensen's letter when he asserts that Stensen "holds as «true»" what the "particular letters" claim. Ibid., n. 86 he explains Stensen's "naïve attitude towards «miracles»" in the question of deformities with the latter's new-found perspective of faith. This explanation, however, is inconsistent with Stensen's levelheaded natural scientific investigation of the authenticity of miracles during his time as bishop (see note 711, in Chap. 2 below).

and mental defects of another person" served as "food for disdainfulness" for other people.[600] This attitude expressing Stensen's empathy and, from a medical point of view, his sympathy with the patient suffering in body and soul can be assumed to have characterized him in regard to mother and child in cases of severe birth defects as well.

2.3.2 In the Face of Suffering and Death

Human life in the seventeenth century was exposed in myriad ways to imperilment by disease and epidemics, eg the plague, which Stensen was confronted with as early as 1654 through the deaths of several classmates from his Latin school.[601] From age three to age six, "hindered by a rather severe illness", Stensen had "always been under the hands of [my] parents and older friends"[602] and was thus dependent on their care. He presumably also survived an infection with smallpox, as can be deduced from his statement as suffragan bishop of Münster to the Baroness of Galen around May 1681.[603] Although these were harrowing experiences for Stensen, his parents appear to have cared well for him as he speaks of their "hands" – the hands had a fundamentally positive connotation in Stensen's spirituality. And although he was never a practicing physician, it is documented that in the days before his own death, Stensen cared for the Viennese priest of the order of Hermits of St. Augustine and Schwerin court chaplain Jakob August Steffani OESA (†1686), in part by pre-scribing the latter's medication.[604] What was Stensen's attitude as anatomist and bishop toward the questions of suffering and death, in particular in terms of the views taken by him as a theologian?

2.3.2.1 Patient and Creator: On the Function of Anatomy

As a student in Leiden, Stensen had described the common glandular diseases, which frequently posed a great threat to the patient's life, as an evil, which is why he considered the examination of healthy and diseased glands advisable for improv-ing the understanding of the human body and for achieving and maintaining good health.[605] Ole Peter Grell argues that Stensen's concept of the role and importance of anatomy had changed between the times of his Copenhagen study diary, in which

---

[600] Sermo 27, p. 284, lines 31–33, here 32–33: "cibus superbiae [...] alieni defectus corporis et animae".

[601] See details in Kardel/Maquet, *Biography and Original Papers*, p. 30.

[602] DefConv., p. 394, lines 16–18: "morbo satis difficili impeditus semper inter manus parentum et amicorum aetate provectorum".

[603] E 229, p. 485, lines 7–8.

[604] Details in Add. 47, p. 993, lines 35–40. On Steffani: Scherz, *Epistolae*, vol. 1, pp. 125–126.

[605] GlandOr., pp. 12–13, 18, lines 16–21.

he had defined the "benefit of one's neighbor" as the goal of medicine,[606] and that as "Anatomicus regius": Stensen had later "excluded the beneficial aspect of such undertakings for the Godly community", and anatomy now served "a purely religious purpose", having become "something akin to a Catholic, baroque play" to him in which the anatomist resembled an actor who was "manipulated" by the playwright God.[607] The medical historian Grell, hereby touching on the theological question of human freedom and divine predestination in a less than constructive manner, alludes to the contemporary metaphor of the 'World Theater' with its notion of the public dissection as a 'theatrum mundi', according to which "the human existence is determined as a role in a play presaged by God and fate".[608]

Grell bases his assertion mainly on the end of Stensen's Copenhagen inaugural address. Here, following his words on the function of the sensory organs which according to him lies in recognizing the Creator in His creation by use of the questions asked by science and, as a result, in transferring onto the Creator the love hitherto only shown for the creation, Stensen says: "Thus it is those who are in error, and make use of anatomy below the dignity of its concern, who make it their servant with the goal *only* of preventing or treating diseases."[609] Hence Grell's conclusion is invalid already from a linguistic point of view, as in contrast to Stensen, who merely states that it cannot be anatomy's task to "only" (solis) prevent and treat disease, he incorrectly claims that Stensen had "excluded" disease prevention. Stensen's stance on the function of anatomy will be further elucidated by the following six points:

1. Disregarded by Grell,[610] Stensen differentiates immediately following his words mentioned above: "It [ie anatomy] surely has its application there, though not to the extent that we assume, for the knowledge of the preternatural [= pathological] condition cannot outreach the knowledge of the natural [= healthy] condition; as the latter is still very limited, the former will likewise not be able to advance its boundaries very far."[611] Stensen was therefore not redefining anatomy, but

---

[606] See note 305, in Chap. 2 above.

[607] Grell, "Conversions", pp. 208–209, 216, here 216. "Godly community" is a Reformation term inapplicable in this context. In Casella/Coturri, *Opere scientifiche*, pp. 1–107 (Biografia e introduzione alle opere), here 3–4, the two passages are merely parallelized. Similar to Grell, Kaiser/Völker, 17. Jahrhundert, p. 233 spoke 18 years earlier at Martin-Luther-Universität Halle-Wittenberg (GDR) of the Lutheran Stensen as a "passionate 'Stürmer und Dränger'" (likening him to the artists of the Sturm und Drang period), whereas the convert Stensen had been an "occasional anatomist for whom the demonstration of results was primarily the explanation of the divine wisdom of order".

[608] Stockhorst, "Leichensektion", p. 278.

[609] Prooemium, p. 255, lines 14–21, here 19–21: "Frustra itaque sunt, & infra rei dignitatem cum anatome agunt, qui solis morbis praecavendis aut curandis eam famulam faciunt".

[610] This may have to do with Grell's use of the non-literal translation "It does have a certain use there but not as much as we think since […]" of ibid., lines 21–25 in Kardel, *Steno*, p. 125. This translation should be corrected as follows: "It certainly has its application there, though not as much as we think, since […]".

[611] Prooemium, p. 255, lines 21–25: "habet quidem illa suum ibi usum, non tamen quantum credimus, cum status praeternaturalis agnitio non possit ultra cognitionem status naturalis sese extend-

merely lamenting a lack of fundamental research and the resulting ineffectuality of medicine; he held no illusion concerning a brisk advancement of medical science. He was not denying the working of medical science for the "benefit of one's neighbor", but in fact beheld in it a deeper spiritual meaning. Reinhard V. Putz, who attributed to the gross anatomy practical an "important reflection on the foundations of one's own life", should be remembered in this context.[612] It is precisely this moment, taking pause in a stream of medico-technical development and simultaneously orientating towards the Creator, that conveys the fascination of Stensen's inaugural address and has allowed it to remain topical to this day with a view to man's finiteness and the development of medical science with all the problems accompanying it.

2. In the study note from 1659 cited by Grell, Stensen also places the "pleasure of the joyous intuition of the marvels of God"[613] on an equal level to the "benefit of one's neighbor". As argued by Helge Kragh, this statement by Stensen represents a "different and slightly later variant" of the "natural theological thinking" of the professor of physics and later Lutheran bishop Dr. theol. Peder Winstrup (1605–1679).[614] In the disputation *Disputatio philologica et philosophica de usu linguarum et disciplinarum philosophicarum in theologia [...]* (Philological and Philosophical Disputation on the Use of Languages and the Philosophic Disciplines in Theology [...]; Copenhagen 1633) over which he presided, Winstrup stated that physics and mathematics supported theology in its battle against heretic ideas and were also capable of increasing Bible knowledge by providing precise data from nature.[615] Stensen's intent with his inaugural address, which also had pastoral goals in mind, was a similar one.

3. Stensen's criticism of the mere orientation of anatomy towards prevention and treatment of disease also calls attention to his methodical criticism of medicine. Many years earlier, Stensen had emphasized the need for autonomy of science independent of any application in the sense of fundamental research. His thoughts on the relationship between science and practice, voiced at the end of the letter to Thévenot attached to his Myology, are as follows:

> That this same, which remains to be examined under great effort, deserves to be known, will hardly be denied by he who loves the truth and does not disdain health. It is therefore not necessary to borrow arguments from orators to prove incontrovertibly that to sweat during this work is not the idiosyncrasy of a person who abuses his leisure time, as our critics are not ashamed to frequently assert in the presence of famous men. The elegance of the artwork alone, while providing the most obvious proof of intelligent nature, would deserve the effort of investigation, even were it thousandfold greater. In

---

ere; haec autem cum etiamnum sit admodum limitata, nec illa fines suos multum promovere poterit."

[612] See note 93, in Chap. 1 above.

[613] See note 305, in Chap. 2 above.

[614] On Winstrup: Bjørn Kornerup, "Winstrup (Vinstrup), Peder Pedersen"; in: DBL³ 15 (1984), pp. 609–611, here 609–610.

[615] Winstrup/Ulmivallius, *Disputatio* and on this Kragh, "New Science", p. 76: "a different and slightly later variation [...] natural theological thinking".

addition, it is about the movement-inducing fiber, about the part that sets the limbs swinging, that injects the air, that sets the blood [in its circulation] in motion, in short, that upon which the indicators of life and of death depend. But who would betoken it the idiosyncrasy of an idle man to wish to examine the nature of this part if he recognizes that it was hitherto practically unknown and sees that something can be achieved in its study: But this does not strike our critics. You may easily remember that not only I, but even you, who supported my side, were often reproached: But what use is it to want to know this? In what way does it affect the practice? By repeating this question incessantly and embellishing it with various figures of speech, they attempt in front of everyone to ridicule, not to say to make abhorred, those who are intent on new facts. It will suffice to show elsewhere what the practice of this century owes to anatomical experiments, not least for the reason that it has exposed countless mistakes found in the explanation of the causes (of diseases), and at the same time proved erroneous many reasons which they [= Stensen's critics] put forth in the use of remedies. In place of an answer, I wish to confront them at this point with the question that they may examine their conscience and see what consistency is at the basis of what they put forth with audacious eloquence in their explanations of apoplexy, paralysis, pathological tension, convulsions, languishing strength, fainting and other symptoms of animated motion and on what foundation is built that which they administer in terms of remedies to rectify these ills, and this not in regard to *the* paralysis and not in regard to *the* convulsion, but concerning *this* paralyzed person and *this* patient with convulsions. But if they realize that nothing but words are contributed to the cognitive process, and that mere assumptions govern the healing process, they will even reluctantly admit that there is for the use of good to want to examine what truth and certainty there is in this area of anatomy. And there is also no reason for them to object that over the course of so many centuries everything had remained in the same state. The answer is obvious: Everyone has searched for the remedies, [but] only very few have worked towards becoming acquainted with the realm that the remedies are applied to. But indeed the construction of an automaton created by another must be precisely researched by the one who is to reestablish the damaged motion of that automaton, and the nature of the blood, the nervous fiber, the motion-inducing fiber must, as far as it is possible by human assiduity, be investigated by him who wishes to cure not merely by chance the symptoms which damage natural motion. And because much is concealed from us in myology that could be known, and because it is not only in the interest of truth but also in that of health to know these same things, anyone can see what right our critics have to call the efforts of anatomists the trifles of an idle man by deriding their new experiments. And it is this which I thought necessary to state publicly, so that my friends may realize how to answer those who speak in unfriendly fashion about my work.[616]

---

[616] Myol., p. 105, line 27–p. 106, line 37 = Kardel, *Steno on Muscles*, p. 220, lines 19–29 & p. 222 & p. 224, lines 1–23: "Quod eadem illa, quae magno labore restant investiganda, sciri mereantur, qui veritatem amat, nec sanitatem spernit, vix unquam negaturus est. […] Non itaque opus, a rhetoribus argumenta mutuari, ut evincam, non esse hominis otio suo abutentis isti labori insudare, quod in illustrium virorum praesentia saepius asserere non erubuerunt censores nostri. Sola artificii elegantia, dum evidentissimum intelligentis Naturae argumentum exhibet, indaginis laborem, etiam millies majorem, mereretur. Adde, quod de fibra motrice agatur, de parte, quae membra agitat, quae aërem inspirat, quae sanguinem movet, paucis, unde vitae mortisque signa dependent. At quis otiosi dixerit, velle istius partis naturam indagare, cum eam hactenus quasi ignotam deprehendat, videatque, posse quid in ejus indagine praestari; sed haec censores nostros non tangunt. Memineris facile, non mihi tantum, sed ipsi tibi meas partes suscipienti objectum saepius; at cui bono, haec scire velle? quid haec ad praxin? quam suam interrogationem identidem repetendo, variisque figuris exornando, apud omnes id agunt, ut novis rebus invigilantes ridiculos, ne dicam invisos, reddant. Licebit alibi fusius demonstrare, quantum hujus seculi experimentis anatomicis

In his confession notes from Hamburg as a bishop, Stensen later gave thought to the opponents of his research with regard to the factual, yet somewhat harsh judgments he had passed on them – Stensen was sensitive and did not wish to offend anyone.[617] These opponents presumably included egotistic 'practitioners', whom he addressed at the end of his Myology and whom he also had in mind when he stated introductorily in the attached letter to Thévenot that he had no doubt the less than objective critics of his works, as was customary for them, would be intent on pushing forward their gums, ie attacking him verbally.[618] No one who loved the truth and did not disdain health would deny that that which remained to be examined under great effort deserved to be known.[619] Purely practice-oriented pursuing of medicine therefore led to the opposite of what was desired, and humble basic research, as Stensen asserted in his inaugural address, recognized God as creator.

4. Holger Jakobsen's *Exercitia academica*, which characterize Stensen as a critical scientist of the highest factual standards and concerned with the well-being of the patient, also paint a picture differing from Grell's.

5. A further indication can be found in Stensen's letter to Heinrich Meibom Jr., whom he informed on April 5, 1673 st.v. – 2 months before his inaugural address in Copenhagen – that it was beyond any doubt that the discoveries made daily in anatomy were worthy of praise, for through them "more proof of the essence of God and an easier method of treatment" became known.[620] Here again, dissection and faith are not viewed as counterparts, but in relation to each other.

---

debeat praxis, vel eo solo, quod innumeros errores, qui in causarum explicatione occurrunt, detexerit, simulque rationes plurimas, quas in remediis applicandis afferunt, erroneas demonstrarit. Hic responsi loco rogatos eos volo, suam ipsi excutiant conscientiam, videantque, quid solidi subsit omnibus iis, quae in apoplexia, paralysi, contractione, convulsionibus, virium prostratione, syncope, aliisque motus animalis symptomatis explicandis audaci facundia pronuntiant; cui fundamento innitantur, quae iisdem malis tollendis applicant remedia, idque non paralysin, nec convulsionem, sed hunc paralyticum, hunc convulsum sumendo. Quod si videant, in cognitione, praeter verba, nihil afferri, in curatione solam conjecturam principatum obtinere; vel inviti fatebuntur, esse alicui bono, velle veri certique quid in hac anatomes parte indagare. Nec est, quod objiciant, tot seculorum decursu in eodem statu mansisse omnia. In promptu responsio est: remedia quaesiverunt omnes, partem, cui remedia applicant, cognoscere, pauci allaborarunt. At vero automati ab alio confecti constructio illi exacte investiganda est, qui ejusdem automati motum laesum restituere debet, & sanguinis, fibrae nervosae, fibrae motricis natura, quantum humana industria fieri poterit, illi indaganda est, qui motum naturalem laedentia symptomata non solo casu curare desiderat. Cum itaque in myologia multa nos lateant, quae sciri possunt; cum non veritatis tantum, sed sanitatis intersit, ut eadem sciantur: cuilibet manifestum est, quo jure censores nostri, nova anatomicorum experimenta ridendo, illorum labores otiosi hominis occupationes clamitant. Et haec illa sunt, quae in medium afferenda judicavi, ut pateat amicis, quid illis respondendum sit, qui parum amice de meis laboribus loquuntur."

[617] Op. 4, p. 436, line 23–p. 437, line 1.

[618] Myol., p. 94, lines 29–30 = Kardel, *Steno on Muscles*, p. 184, lines 1–3.

[619] Myol., p. 105, lines 27–29 & p. 106, lines 30–31 = Kardel, *Steno on Muscles*, p. 220, lines 19–21 & p. 224, lines 15–17.

[620] Bruun, "Fem nyfundne Niels Stensen-breve", p. 150, lines 16–18, here 17–18: "plura [...] divinitatis argumenta et facilior medendi ratio".

6. The final reference in this context is to Dr. med. Charles Drélincourt (1633–1697),[621] successor to Van der Linden in Leiden from 1668/1669, professor of anatomy from 1670 after Van Horne's death, and opponent of Sylvius' iatro-chemical theories. In his inaugural address as Van Horne's successor, Drélincourt emphasized that the constituent parts of the human body bespoke the "finger of God".[622] For him, as Tim Huisman points out, the admiration of the human body as the work of God was "the true fruit of anatomy".[623] Huisman puts forth the theory that as compared to Van Horne, who was completely occupied with his research and in whose writings only few theological considerations can be found, Drélincourt represented a new type of anatomist in whom, in contrast to the emblematic or allegorical point of view of the so-called Book of Nature in Renaissance anatomy, a "reconciliation of the metaphysical and theological notions of life with the mechanistic way of thinking about nature" seemed to be taking place resulting from the concept that the complex beauty and ingenuity of nature was in itself a sign of the divine.[624]

With his metaphysical statements during the Copenhagen inaugural address, Stensen can in fact be positioned within this development outlined by Huisman. The true task of the anatomist, Stensen says in the address, consists in guiding the spectator by way of the astonishing artwork of the body to the dignity of the soul, and through both of these to knowledge and love of the Creator (cf. Fig. 1.5).[625] This is likewise expressed in his coat of arms showing a cross atop a three-dimensional and anatomically correctly shaped heart with the cardiac apex (Apex cordis) (Figs. 1.1 and 1.8).[626] Thus the Copenhagen inaugural address delineates the quintessence of Stensen's concept of the anatomist's profession.

2.3.2.2   The Spiritual Dimension of Disease

Like his assessment of deformities, Stensen considered the diversity of diseases a consequence of original sin in his Christian religious instructions as a bishop.[627] In a letter to Cosimo III from Münster dated March 19, 1683, Stensen described it as the root of all evil that we did not get around to learning from an early age that it was our duty to suffer.[628] Van Rensselaer Potter likewise accepted the "inevitability of some human suffering" originating from the natural disorder innate in humans and

---

[621] On Drélincourt: DDPhS, s.v. "Drélincourt (Drelincurtius), Charles (Carolus)", cols. 488–490.

[622] Drélincourt, Praeludium anatomicum, p. 22: "Dei digitum", & p. 65: "Dei [...] digito".

[623] Huisman, Finger of God, pp. 104–105, here 105.

[624] Ibid., pp. 106–107, here 106.

[625] Prooemium, p. 254, line 35–p. 255, lines 5, 25–28.

[626] See details in Sobiech, Herz, Gott, Kreuz, pp. 234–240.

[627] Op. 5, p. 454, line 25.

[628] E 291, p. 578, lines 15–16.

the environment,[629] albeit from his mechanistic view of the 'free' will of man.[630] Stensen perceived the physical abilities given to man by God as an undeserved gift; already in one of his spiritual records written after his conversion in Florence he asked himself how it would be if he lost his eyesight, his bodily health, his mind, his memory or his life in ignominy and pain.[631]

Stensen thought the spiritual value of illness to be high: that which the physicians or the underdeveloped art of medicine lacked should be replaced through the awakening of acts of Faith and through praying for strength to endure the illness so as to increase the grace given by God and in God's honor; everything should be expected from God's providence. A day spent in sickness could therefore be as significant as an entire month of health; if we walked God's paths and did not place obstacles in his way, he would provide more signs of his love and more abundant consolation than in times of health.[632] With this in mind, Stensen advocates consciously viewing illness as a sacrifice to God.[633]

The experience of the Viennese Jewish professor of neurology and psychiatry Dr. med. Viktor E. Frankl (1905–1997)[634] in spring of 1945 while in the Dachau satellite camp Kaufering VI where he, himself a detainee, was allowed to work as a physician, may convey an idea of what Stensen meant. Under the title "Fate – a Gift", Frankl writes about the last days of a sick female detainee: "This young woman knew that she was going to die during the next few days. When I spoke to her, she was nevertheless cheerful. 'I am thankful to fate for treating me so harshly', she said to me literally; 'for in my previous, civic life I was too spoilt and not very serious about my intellectual ambitions.' During her last days she was very much within herself. 'That tree there is the only friend in my solitude', she said and pointed through the barracks window. Outside there was a chestnut tree in full bloom, and if one leaned down to the woman's bed one could just see a single green branch with two blooming shoots on it through the small window. 'I often talk to that tree', she then says. I am taken aback and do not know how to interpret her words. Could she be delirious and occasionally hallucinate? Thus I ask inquiringly whether the tree answers her as well – yes? – and what it says to her. And she answers me: 'He has told me: I am here – I – am – here – I am life, eternal life …'"[635] In this moving portrayal of the unknown young woman, which breathes a deep spirituality, two parallels with the Bible can be recognized: on the one hand the story of the burning bush (Exod 3:1–15), and on the other, although one may assume the young woman

---

[629] Potter, *Bioethics*, interior of front cover page & p. 196 ("A Bioethical Creed for Individuals"), here "4. [= Fourth] Belief".

[630] See note 41, in Chap. 4 below.

[631] Spir. 11, p. 130, lines 14–15. See details in Sobiech, *Herz, Gott, Kreuz*, pp. 197–198, 320.

[632] Op. 11, p. 519, lines 6–19.

[633] Op. 3, p. 427, lines 18–23.

[634] On Frankl: Joseph B. Gerwood, "The Legacy of Viktor Frankl: An Appreciation upon His Death"; in: PsRep. 82 (1998), pp. 673–674.

[635] Frankl, *Leben*, pp. 112–113, here 113; on place and time of Frankl's practice and the mentioned event see Epple, *Türkheim*, p. 118, no. 1 & p. 119, no. 6.

was of Jewish faith, the self-description of Jesus in response to Martha (John 11:25–26).

This notion besetting the young woman in Frankl's narrative bears similarity to a note by Stensen from his time as a bishop stating that the human soul hoped for the health of eternal life.[636] From the perspective of his spirituality of the 'circulation of love', in which the human senses act as "interpreters" of divine love, Stensen advises to experience the love of the caring and healing God precisely in the pleasures and hardships perceived with the senses.[637] This spirituality so close to reality, practiced by Stensen since his time in Florence, was channeled by the young concentration camp detainee when she concluded what was sensually not perceivable from something she was able to experience with her senses.[638]

2.3.2.3 Asceticism as Vicarious Atonement

The American medical historian and gynecologist Dr. med. Kate C. Hurd-Mead (1867–1941) said the following – probably referring to the portrait showing Stensen as so-called "Anatomicus regius" (Fig. 2.6): "Stensen's face, as we see it in his portraits, is that of an ascetic, with an expression of intense concentration."[639] From a medical point of view, this raises the question of the significance of the ascetic lifestyle practiced by Stensen already during his time as "Royal Anatomist", which included material poverty, fasting and exercises in penance and, during the last years of his life, frequent vigils and increased fasting. Stensen emphasized, however, that his way of living should not be generalized, and fed the members of his episcopal household well.[640]

One may begin examining this question by looking to the end of Stensen's life: a Czech medical expertise published in 1971 states that the cause of his death was a "chronic disease of the abdominal cavity organs with recurring obstructions (convolvulus)" which extended over a period of 2 years and "ended with a [paralytic] ileus".[641] Its origin cannot be determined in retrospect. In his pastoral care for the sick as suffragan bishop of Münster Stensen had mandated "obedience toward the physician",[642] and he complied with his own instruction: according to his chaplain

---

[636] Op. 3, p. 424, lines 10–11.

[637] Sermo 12, p. 229, lines 10–11.

[638] See note 399, in Chap. 2 above.

[639] Hurd-Mead, Women, p. 440. On Hurd-Mead: Philip K. Wilson, "Hurd-Mead, Kate Campbell"; in: ANB 11 (1999), pp. 557–559. A similar opinion is voiced by Tjomsland, "Stensen", p. 491. On the portrait of Stensen as a Florentine anatomist cf. Sajner/Pačes, "Krankheit", p. 56. As point of departure for comparative placement of the portraits cf. Wegner, *Anatomenbildnis*, pp. 77–78 with fig. 44 before p. 77 (copy) & p. 175, here no. 44. On Stensen's emaciated face in Hamburg cf. Sobiech, *Herz, Gott, Kreuz*, pp. 344–345.

[640] For details see Sobiech, *Herz, Gott, Kreuz*, pp. 86, 313–325.

[641] Sajner/Pačes, "Krankheit", pp. 53, 55, 58–60, 63, here 59. On Stensen's own diagnosis, a concrement lodged in his bladder, cf. ibid., p. 60 as well as E 477, p. 895, lines 29–36.

[642] Op. 11, p. 520, line 8: "obedientiae in medicum".

Kaspar Engelbert Schmael (†1692) he followed all directions by the (unknown) physician treating him at his Schwerin sickbed, and no sound of pain crossed his lips.[643] When the physician ascribed Stensen's approaching death to his overly ascetic lifestyle, the latter answered that if that were truly the cause, he did not regret it as he was convinced of having practiced his lifestyle with the intent of serving God. Some people had lived far more ascetically than he had, consuming only bread and water, and had lived to an old age. Why should he not be able to do so, if it was the holy will of God that he should live?[644] Stensen had no doubt that he had acted in a medically responsible manner,[645] and he was aware of the value of his own health also as bishop.

A pertinent work on the health of clerics in the seventeenth century was the *Hygiasticon seu vera narratio valetudinis bonae et vitae [...]* (Book of Health or True Instruction for Good Health and Conduct of Life [...]; Antwerp 1613) by Leonard Lessius SJ (1554–1623),[646] which featured an endorsement[647] by Dr. med. Gerard de Vileers (c. 1565/1566–1634),[648] a professor at the medical faculty of the University of Leuven, and a poem of dedication by the physician Dr. med. Franciscus Sassenus (1578–1620),[649] and was reprinted multiple times. In this treatise, Lessius criticized that man wanted to generously indulge his palate and appetite: "It is

---

[643] Add. 26, p. 952, lines 38–39; Add. 42, p. 983, lines 8–10; Add. 43, p. 986, lines 10–15. Stensen produced no sound of pain on Wednesday 11/24/1686 st.v. when five enemas were administered to him without success (Meyer/Schwarz, "Vita", p. 65 r. col. ad l). On Schmael: BAM, Coll. Kleruskartei, s.v. "Smale, Kaspar Engelbert"; E 297, p. 588n3.

[644] Add. 26, p. 953, lines 1–6; for details on Stensen's fasting see Sobiech, *Herz, Gott, Kreuz*, pp. 321–325.

[645] This is supported by Stensen's letter to Malpighi on 11/24/1671, in which he recommends from a dietetic perspective a change in Malpighi's diet, which had an effect on the juices of the body and was therefore more important for one's health than air, activity and emotional agitation. Stensen mentions, among others, three persons known to him of whom two, given up by the doctors, had regained their full health exclusively by means of an appropriate diet, and the third was still walking without a cane at well over 90 years of age due to his transitioning to a more moderate lifestyle. Stensen adds that he would not dare to speak as a physician, however, since he lacked appropriate practice. Particularly in light of the few things to which humans could contribute in matters of nature, the extraordinary help of the "doctor sovereign", ie God, played a role here if it was useful for the salvation of the soul (E 65, p. 248, lines 15–25, 30–32 = Adelmann, *Correspondence*, vol. 2, no. 271, p. 597). That Stensen recognized the danger of overestimation of one's own capabilities is evidenced in Sermo 23, p. 273, lines 24–25, where, in order to illustrate the necessity for prayer, he makes the comparison that he who considered himself strong enough to subsist without nourishment would die.

[646] Due to an infectious disease incurred during the time of his education, Lessius had engaged in the study of medicine and dietetics in private. From 1585 to 1600 he was a professor of theology at the Leuven Jesuit College; on Lessius: Albert Ampe, "Lessius (Léonard)"; in: DSp. 9 (1976), cols. 709–720, here 710.

[647] Lessius, *Hygiasticon*, final three pages before the beginning of pagination.

[648] Villeers taught at Leuven University from 1593 to 1634; on Villeers: DHM 4 (1778), s.v. "Villeers (Gerard de)", p. 536.

[649] Professor at Leuven University from 1618 to 1620; on Sassen: Dauwe, "Van Sassen", pp. 271–272.

viewed as miserable to live by the instructions of physicians, as per the common adage: He who lives medically lives wretchedly."[650] Through accumulation and decay of harmful humors they contracted fatal diseases, which he had observed in many men who had wasted away at a young age. Thus they were no longer able to be of any help to their environment and forfeited their chances for a greater crown in heaven. Even many religious and secular priests were unable to meet the obligations of their calling due to ignorance regarding the requirements of their own health.[651] This need for clerics to be in good health was also emphasized by Johannes Sterck SJ (1630–1692),[652] Stensen's father confessor during his time as "Anatomicus regius". While Lessius also mentions the value of ascetic exercises, he advocates moderation in all regards.[653] Sterck appealed to Stensen's medical education with respect to his asceticism, which included sleep deprivation and in Sterck's opinion exceeded the appropriate extent, thereby causing detriment to his pastoral work, in an emphatic letter quoting Ovid: it was not always in the physician's powers to deliver the patient from his illness.[654] The "patient" he was referring to were the church-internal grievances which Stensen had experienced particularly within the Münster diocese, and which he condemned. Sterck however was also indirectly accusing Stensen of having his own perfection in mind with his asceticism and not the salvation of his entrusted souls.[655] Stensen however viewed the atonement he was exercising vicariously for others as supporting his pastoral efforts.[656]

In one sermon, Stensen mentions how the faithful who had made the decision eg to fast were confronted with the talk of others who sought to dissuade them from their resolution with ungrounded warnings about melancholy and the shortening of life.[657] This is further evidence that Stensen did not intend to harm his own health with his physical asceticism. In his final years, he was increasingly isolated due to his aforementioned criticism of grievances in the pastoral work of the Münster diocese as well as due to the diaspora situation in Hamburg and Schwerin. This isolation weighed him down greatly, leading to intensive self-observation and self-criticism which resulted, among other things, in frequent lamenting about his own inadequacy and profession of his "unworthiness" for the episcopal office in

---

[650] Lessius, *Hygiasticon* (no. 1), pp. 2–3, here 3: "Miserum putant, ex medicorum praescripto vivere, juxta dictum vulgare, *Qui medice vivit, misere vivit*". Cf. PSLMA 4 (1966), p. 199, no. 24238. On the *Hygiasticon* cf. De Nave/De Schepper, *Geneeskunde*, pp. 54, 234–235 (no. 80), here 234 with title page of the first edition.

[651] Lessius, *Hygiasticon* (nos. 1 & 2), pp. 3–5.

[652] E 253, p. 508, line 23–p. 509, lines 3, 8–14. On Sterck: Scherz, *Epistolae*, vol. 1, pp. 77–78.

[653] Lessius, *Hygiasticon* (no. 60), p. 140; ibid. (no. 36), p. 89; ibid. (no. 66), p. 151.

[654] E 253, p. 510, lines 7–12, here 7–8 as a quote from the *Epistulae ex Ponto* I 3,17 by Ovid, cf. Holzberg, *Tristia*, p. 312. Details in Sobiech, *Herz, Gott, Kreuz*, pp. 324–325.

[655] E 253, p. 509, lines 1–3, 12–14.

[656] For details on the "vicariousness" practiced by Stensen, which represents an essential element of his spirituality oriented around the well-being of his neighbors, see Sobiech, *Herz, Gott, Kreuz*, pp. 20, 41, 200, 215–216, 248, 261–263, 271, 288, 295, 305, 313, 321–325, 332–333, 341–342, 344.

[657] Sermo 24, p. 275, lines 16–20, 23.

various letters.[658] On July 16, 1685 the Congregation for the Propagation of the Faith gave Stensen permission to retreat to Italy for several months.[659] Stensen, who continued to think and act rationally until his early death despite the extreme strain he was subjected to, conducted anatomical studies on the nervous system in Hamburg before leaving for Italy, and postponed[660] the journey entirely when the opportunity for missionary work presented itself in Schwerin, where he laid the foundations for a Catholic community.

2.3.3   The Priest as Physician of the Soul

It could be said that Niels Stensen only truly lived the profession of medical practitioner as a priest; just like he wanted to serve the health of the body with his research, he later wished to serve the soul's health with his pastoral work. In his letter to Cosimo III from Hanover on March 16, 1679, he identifies himself for the first time as a "physician".[661]

2.3.3.1   Care for the Body and Care for the Soul

As the otherworldly well-being of his fellow men became Stensen's priority as a priest and bishop, the disparity between medical care for the mortal body and spiritual care for the immortal soul, which he observed in many people, burdened him: "What does man endure to keep his body in health and beauty; how many worry about breaking a stone,[[662]] how many about their arm and leg; and this they are able to endure for their mortal body, [but] what for their soul? If we had the same love for the glorified body [of eternity] that we have for the mortal body, how much would we endure for that concern [= caring for one's own soul]? How many physicians, surgeons, and pharmacists, how many medications [would then be available

---

[658] See in detail Sobiech, *Herz, Gott, Kreuz*, pp. 88–89 – The Münster episcopal judicial vicar Prof. Dr. theol. Dr. iur. can. Max Bierbaum (1883–1975) (on him: Thomas Flammer, "Bierbaum, Max"; in: BBKL 28 = supplemental vol. 15 [2007], cols. 122–127) noted in his biography of Stensen (1959) with regard to the "excessive rigor" and "anxiety" in Stensen's ascetic religious life during his last years (Bierbaum, *Stensen*, pp. 140–141) that one need not "speak of pathological traits" as it was "nothing extraordinary" in the history of Christian piety if Stensen "at times almost self-tormentingly suffered under the consciousness of his own imperfection." The professor of anatomy Adolf Faller removed this statement by Bierbaum from the second edition (1979) of the biography, which he edited (see Bierbaum/Faller, *Stensen*, p. 105). For him, Stensen was "the anatomist of spiritual life" who "disassembled and separated the subject matter he dealt with into its smallest parts", which explained the strictness of judgment in his spiritual life (ibid., p. 120).

[659] E 414, p. 792.

[660] E 425, pp. 810–811.

[661] E 170, p. 405, line 17: "medico".

[662] A concrement like a urinary stone, which many contemporaries of Stensen suffered from.

for that end]?"[663] Only the supernatural beauty counted for the pastor Stensen. A parallel to this can be found in *Medicus Christianus [...]* (Christian Physician [...]; Antwerp 1665) by the theologian Ludwig Bertha OP (1620–1697).[664] In the introduction to the section on the sacrament of confession, the author asks the following question: "If there were a physician in the world able to cure all the different wounds and illnesses with a single powder at a low price, how great an onrush would there be to him from everywhere?"[665] Stensen likewise emphasizes in a sermon that, just like people would appreciate one who had a "universal medicine" (medicina universalis) – the so-called panacea –, so should God be appreciated who had given us the remedy of eternal life.[666]

Stensen took the precedence of the salvation of the individual's soul very seriously, eg during the 40 day Lenten period before Easter. In a sermon on Ash Wednesday he emphasized that, if there was no "sensible reason" to do otherwise, the fast should be maintained and one was not allowed to grant oneself dispensation from it, adding with respect to those who did so: "They think the consent of the physician suffices, and they are both sinning."[667] As Stensen's sermons are not fully written out drafts, the concrete background against which this particular statement should be interpreted – and which is unknown – can only be speculated upon. Michael Boudewijns supported dispense from fasting by a medical practitioner in cases of "urgent necessity".[668] As Stensen accepted only "sensible reasons", and in another sermon on fasting mentioned that many people unwilling to fast would falsely purport danger of illness,[669] he was apparently referring to "consent" given by medical practitioners as a result of carelessness or conscious abetment.

---

[663] Sermo 12, p. 226, line 35–p. 227, lines 3, 5–6, here p. 226, line 35–p. 227, line 3: "Quid non patitur homo, ut conservet corpus in sanitate, pulchritudine, quot curant scindi calculum, quot brachium, pedem; et hoc possunt pati pro corpore mortali, quid pro anima? Si nobis ille esset amor corporis gloriosi, qui est amor corporis mortalis, quanto in id studio incumberemus? Quot medici, chirurgi, pharmacopei, quot medicamenta?"

[664] On Bertha: SOP 2 (1721), s.v. "F. Ludovicus Bertha", p. 745.

[665] Bertha, *Medicus Christianus* (Pars tertia, Tractatus II, Caput I), p. 347 l. col.: "Si quis foret in mundo chirurgus, aut medicus, qui singula vulnera, & morbos, unico parvi pretii pulvere sanare novisset, quantus ad eum undequaque fieret concursus?"

[666] Referring to the Communion in Sermo 26, p. 280, lines 6–8. In Examen I, p. 331, lines 4–10, here 4–5 Stensen described to Leibniz in 1679, "if it were allowed to be joking, with a nevertheless non-jocular joke" (Si jocari liceret, joco tamen non jocoso), the power of Jesus to heal all pathological enfeeblements as "true panacea" (vera panacaea) and the providence of God as the "philosophers' stone" (lapis philosophorum). Examples for contemporary 'panaceas' are: (1) the "panacea antimonialis" of the chemist Johann Rudolph Glauber (1604–1670), cf. Kathleen Ahonen, "Glauber, Johann Rudolph"; in: DSB 5 (1972), pp. 419–423, here 419; (2) Jean Pecquet's 'Panazee', through which he became an alcoholic, cf. Sylvie Taussig, "Pecquet, Jean"; in: DFPh. 2 (2008), pp. 969–971, here 970.

[667] Sermo 8, p. 210, lines 30–33, here 31–32: "rationalis causa [...] Putant sufficere assertionem medici, et peccant uterque."

[668] Boudewijns, *Ventilabrum*, pp. 258–264 (Pars prima, Quaestio XLIV. An medicus poßit relaxare ieiunia, & concedere ab Ecclesia vetita?), here 259: "urgente neceßitate".

[669] Sermo 10, p. 217, lines 5–7.

Stensen continues the above-mentioned sermon by quoting St. Ambrose of Milan (339–397): "Opposing the divine judgment are the rules of medicine; they restrain from fasting, they disallow night vigils and prevent any diligence in contemplation."[670] This sentence becomes comprehensible in the context of its origination: Ambrose was referring to pagan body culture.[671] Only few critical statements are documented from the time of early Christianity concerning a pagan medicine viewed as fixated on the body; rather, the "physician" (medicus/ἰατρός) was used "from a relatively early time onward on a more abstract level as a positive and frequently used comparison image", especially for Christ (cf. Mark 2:17), in a metaphorical fashion – as done here by Ambrose.[672] Hence Stensen presumably intended to criticize excessive care for the body, or care for the body under disregard or neglect of care for the soul, with his statement.

In his pastoral care for the sick, Stensen suggests that persistent thinking about illness and its cures brought with it only impatience, sadness about the various concomitants of the illness, and a "too great trust in physicians and medicine".[673] He was aware of the "deficiency of the physicians" of his time which, as he noted as suffragan bishop on behavior in times of sickness, could be compensated by God.[674] This is nothing else than a theological comment on his study excerpt from 1659: "and so one calmly awaits the restoration of good health, in which nature plays its part",[675] whereby he now trusted in the power of prayer.

In his "Rule for Visitations" written as suffragan bishop in Münster, Stensen admonishes that the priest performing a visit at the sickbed was not allowed to bring accusations against the physicians who had called on the patient or were yet to do so, nor against other persons, but instead only against the sins, and herein against his possibly worse *own* sins (cf. Phil 2:3); he was to instill hope in the patient under reference to God's omnipotence and charity and instruct him to accept everything with "equanimity" from the hand of God, as God may be intending to lead the patient to holiness in this way.[676] Esteem for the physician's profession, but also a reference to the finiteness of the worldly physical existence are expressed in his sermon note stating that in sickness, solace came from the physician and his assis-

---

[670] Sermo 8, p. 210, line 36–p. 211, line 1: "Contraria sunt divinae cognitioni praecepta medicinae; a jejunio revocant, lucubrare non sinunt, ab omni intentione meditationis abducunt." Cf. Petschenig/Zelzer, *Sancti Ambrosi opera*, p. 500 (Expositio psalmi CXVIII, 22,23): "contraria autem studiosis divinae cognitionis praecepta medicinae sunt; a ieiunio revocant, lucubrare non sinunt, ab omni intentione meditationis abducunt."

[671] Rothschuh, *Konzepte*, p. 53.

[672] Schulze, *Medizin*, pp. 159–160, 162, 183, here 162; cf. Rothschuh, *Konzepte*, p. 53.

[673] Op. 11, p. 520, lines 9–10, 20–23, here 23–24: "nimia fiducia in medicos et medicinam." Around November 1680 Stensen wrote to Hortensio Mauro in E 199, p. 441, lines 37–38 in regard to the illness of Prince-Bishop Ferdinand II that the almighty God alone could compensate the shortcomings of the medications.

[674] Op. 11, p. 519, lines 11–14, here 13: "defectus medicorum".

[675] See note 198, in Chap. 2 above.

[676] Op. 9, p. 484, line 30: "Regula visitationum"; ibid., p. 485, lines 3–8. On the "conformity" (conformitas) with the will of God see Sobiech, *Herz, Gott, Kreuz*, p. 136 & passim.

tant, while in death it came from Jesus alone.[677] Stensen appears to have considered harmonious cooperation between physician and priest to be the ideal case. In his letter to the Baroness of Galen from Hamburg on November 19, 1683, he writes with reference to the demand in Luke 14:26 to love one's relatives and one's own life less than Jesus: "We have only one [single] soul, child of God, [and] when it is lost, all is lost."[678] To him this represents the central criterion of human life.

2.3.3.2   Therapeutic Effect of the Sacraments

In a letter written on December 29, 1682 to Odoardo Cibo in Rome, Stensen asks for the authority to provide the physician Dr. med. Johann Georg Rötenbeck, who had converted under his spiritual guidance and been ordained a priest, with the necessary canonical faculties to work in Mecklenburg. In so doing he states that Rötenbeck would be able to "conceal the practice of spiritual medicine under the title of Doctor of Medicine."[679] Apparently Stensen considered the Catholic faith and the sacraments to be medicine for the soul. In this sense he also described God as "most loving father and wisest physician" in his notes created in Münster on pastoral care for the sick.[680] Already as a priest in Florence he had noted down that one should seek God's presence like a sick person seeks the physician,[681] thus continuing the long tradition of medical speech in theology, eg in the therapeutic theology of the early Church Fathers.[682] How did he see the role of the patient in this respect?

It was not surprising, Stensen stated in one Ash Wednesday sermon, that a sick person should die, even if he had the best physician, if he had ignored the latter's instructions and chosen to do the opposite.[683] Precisely the people suffering from very complicated diseases spurned the physician and his medications the most, which could go so far that those patients who were most dangerously ill would fantasize in a delirium that they were perfectly well. It was certain, however, that it could be expected least of all from the sick person himself to call a physician, but that this was the duty of those affiliated with the patient and the physicians them-

---

[677] Sermo 41, p. 354, lines 4–7.

[678] E 328, p. 647, line 30–p. 648, line 2, here p. 648, lines 1–2: "Wir haben nur eine Seele, Kindt Gottes, wan selbige ist verlohren, ist alles verlooren."

[679] E 290, p. 574, line 30–p. 575, line 12, here p. 574, lines 32–33: "sotto titolo di dottore di medicina puo nascondere l'esercizio della medicina spirituale." The authorizations requested by Stensen for Rötenbeck were granted on 04/23/1683, cf. Scherz, *Epistolae*, vol. 1, p. 49.

[680] Op. 1, p. 407, line 10: "patre amantissimo et medico sapientissimo".

[681] Spir. 5, p. 92, lines 10–12.

[682] E.g. by St. Jerome (c. 347–419/420), cf. Dörnemann, *Krankheit und Heilung*, pp. 339–340. – A "theologia medicinalis" likewise existed in the Lutheran orthodoxy of the Baroque, see Steiger, *Medizinische Theologie*, here overview on pp. 51–53.

[683] Sermo 7, p. 204, lines 22–25.

selves.[684] The same should occur in the parish, which was more akin to a house for the mentally ill whom one could likewise not expect to ask the physicians for medication of their own accord in order to restore the faculties of their reason.[685] Thus the 'patient' must be sensitized for the necessity of medication. To afford the confessor in the sacrament of penance the possibility of recognizing, like a physician, the nature of the mental illnesses and the severity of the sins and thereupon prescribe appropriately, the penitent had to have spent at least as much time and effort beforehand, in his examination of conscience, on recognizing the blemishes of his own soul as he did on recognizing the flaws of his own body; if it were not painful to care for one's own body then one could not assert anything else for the care of the soul.[686] It was furthermore possible that God would grant good health by merit of a good confession.[687] The early Church Fathers had emphasized the therapeutic effect of penitence in much the same way.[688]

The Eucharist, which to Stensen was primarily associated with the love of one's neighbor, also possesses therapeutical effects according to the tradition of the Church Fathers.[689] One incitement for the Holy Communion, Stensen said, could be that we wished to encounter Jesus as a physician who could cure every weakness in order to present to him, in the sacramental union with him until the disappearance of the Eucharistic species, with the help of all our senses the concern for our salvation and that of other people, our sins and the imperfections of our body and soul.[690] This advice by Stensen in regard to Communion is an expression of his spirituality of the human senses in the 'circulation of love' between God and man.

2.3.3.3 Pastoral Border Cases

That Stensen set great store by the care of the sick since the time of his conversion to the Roman Catholic Church is apparent in the fact that the expertise composed by Cardinal Francesco Nerli Jr. (1636–1708),[691] archbishop of Florence from 1670 to 1682, pertaining to Stensen's episcopal ordination lauded him for visiting the hospitals and prisons in Florence following his conversion, assisting the patients and inmates in their physical and spiritual needs.[692]

---

[684] ParHAg., p. 44, lines 7–13 (= Altoviti, *Obbligo* [Cap. XXXV], p. 54, lines 14–22 = Attavanti, *Obbligo* [Cap. XXXV], p. 45, lines 5–13).

[685] ParHAg., p. 44, lines 13–17 (= Altoviti, *Obbligo* [Cap. XXXV], p. 54, lines 22–27 = Attavanti, *Obbligo* [Cap. XXXV], p. 45, lines 13–17).

[686] Op. 3, p. 423, lines 17–25. Cf. Spir. 8, p. 102, lines 5–6; Sermo 17, p. 242, lines 32–34.

[687] Op. 1, p. 408, lines 32–33.

[688] Schulze, *Medizin*, pp. 160 (Ambrose of Milan), 162.

[689] Dörnemann, *Krankheit und Heilung*, pp. 340–341.

[690] In more detail in Op. 2, p. 415, lines 26–28, 33–p. 416, line 4; ibid., p. 422, lines 4–5.

[691] On Nerli: RIIK-P, s.v. "Francesco Nerli junior", pp. 887–889.

[692] Add. 15, p. 932, lines 37, 42–p. 933, line 1. This Christian labor of love was also practiced by other laymen at the time, eg by Count Filippo del Pozzo and "other pious souls" (altre anime pie)

Later, as suffragan bishop, Stensen recommended to his parish priests the third part of the *Regula pastoralis* by St. Gregory the Great ([* c. 540] 590–604) as an exemplary testimony of individual pastoral care.[693] In this work, Pope Gregory I depicts the art of medicine as a template for pastoral care, emphasizing the description of therapy and curing.[694] According to Achim Thomas Hack, Gregory viewed the "medical ethos" as equally binding for "physicians of the heart" and "of the soul" ("medici cordis" and "mentis"), ie for pastors.[695] The rule found in the third part of the *Regula pastoralis* stating that he who incurred greater guilt could also become the one who loved more, which – following a long-standing theological tradition – is applied in particular to sexual aberrance and entanglement (Gregory cites Luke 7:47; 15:7),[696] was visibly expressed and confirmed in the eyes of Stensen eg in his personal acquaintance, the Florentine Knight Hospitaller Ferdinando Bonaccorsi (†1685) who ceased frequenting brothels, instead practicing acts of Christian charity, eg the "visiting of hospitals", and eventually becoming a priest.[697]

Another case characteristic of Stensen's medical-sacerdotal handling of drastic changes in the lives of others is his concern for the apostatized Neapolitan Franciscan Fr. Antonio Nepeta (OFM)[698] – whom he referred to as "the unhappy soul" (l'infelice anima) –, who had left his order and the Church and practiced medicine in Hamburg in the early 1680s, especially to treat "the French disease" (il mal Francese), ie syphilis, as Stensen reported to Cosimo III from Hamburg on March 15, 1684.[699] Stensen's unexcitedly factual and therapeutic approach to his pastoral support of Nepeta is striking: at the beginning of June 1678, Stensen had written to Cosimo from Hanover that, if Nepeta's wife who was pregnant at the time gave birth, "the third soul" (la terza anima) – ie that of the child – would also be won for the Church.[700] In the letter on March 15, Stensen emphasized that Nepeta's wife had informed him that she intended to abstain from marital intercourse (cf. John 8:10–11). She strongly wished, Stensen added, to become Catholic, and was teaching her son to pray in Catholic fashion; during the 6 months of his own stay in Hamburg, he had seen nothing but "a good and honest life" (una vita buona ed honesta) by Nepeta and his wife. Stensen characterized her benevolently: "She seems to be a good

---

who visited the sick in Roman hospitals, see E 290, p. 575, lines 13–17.

[693] ParHAg., p. 15, lines 13–18 (= Altoviti, *Obbligo* [Cap. IIII (sic!)], p. 11, line 35–p. 12, line 4 = Attavanti, *Obbligo* [Cap. IV], p. 5, lines 22–30); Sobiech, *Herz, Gott, Kreuz*, p. 275 with n. 36.

[694] Hack, *Krankheit*, pp. 151–152; on therapy and cures ibid., pp. 163–176.

[695] Ibid., pp. 181, 183.

[696] Judic, *Règle pastorale*, pp. 456–469 (Pars III, Caput XXVIII), here 464 & 466 (with the cited verses of Luke). Cf. Weber, *Allgemeine Moraltheologie*, p. 316n501.

[697] Scherz, *Epistolae*, vol. 1, p. 67n6. On Bonaccorsi: BNCF, Magl. XXV, 42, pp. 399–401 (no. 607), here 399: "frequentare spedali".

[698] On Nepeta: Scherz, *Epistolae*, vol.1, E 160, p. 391n1.

[699] E 342, p. 664, lines 34–35 with citation in line 34; ibid., p. 665, lines 11–12. He showed however, Stensen said, a desire to return to the Church and had already been abstaining from "any practice of the other religion" (da ogni esercizio d'altra religione) for several years (ibid., p. 664, lines 35–37 with citation in line 37), which in contemporary diction could also refer eg to Lutheranism.

[700] E 160, p. 389, lines 32, 37–39.

Fig. 2.9 Wax seal with
heart-and-cross coat of
arms on a letter by Niels
Stensen (Hanover, April
19, 1678 st.v. = April 29,
1678 st.n.) to the Grand
Duke of Tuscany Cosimo
III de' Medici
This is Stensen's seal
emblem from the time after
his consecration as bishop
in 1677, cf. E 150,
p. 376–377
APUG 576, fol. 44ʳ–45ᵛ,
here 45ᵛ

woman, they say she is of good nobility."[701] As he wrote to Cosimo on October 11, 1684 from Hamburg, Stensen spent much time on the spiritual care of Nepeta, whom he believed to be close to atheism, and even attempted – in keeping with the symbolism of heart and cross in his episcopal coat of arms (Fig. 2.9) – to demonstrate to him the wisdom of God in the processes of nature by way of the dissection of an animal heart.[702] Stensen nevertheless feared for the salvation of Nepeta, who in his opinion appeared to have adopted "some of the principles of atheism".[703] Stensen seems also to have been cautious with his words in his episcopal pastoral care for the faithful whose marriages faced deep crises, always examining the individual case as he was sensitized by Cosimo's tragic marriage, the result of a political wedding. While he never mentioned it in his correspondence, he was presumably thinking of it when in early 1678 he admonished as censor of the book *Philanthon vindicatus [...]* (Philanthon, vindicated [...]; Hanover 1678) by his father confessor Dionysius von Werl OFMCap. (c. 1640–1709)[704] that the Helmstedt Lutheran professor of medicine and philosophy Dr. iur. Hermann Conring (1606–1681) could

---

[701] E 342, p. 665, lines 3–7, here 5–6: "Ella pare buona donna, si dicono di buona nobiltà." – Nothing more is known about her.

[702] E 377, p. 729, lines 8–15.

[703] Ibid., line 9: "qualche principii dell'atheismo"; ibid., lines 14–15. On this and on the (lost) treatise composed for Nepeta about passages in the Bible which appear to contradict nature, see in more detail Sobiech, *Herz, Gott, Kreuz*, pp. 147–148, 235–236.

[704] On Dionysius: Johannes Friedrich Werling, "Dionysius v. Werl"; in: LThK³ 3 (1995), col. 249.

not be blamed for the "sins of his wife" Anna Maria née Stucke (1616–1694).[705] Blanket judgments clashed with Stensen's analytical mindset.

Stensen claimed on May 7, 1680, while performing a public dissection in Celle in the course of a colloquium on 'controversial theology' which he was attending as vicar apostolic, that he himself during his studies in the Netherlands had come close to an "atheism" for some time; in the reference framework of the seventeenth century, this means a position which denies a personified creator god and his providence. Stensen's endeavor to use dissection for 'pastoral medical' practice demonstrates his wish to find God even in something as 'profane' as dissection.[706]

In regard to his positions on medical recoveries inexplicable through natural science and on possession,[707] Stensen followed his experience as anatomist and based his thoughts on his observations: in his letter to Cosimo from Münster on July 22, 1681, he reports having seen three persons, one possessed, one hydropic and one suffering from malignant fever, ie to the point of endangerment of her life, be healed practically instantly following a prayer by the Capuchin priest Marco d'Aviano OFMCap. (1631–1699), who was on a journey through the German territories, during his visit to Münster from July 12 to 13 at which occasion, Stensen says, Marco "allowed us to see various effects of the divine blessing".[708] The Lutheran rulers over the Brunswick territories had presented to Stensen, presumably with a view to his reputation as anatomist and also as convert, two reports by the Lutherans of Regensburg on the work of this priest. Hereupon Stensen performed a detailed inquiry in order to be able to pass proper judgment, including in particular the exemplary comparison of the condition of a person before and after the reputed healing he was to examine.[709] Since August 1, 1681, Stensen had been a member of the commission appointed by Ferdinand II (1626–1683), prince-bishop of Paderborn and Münster, for the examination of the miracles effected by Marco d'Aviano's prayers. On October 10, 1681 Stensen reported the following to Cosimo from Münster: "I am among the number of those instructed to investigate the miracles which are said to have occurred due to the blessing by Fr. Marco. We have already

---

[705] E 148, p. 374, lines 24–27, here 26: "peccata uxoris"; on Anna Maria Conring: Herberger/ Stolleis, Conring, p. 89 & passim. On the marriage of Cosimo III see Scherz, Epistolae, vol. 1, pp. 38–39. Stefano Lorenzini and his brother were roped into Cosimo's marriage problems in 1681, see Altieri Biagi/Basile, Scienziati del seicento, s.v. "Stefano Lorenzini", pp. 791–822, here 794–795.

[706] See in more detail Sobiech, Herz, Gott, Kreuz, pp. 51, 61–62, 82, 234–235. Hübener, "Praxisbegriff", p. 55 and ibid., Prämoderne, p. 39 claim that Stensen was here attempting to demonstrate "the creative power of God ad oculos", ie in a palpable way.

[707] Cf. Schulz, "Besessenheit", pp. 20–48, where cases from the twentieth century and opinions on possession from the theological as well as the medico-psychiatric point of view are to be found.

[708] E 235, p. 490, lines 16–20, here 17–18: "ci ha fatto vedere diversi effetti della divina benedizione." On Marco: Volker Press, "Marco d'Aviano"; in: NDB 16 (1990), pp. 128–129; Marco was beatified on 04/27/2003, cf. AAS 96 (2004), pp. 730–732.

[709] E 205, p. 449, line 31–p. 450, line 6 (letter by Stensen to Hortensio Mauro c. November 1680); further information in Scherz, Epistolae, vol. 1, pp. 72–74; Scherz, Biographie, vol. 2, pp. 116–118.

found several cases in which miraculous changes have happened, but one that can be subjected to the objections of politicals [710] and non-Catholics we have hitherto not found; we have however not investigated all cases that have occurred."[711]

Mention should be made in this context of Stensen's careful handling of the case of Sr. Elisabeth Bisping (* c. 1649) from the Augustinian convent in Rosenthal, which he visited in the spring of 1683. Quarreling and conflicts had been daily occurrences there for 13 years, and all discipline had collapsed some years before his visit; one of the sisters, a source of the permanent strife, had accused herself in Stensen's presence of having a pact with the devil which included a sexual component.[712] Stensen took this case, which he was unable to conclude due to the prince-bishop's death, quite seriously as one of involvement with the devil, and attempted to remedy it by prayer and working towards voluntary reform of the sister in question. His final words in a letter to the convent sisters from April 30, 1683, which provisionally concluded his visitation, are as follows: "You should all apply yourselves to the true unity and be aware that where there is love and peace, there is God, and where there is strife, abuse and quarreling, the devil reigns. Jesus' peace be in all of your hearts!"[713] Hence the aforementioned assumption by Anne Tjomsland that Stensen had in extraordinary fashion been free of all superstitious notions prevailing in his time is once again corroborated.[714]

Stensen's pastoral activity, presented with the help of selected border cases, thus also confirms that he retained as a bishop the scientific ethos built around autopsy and experience and characterized by empathy and caution which he practiced as an anatomist, and that he examined the cases he occupied himself with not only theologically, but also from a rational and natural scientific point of view.

---

[710] See note 246, in Chap. 2 above.

[711] E 243, p. 498, lines 22–27: "Sono nel numero de' commissarii, che devono esaminar i miraculi, che si dicono fatti per la benedizzione del Padre Marco. Troviamo gia diversi casi ne' quali si sono fatte mutazioni maravigliose, ma tali che si possono esporre all'objezzioni de' politici ed acatholici non troviamo per ancora; non abbiamo però esaminati tutti li casi, che si sono passati." On Stensen's activity in the commission see Scherz, *Epistolae*, vol. 1, p. 74; Scherz, *Biographie*, vol. 2, pp. 118–119; supplementarily Ernesti, *Ferdinand*, p. 318n100.

[712] In a letter from Hamburg to the cardinals of the Congregation for the Propagation of the Faith c. 10/02/1683: E 321, p. 626, lines 22–28. In detail in Sobiech, "Hexenverfolgung", pp. 110–123, here 116–123 on the mentioned case.

[713] E 294, p. 583, lines 30–33: "Sie wollen sich doch alle befleissigen der wahrer Einigkeitt undt gedencken, wo Liebe undt Friede ist, da ist Gott, undt wo Unfriede, undt Tros undt Zanck ist, da regieret Sathan. Der Friede Jesu sey in aller ihrer Hertzen!"

[714] See note 39, in Chap. 1 above.

Chapter 3
Niels Stensen's Character Sketch in History (Seventeenth to Twenty-First Century)

In the more than 325 years since his death, various aspects of Niels Stensen's personality and work have been appreciated by scientists engaging in medical research, physicians, philosophers, theologians and representatives of the natural sciences. Several authors discussing Stensen's research and ethos have already been mentioned in the course of this study.[1] What is still missing, however, is a synopsis of significant references to his scientific and personal character sketch in history and his veneration as one of the blessed.[2]

From a bibliometric perspective, i.e. with regard to citation frequency, Stensen's anatomical writings in scientific literature published during his lifetime presumably are to be ranked highest. His discoveries were also cited during his time as a priest and bishop: Redi, for example, mentions in his microscopic parasitological magnum opus of 1684 – at which time Stensen was serving as vicar apostolic in Hamburg – the embryological research on a shark published in the *Acta Hafniensia* in 1675 which Stensen, "that most learned prelate", had conducted in Redi's private chambers in Livorno in spring 1667 while the court of the Medici was residing there.[3] It is noteworthy that, when Stensen is not only cited but also characterized,

---

[1] The reader is further referred to the *Dissertatio de anthropologia* (Frankfurt 1737) by Prof. Dr. med. Friedrich Christian Cregut (1675–1758) mentioned in De Angelis, *Anthropologien*, pp. 398–400, here 399n140, in which the various authors who had written about the natural and healthy state of the body are compiled and in which Stensen is mentioned several times.

[2] The *Cimbria literata [...]* (Learned Jutland [...]; Copenhagen 1744), belonging to the early modern genre of the "Historia Literaria" and written by the rector of the Flensburg Latin school Johannes Moller (1661–1725), who was staying in Hamburg from 1681 to 1684 – Stensen's last sojourn there lasted from mid-September 1683 until 12/11/1685 – , contains a compilation of eulogies of Stensen by contemporary authors: Cimbria lit. 2 (1744), s.v. "Nicolaus Stenonis", pp. 867–869, here 868–869. On Moller: Hendrik Andreas Hens, (Bjørn Kornerup): "Moller, Johannes"; in: DBL[3] 9 (1981), pp. 619–620.

[3] Redi, *Osservazioni*, p. 159, lines 5–11, here 10: "quel Dottissimo Prelato". The treatise mentioned by Redi ibid., lines 7–8 is OvaViv. II.

© Springer International Publishing Switzerland 2016
F. Sobiech, *Ethos, Bioethics, and Sexual Ethics in Work and Reception of the Anatomist Niels Stensen (1638-1686)*, Philosophy and Medicine 117, DOI 10.1007/978-3-319-32912-3_3

he is furnished with a multitude of honorable epithets in the superlative (resp. the elative).[4] When Stensen took his final private anatomical notes on the brain and nervous system in 1684, the Amsterdam physician Dr. med. Theodor Jansson van Almeloveen (1657–1712),[5] professor of medicine at the University of Harderwijk from 1701/1702, lauded Stensen as an "exceedingly perspicacious researcher of the structure of the human body".[6] The *Lindenius renovatus [...]* (Modernized Linden [...]; Nuremberg 1686), the medico-bibliographical standard work corrected and revised by Dr. med. Georg Abraham Mercklin Jr. (1644–1702)[7] and published in the year of Stensen's death contains the last mention of the latter's writings during his lifetime.[8] Stensen's name, his reputation as anatomist and some of his research results were handed down in medical literature – despite considerable fluctuation in the frequency of publication[9] – beyond the first half of the eighteenth century.

3.1   Talent and Originality

Stensen's scientific aptitude was attentively noted by his compatriots since his study days: Ole Borch, for instance, wrote to Thomas Bartholin from Leiden on January 9, 1662 with regard to "Stensen's short treatise",[10] meaning the miscellany presumably published on January 1, 1662 whose third item was dedicated to Borch,[11] that Stensen indeed possessed a competent talent. Should Stensen be entrusted with the public concern for anatomy in Copenhagen, i.e. be given a professorship, after having spent some more years "in these holies" (in his sacris) – meaning anatomical dissections – then he, Borch, would be hopeful for the further accretion of discoveries in Stensen's hitherto so rewarding studies. Borch characterizes him: "From close up he possesses incredibly lively eyes, indefatigable laboriousness and providential judgment, and is also practiced and educated in the humanist disciplines."[12] Two years later, however, Stensen was passed over during the appointment of new

---

[4] Examples of this already in Walsh, *Churchmen in Science*, p. 146.

[5] On Van Almeloveen: HGK, s.v. "Almeloveen, Theodoor Jansson van", pp. 24–25.

[6] Van Almeloveen, *Inventa nov-antiqua*, p. 29: "sagacissimus structurae corporis humani perscrutator".

[7] On Mercklin: Julius Pagel, "Georg Abraham Mercklin junior"; in: BLÄ 4 (1932), pp. 170–171.

[8] LindR, s.v. "Nicolaus Steno", pp. 843–844. On *Lindenius renovatus* cf. DDPhS, s.v. "Linden (Lindanus), Johannes Antonides van der", cols. 1200–1203, here 1203.

[9] A bibliographical overview of publications from this time mentioning Stensen in their titles can be found in Koch, "Bibliografia", pp. 138–141. A further bibliographical groundwork is the librarian final thesis BNSten., which compiles the literature published by and on Stensen until 1986. Neither of these two bibliographies (both of which feature limitations and gaps) register literature dealing with Stensen without mentioning him in the title, however.

[10] Bartholin, *Epistolarum centuria III* (Epist[ola] XCVII. [...]), pp. 416–425, here 417: "Stenonii tractatulus".

[11] GlandOc. The dedication to Borch can be found ibid., p. 77, lines 7–11.

[12] Bartholin, *Epistolarum centuria III* (Epist[ola] XCVII. [...]), p. 417: "in his sacris [...]. Vivacissimis est e propinquo oculis, indefessus labore, judicio non infelix, etiam in humanioribus exercitus, cultusq[ue]."

chairs at the University of Copenhagen, forcing him to begin his Peregrinatio erudita by going to France.

While Stensen was in Florence, the Copenhagen professor Simon Paulli Jr. emphasized in an extended reprint (Strasbourg 1667) of his work *Quadripartitum botanicum de simplicium medicamentorum facultatibus [...]* (Fourfold Botanical Book on the Effects of Simple Medications [...]), originally published 1639 in Rostock, that he wished for his former student Stensen, who was "born for the sciences", that God might preserve him for them and support his most honorable endeavors.[13] In his speech *De officio medicorum* (On the Profession of Physicians), held on June 1, 1665 and attached to the above-mentioned publication, Paulli had rendered this praise more precisely by stating that Stensen, who had excelled in anatomical studies under his and Thomas Bartholin's preceptorship, concerning these studies "could be regarded not so much as learned but rather as completely by nature to be destined to them".[14] Paulli's original wish was never to be fulfilled, as Stensen's conversion rendered his assignment to the University of Copenhagen – which was bound to the Lutheran confession – impossible.

In the months preceding Paulli's address, Stensen had already been attracting attention in Paris. For example, the Calvinist country physician Dr. med. André Graindorge (1616–1676)[15] reported on May 19, 1665, during his sojourn in Paris lasting from May 1665 to April 1666, to the amateur scientist, critic of Descartes, and apologist – and later priest (ordained in 1676) and bishop of Avranches – Pierre-Daniel Huët (1630–1721)[16] in Caen about Stensen, who at the time was conducting anatomical demonstrations at the private academy of Thévenot. Graindorge, who had associated with Huët in the founding of a very informal and unsystematically operating amateur assembly for dissections in Caen in 1662,[17] had the following to say about Stensen's tireless and passionate dedication: "This Mr. Stensen works like a whirlwind. We have seen [him] this evening after dinner [dissect] the eye of a horse. To tell you the truth, in comparison to him we are but beginners. He is always working. He possesses an unimaginable patience, and with practice he has acquired a skillfulness above the average."[18] On the following May 30, Graindorge explained himself to Huët, who was apparently surprised by the comparison, with regard to Stensen's dissection of a horse's eye: "When I betokened us beginners in compari-

---

[13] Paulli, *Quadripartitum botanicum*, p. 313: "publicis studiis natum". On the *Quadripartitum botanicum* cf. Egill Snorrason/Anne Fox Maule, "Paulli, Simon"; in: DBL³ 11 (1982), pp. 181–183, here 181.

[14] Paulli, *Quadripartitum botanicum*, pp. 628–634 (De officio medicorum), here 628: "non tam eductus quam totus natus iis esse videri possit".

[15] On Graindorge: Nicholas Dew, "Graindorge, André"; in: DFPh. 1 (2008), pp. 559–560, here 560 on Stensen.

[16] On Huet: Dinah Ribard, "Huet, Pierre-Daniel"; in: DFPh. 1 (2008), pp. 603–610.

[17] Lux, *Patronage*, p. 27.

[18] Tolmer, "Lettres", pp. 269–272, here 269–270: "Ce Mr. Sténon fait rage. Nous avons vu cet après-dîner un oeil de cheval. A vous dire le vrai, nous ne sommes que des apprentis auprès de lui. [...] Il est toujours en exercice. Il a une patience inconcevable, et par routine il a acquis une adresse au-dessus du commun." Also in Tolmer, *Huet*, p. 330. Cf. Lux, *Patronage*, p. 29.

son with Mr. Stensen, I had reason to, as I have never seen such dexterity; for, without applying his eye, scissors or any [other] small instrument to anywhere else but his hand which he held continually facing the audience, he let us see everything that can be observed about the structure of the eye."[19] Graindorge was obviously very impressed by Stensen's proficiency as anatomist. After his return from Paris, he reported to Huët from Caen on May 14, 1666 how he had been surprised here by a large number of women keen on seeing the heart unraveled in the manner of Stensen, so that he, who had attended two heart dissections by Stensen, conducted a first public dissection of a heart which taught him, in his own words, that spectating during such a performance, mastering the technical terminology, and practical dissection were completely different things.[20] His stay in Paris and Stensen's dissections which he observed there led Graindorge, who from his perspective as a natural philosopher had until then viewed it as below his dignity to dissect 'lowly creatures' like frogs, rabbits or fish, to a new understanding of natural science in autumn 1665; through his letters to Huët, their informal assembly was eventually transformed into a scientific research academy, the *Académie de Physique de Caen*, which existed until 1672.[21]

The physician Licent. med. Johann Valentin Wille (1651–1677), royal Danish field medic in Copenhagen who died aged only 25, undertook a study trip through Europe following the completion of his study of medicine in Strasbourg and came to Copenhagen no later than autumn 1673, where he continued his studies under Stensen (among others) and stayed until summer 1674.[22] Wille ends an article published in the *Acta Hafniensia* for the year 1673 with a comment stating that he had been taught the anatomical "methods" (viae) exceptionally well by "the inimitable Stensen, my preceptor here in Copenhagen".[23] In his posthumously published (1678) *De philiatrorum Germanorum itineribus dissertationes tres [...]* (Three Treatises on the Travels of the True Friends of the Medical Craft [...]),[24] Wille places Stensen among the youthful minds seduced by the "religious precipices" (religiosi scopuli) of the Peregrinatio academica, but voices his admiration for Stensen's ethos in the despairing exclamation: "What shall I say of Stensen, the name most endeared to me?"[25] Although Wille would never know the circumstances

---

[19] Tolmer, "Lettres", pp. 272–274, here 272: "Quand je nous mettais comme apprentis en comparaison de Mr. Sténon, j'avais raison, car je n'ai jamais vu tant de dextérité; car, sans mettre ni l'oeil, ni les ciseaux, ni un petit instrument autre part que dans sa main qu'il tint toujours exposée à la compagnie, il nous fit voir tout ce que l'on peut remarquer dans la constitution de l'oeil." Also in Tolmer, *Huet*, p. 330. Cf. Lux, *Patronage*, pp. 40–41.

[20] Lux, *Patronage*, pp. 68–69. On Graindorge's spectating during Stensen's heart dissections cf. Tolmer, "Lettres", pp. 270, 305.

[21] Lux, *Patronage*, pp. 1, 27–30, 38–42, 46–47.

[22] On Wille: Høeg, *Wille*, pp. 19–21, 48; Scherz, "Wille und Stensen".

[23] AMPH 2 (1673), pp. 301–304 (CXX. Unguentum ophthalmicum efficax), here 303: "incomparabili Stenonio, praeceptore hic Hafniae meo".

[24] On this, see Høeg, *Wille*, pp. 40–41 & 55, no. 10.

[25] Willius, *De philiatrorum Germanorum itineribus* (Dissertatio prima), p. 5: "Quid de Stenonio dicam, dulcissimo mihi nomine?"

of Stensen's conversion, he was nevertheless convinced of the sincerity of the latter's motives and characterized benevolently how he experienced him in Copenhagen: "He returned from Italy Catholic by confession, highly Christian in his lifestyle and attitude."[26] He also mentions Stensen with regard to the educational value of dissections performed with one's own hands: "Whom the preceptors Bartholin and Paulli profess to subordinate to, Mr. Niels Stensen, ordered him [i.e. Wille], who asked for friendly advice for his anatomical studies, to observe meticulously with sharpness of the knife, the eyes and the mind, far from any companions, where each of the fibers leads, that one was not to cut one of them before its orifices, components and connections were certain, and that one must not choose as teacher anyone who applied that method."[27] Stensen had presumably advanced this "codex of precautionary measures"[28] (Georges Canguilhem) not least because during dissection of the liver of a female ray in 1664, he had unintentionally severed the ovarian end of the oviduct and therefore had been unable to discern the opening of the oviduct into the abdominal cavity, resulting in his not discovering the function of the ovary in the vivipara until 1666/1667.[29] Already in his Parisian *Discours*, Stensen had criticized cutting dissection objects strictly by the book during the study of medicine and advocated a plurality of methods.[30] His methodical precision is also attested to in Holger Jakobsen's notes.[31] Wille adds: "Denmark would not be forced today to dispense with Stensen, that famous inimitable oracle of nature, whose knife even Hippocrates [of Kos (c. 460 B.C.–c. 370 B.C.), *archetype of every physician*], if he could come back to life, would admire, and it [= Denmark] [would not be forced] to forfeit so suddenly the glory of its felicitousness in anatomy, had not first the superstition of the Roman Religion and then the benignity of the Grand Duke of Tuscany won him over, under the protection of whose majesty he now already shines in a position commensurate with his merits."[32] As proof that too much time was wasted in the domestic study of anatomy without guidance and correction by a teacher, Wille refers to a statement by Stensen himself: "In written form as well as orally, Stensen often professed that he could teach as well as strikingly demonstrate within

---

[26] Ibid. (Dissertatio secunda), p. 96: "Rediit ille ex Italia professione Catholicus, vita & moribus maxime Christianus."

[27] Ibid. (Dissertatio prima), pp. 44–45: "Cui praeceptores Bartholinus & Paulli se cedere fatentur, d[ominus] Nic[olaus] Stenonius amicum consilium in anatomico studio rogantem iussit sedulo cultri, oculo[45]rum & mentis acie, remotum a congerronibus, observare, quo quaeque fibrarum ducit, nullam rescindendo, antequam de ejus insertionibus, partibus & connexionibus constiterit, nullo usurum hodego, qui hac methodo uteretur."

[28] See note 230, in Chap. 2 above.

[29] Lesky, "Säugetierovar", pp. 238–239.

[30] Discours, p. 18, lines 13–30.

[31] E.g. ExAc., p. 294, lines 21–26; see also Scherz, *Vom Wege*, p. 121.

[32] Willius, *De philiatrorum Germanorum itineribus* (Dissertatio prima), pp. 76–77: "Stenonio suo, incomparabili illo naturae oraculo, cujus scalpellum, si revivisceret, ipse Hippocrates admiraretur, non hodie Dania carere cogeretur, nec ita repente anatomicae felicitatis gloria cadere, nisi itineribus Italicis intentum primum superstitio Romanae Religionis, post Magni Hetruriae Ducis clementia cepisset, in cujus numinis tutela, [77] condigna meritis sorte, jam ille lucet."

a matter of hours what had been discovered not over a period of months or years, but over the period of Olympiads [= 4 years each] and lustra [= 5 years each]."[33] Yet Stensen's outstanding giftedness never misled him to hauteur.

3.2 Science and Humility

In 1661, Ole Borch reported that Stensen, quite modest and until then inexperienced with regard to "ambition", indicated rather clearly that he was willing to defend himself against Blasius if the situation demanded it.[34] As Stensen's credibility was at stake in this case – and he was left to his own devices concerning his defense – , he availed himself of all permissible scientific arguments without resorting to insults like Blasius. Even his Copenhagen study mentor Thomas Bartholin found the altercation between his friend Blasius and Stensen discomforting. Stensen was also forced to defend himself against Deusing, who flat out ignored his person. This proves that he knew how to stand up for himself. Having taken a deist stance 'sui generis' starting in about 1662/1663, influenced by the multitude of confessions he experienced in the Republic of the Seven United Netherlands, Stensen strived from about 1664 onwards to avoid anything that even remotely smacked of "human, not gospel-compliant" wisdom.[35] From this time on, Stensen was known among his contemporaries for the humility he lived out consciously. This humility was paired with his sensitive approach, which was probably connected to his 3 years of illness as a child and supported the development of his research ethos characterized by empathy and caution. In the months prior to his conversion, in particular following the Corpus Christi procession he had witnessed in Livorno on June 24, 1666,[36] Stensen found himself in permanent spiritual turmoil, which may explain his somewhat irritated tone in parts of the letter to Thévenot written in this time and attached to his Myology. A witty statement Stensen made to Antonio Magliabechi, mentioned again in context in the following, may also be owed to this agitation. That Stensen's ethos was deepened even further particularly as a result of his heart dissections in 1662/1663, which left him deeply doubtful of scientific positions based on Descartes – and also those of Spinoza – , is evidenced by his letter as a bishop around November 1677 from Hanover to Leibniz: through the anatomical discoveries, God had let him renounce "philosophic pretentiousness" and by and by restrict

---

[33] Ibid. (Dissertatio secunda), pp. 149–150: "Tam scripto, quam ore saepe fassus est Stenonius, se non mensium, non annorum, sed olympiadum, sed lustrorum inventa, horarum spatio & do[150] cere posse & monstrare ad oculum." – The correspondence with Wille is presumably not preserved.

[34] See note 54, in Chap. 2 above.

[35] DefConv., p. 393, lines 10–11, here 11: "humanam prudentiam evangelio non conformem." On this "gospel-compliance" see Sobiech, *Herz, Gott, Kreuz*, pp. 35–36, 64–65, 67.

[36] On that day in Livorno and its consequences for Stensen's spiritual development see Sobiech, *Herz, Gott, Kreuz*, pp. 37–38, 41–42, 66.

himself to the "love of Christian humility" (cf. Eph 4:2), which was the most worthy love of which a rational soul was capable, namely a love for the cognition of what we are in relation to God and ourselves.[37] Stensen's newfound theological freedom due to his conversion, which to him was identical to the grace of God he felt inside himself, presumably caused him to generously accord De Graaf his own research results.

During his journey to France, Stensen came in late autumn 1665 to Montpellier, location of a medical faculty and meeting point of English natural scientists, where he met Martin Lister (1639–1712).[38] A student of medicine at the time, Lister would eventually receive his doctorate in Oxford in 1684 and become a physician and zoologist, and like Stensen set great store by personal observation. He experienced Stensen as an anatomist and conversed with him, later reporting on a lecture in anatomy and several dissections conducted by Stensen in the private chambers of Robert Bruce (1626–1685),[39] member of the House of Lords from 1664 as the 1st Earl of Ailesbury: "the Demonstrations were neat & clever wherein I much admired ye ingenuitie & great modestie of ye Person & w[hi]^ch appeared the rathar, by reason of ye great impertinencie of a French Doctour & Professeur yt assisted alsoe at ye Assemblé. Afterwards I visited Mr Steno, whom I found infinitely taking & agreable in conversation & I observed in him very much of ye Galant & honest man as ye french say, as well as of ye schollar."[40] In his characterization, Lister refers to the French description of the "honest man" (honnête homme), which includes an ethical as well as a gallant and courtly dimension.[41]

Francesco Redi, who conducted anatomical and embryological investigations together with Stensen from January 1667 onwards,[42] wrote on March 21, 1667 from his private rooms in Livorno to Vincenzo Viviani in Florence that he had been able to pass on a "bundle" (fagotto) from Viviani, apparently scientific documents or letters on myology, to Stensen immediately "because, as you know, our Mr. Stensen grants me the honor of accompanying me to my breakfast and dinner table, and I have the joy of delighting in his highly virtuous and highly pleasant conversation; and furthermore, we are not idle and conduct beautiful dissections and beautiful

---

[37] E 143, p. 368, line 38–p. 369, line 5, here p. 369, lines 1–2 = (with exact transcription) Leibniz-Forschungsstelle der Universität Münster, *Philosophischer Briefwechsel*, N. 160a, p. 578, lines 11–15 with citation in lines 12–13: "presumption philosophique [...] amour de l'humilité chrestienne"; see in more detail in Sobiech, *Herz, Gott, Kreuz*, pp. 136–138.

[38] On Lister and his contribution to the question of human procreation: Andrew Pyle, "Lister, Martin"; in: DBPh. 2 (2000), pp. 527–529.

[39] On Bruce: Thomas F. Henderson/Victor Stater, "Bruce, Robert"; in: ODNB² 8 (2004), p. 324.

[40] Scherz, *Indice*, p. 292, Supplement 3 (undated). – The French professor remains unnamed.

[41] Papasoli, "Soggiorno", p. 97.

[42] In Redi, *Esperienze*, p. 121n202, cooperation between Stensen and Redi is mentioned for the first time for 01/20/1667 (corr. from "1666" in ibid.) in Redi's records. In CanCap., p. 117, lines 22–23 & p. 138, line 28, Stensen describes Redi as his "most famous friend" (amicus clarissimus); on Redi's cooperation with Stensen cf. Scherz, *Epistolae*, vol. 1, pp. 21–23.

observations in regard to these saltwater fish every day."[43] In his letter from Livorno to the vicar general of the Prato diocese and poet Valerio Inghirami (†1671) on March 30, 1667, Redi states the following concerning Stensen's future conversion, which he considered a certainty: "Believe me, dear Mr. Valerio, Mr. Niels is truly an angel in his manners, in addition to his nature as the great philosopher, the great anatomist and the great mathematician which he is."[44]

The mathematician Michelangelo Ricci (1619–1682)[45] expresses similar praise in a letter from Rome to Leopoldo de' Medici on May 30, 1667 in regard to the Myology, which Stensen had sent him as well, saying that the latter, whom Ricci places on a level with Galileo Galilei, united "a great proficiency and care in obser-vation, talent and a clear imagination" which was the basis for "such beautiful dis-coveries and the explanations delivered with such great clarity and simplicity".[46] On November 17, 1668 – Stensen had already converted and begun his great geological journey of research which initially led him via Rome to Naples – Ricci wrote to Leopoldo, following a meeting with Stensen in Rome the day before, that he appre-ciated "his modesty and honesty and the intellect which is rich in science and other handsome news".[47]

There are other testimonies besides the report by Tschirnhaus from Rome for the year of Stensen's episcopal ordination that cast a light on his work and personality: the Florentine physician Dr. med. Jacopo Lapi (†1693),[48] who was very popular among the students of medicine and surgery thanks to the open private academy held at his house, described him as "the good and honorable Mr. Stensen, who today is one of the best priests of our city"[49] in a letter to the Luccan physician and writer Dr. med. Mario Fiorentini (1642–1720)[50] on January 30, 1677. The occasion for this letter was the death of Dr. med. Tilman Trutwin (†1693),[51] who hailed from

---

[43] Redi, *Opere*, pp. 213–214, here 213: "poichè, come sa, il Sig. Stenone mi fa l'onore di favorir la mia tavola mattina e sera, ed io ho questa contentezza di godere della sua virtuosissima, e amabilis-sima conversazione, e di più non istiamo in ozio, ed ogni giorno facciamo di belle notomie, e di belle osservazioni intorno a questi pesci di mare."

[44] Redi, *Lettere*, pp. 5–6: "Credetemi caro Sig. Valerio, il Signor Niccolò è veramente un angiolo di costumi; oltre lo essere [6] quel gran Filosofo, e quel gran Notomista, e gran Matematico che egli si è." On Inghirami: DRR I, p. 859.

[45] On Ricci: Luigi Campedelli, "Ricci, Michelangelo"; in: DSB 11 (1975), pp. 404–405.

[46] Fabroni, *Lettere*, pp. 156–157, here 156: "una gran perizia e diligenza nell'osservare, ingegno e fantasia chiara; […] sì belle invenzioni e le spieghi con tanta chiarezza e facilità".

[47] Ibid., pp. 162–163, here 163: "la sua modestia e sincerità, e l'intelletto che ha chiaro e ricco di scienze ed altre belle notizie."

[48] Possibly to be differentiated from Jacopo Del Lapo, cf. Guerrini, "Lettere", p. 196n11. On the other hand, Manni, *Stenone* (Libro II, Cap. XVII), pp. 158–165, here 159–161, reproduces verba-tim a letter (see next note) by the sender "Jacopo Lapi" (ibid., p. 161), but identifies the person as "Jacopo del Lapo" (ibid., p. 159).

[49] Manni, *Stenone* (Libro II, Cap. XVII), pp. 158–165, here 160: "il buono, e venerabile Signore Stenone, che inoggi è uno de' migliori Preti Sacerdoti, che sia nella nostra Città".

[50] On Fiorentini: Lazzarelli, *Cicceide legitima*, p. LXXVIII.

[51] On Trutwin: E 22, p. 190n6; Adelmann, *Correspondence*, vol. 1, no. 2, p. 4n4. Not in DDPhS.

Roermond and had worked as an anatomist in Florence, and to whose conversion some years earlier Stensen had contributed. Similarly, the Florentine Cardinal archbishop Francesco Nerli Jr. pointed out in his expertise from June 1, 1677 on Stensen's suitability for the episcopal office that Stensen had acquired the love and esteem of all after his conversion through his way of life characterized by acts of Christian charity, but without losing his own modest opinion of himself and his great humbleness, although "according to the opinion of the experts" he was the "premier anatomist of Italy" as well as being a polyglot and among the best lecturers on philosophy, which was why he had been appointed by the Florentine grand duke as teacher for his heir to the throne.[52]

Three years after Stensen's death, Dr. med. Jean Bernier (1627–1698),[53] who had been practicing in Paris since about 1674, summed up as follows with regard to Stensen's writings: "Incidentally, I refer the reader to the works of medicine which he has given us as evidence of his capability, leaving this small portrait like a mirror of disinterestedness, of prudence, of education and religion for the physicians of our century, who love nothing more than to make noise, to scheme, and to utter inanities in order to establish themselves and to make money."[54] Though Bernier's characterization of contemporary physicians is intentionally rich in contrast, it nevertheless appears – when compared with Bernard N. Nathanson's descriptions of the worldly temptations facing physicians – that little has changed in terms of the dangers confronting the medical profession.[55] There is similar talk in medical ethics today of the economization of medical science and 'cleverness' as indications of "an erosion of the scientific ethos".[56] Also, according to Heiner Fangerau, "the quest for fame and honor, pressure from superiors, the achieving of a certain 'function' or the 'aesthetics' of a result conforming to the hypothesis" tempt physicians to overstep ethical boundaries in their research. Fangerau goes on to state that this represents an antipode to Robert K. Merton's imperative of the "disinterestedness" of science.[57] As shown above, Bernier rightfully emphasized Stensen's "disinterestedness", thus providing evidence of this aspect of Merton's definition of science in Stensen's ethos.

---

[52] Add. 15, p. 933, esp. lines 8–12, here 10–11: "al parere de' periti il primo anatomista d'Italia".

[53] On Bernier: Michel Prévost, "Bernier (Jean)"; in: DBF 6 (1954), col. 114.

[54] Bernier, *Essais de medecine*, pp. 167–169 ("Jean [sic!] Stenon"), here 168–169: "Au reste je renvoye le Lecteur aux Ouvrages de Medecine qu'il nous a donnez pour preuve de sa capacité, [169] laissant ce petit portrait comme un miroir de désinteressement, de diligence, d'érudition & de Religion aux Medecins de nôtre siecle, qui n'aiment qu'à faire du bruit, qu'à intriguer & à débiter des vanitez pour s'établir & pour gagner de l'argent." – On the income of French physicians in the sixteenth and seventeenth centuries see Brockliss/Jones, *Medical World*, pp. 320–328.

[55] See notes 147 & 148, in Chap. 1 above.

[56] Fangerau, "Ethik", p. 283 following Peter Weingart.

[57] Cf. ibid., and on this Merton, *Social Theory*, pp. 612–614, where the sociological state of "science" versus the "'service' professions" is outlined.

3.3   The Convert as a Role Model

Stensen tended exceptionally well to his students in his function as "Anatomicus regius", as highlighted by Thomas Bartholin in Copenhagen in 1673 and evidenced in the comments by Johann Valentin Wille and Holger Jakobsen. It should further be noted that in the years after his own conversion, Stensen initiated several other conversions and ordinations to the priesthood.

The student of medicine under Stensen and later Jesuit Fr. Johan Didrik Karstensen Atche (*or*: Atke) (1652–1692)[58] has an exceedingly detailed necrology dedicated to him in intra-order reports in comparison with many other members of the Society of Jesus.[59] An abbreviated version of it is found in the "Annual Report of the Lower-Rhenish Province for the Year 1692".[60] The annual report and the necrology relate how Atche had participated in Stensen's dissections as a student in Copenhagen and attempted on the occasion of private conversation with "Mr. Stensen", the "at this time most famous anatomist" and "great man", according to the intra-order reports, "who through the efforts of Ours [= the religious of the Society of Jesus] had converted to the Catholic side", to win him back for the Lutheran faith. In the end, however, it was Atche himself who converted to Catholicism.[61] As stated in his letter of application (for the Chinese or Indian missions), hand-written and sent to the Superior General in Rome, Charles de Noyelle SJ ([* 1615] 1682–1686), Atche continued his studies[62] of medicine in Leiden while supporting converts and people persecuted for their faith there,[63] joined the Society of Jesus on October 1, 1676 – at which time Stensen was already a priest in Florence – and died in 1692 as camp pastor to the imperial soldiers in Peterwardein (*today*: Petrovaradin, a district of Novi Sad) in Serbia.

Towards the end of the year 1677, when Stensen was already serving as bishop in Hanover, the treatise *Dissertatio curiosa* about the giant "Starcuterus", a figure of Danish mythology, was published in Florence.[64] It had been written by the Lutheran

---

[58] On Atche: ARSI, Rh. Inf. 23, fol. 85ʳ–88ᵛ (Catalogus rerum Domus Probationis Socie[ta]tis Jesu Provinciae Rheni Inferioris 1678 […]), here 86ᵛ (no. 28); ibid., Hist. Soc. 49, fol. 254ᵛ; Fejér s. s. 1 (1985), s.v. "P[ater] Atche Theodorus", p. 60; not mentioned in August Ziggelaar, "Stensen (Steensen, Steno), Niels"; in: DHCJ 4 (2001), pp. 3636–3637; with imprecise information Scherz, "Jesuit durch Niels Stensen". On Atche's studies under Stensen cf. Add. 12, pp. 927–928, here 927.

[59] ARSI, Rh. Inf. 46, pp. 787–788: "Elogium P[atris] Theodori Atche."

[60] Ibid., Rh. Inf. 57ᴵᴵ, fol. 686ʳ–742ᵛ (Annuae Provinciae Rheni inferioris de a[nn]o 1692), here 729ᵛ–731ʳ.

[61] Ibid., fol. 729ᵛ–731ʳ, here 729ᵛ: "dominum Stenonium anatomicum eo tempore celeberrimum et ad partes catholicas nostrorum opera transiisse […] magni viri"; cf. ibid., Rh. Inf. 46, pp. 787–788, here 787: "d[omi]num Stenonium anathomicum eo tempore celeberrimum […] magni viri".

[62] Ibid., Rh. Inf. 15, fol. 125ʳ⁺ᵛ (Indipeta autograph by Atche on 10/16/1682 from Hildesheim).

[63] Ibid., Rh. Inf. 57ᴵᴵ, fol. 729ᵛ–730ʳ; ibid., Rh. Inf. 46, p. 787.

[64] Paullini, *ΜΟΣΧΟΚΑΡΥΟΓΡΑΦΙΑ* (appendix [unpaginated]), here no. 10 of the edited works: Paullini, *De starcutero*, retaining the early modern notation: *De Starcutero, famosissimo Ggnante [corr.: Gigante] Boreali, dissertatio curiosa ad Virum Celeberrimum D[ominum] Nic[olaum]*

physician and polyhistor Dr. med. Christian Franz Paullini (1643–1712),[65] who at the time was personal physician to Prince-Bishop Christoph Bernhard von Galen ([* 1606] 1650–1678) at the Imperial Abbey of Corvey, and printed at Stensen's expense. During his time as a priest in Florence, Stensen had encouraged Paullini to accept a professorship in Pisa which he had been offered by Grand Duke Cosimo III upon mediation by Athansius Kircher SJ. Paullini, who did not accept the position, quotes from a letter to himself by Stensen from Florence: "If I had you here, I would not give up the hope of obtaining [for you] a place amongst the professors in Pisa, as his Highness the Grand Duke of Tuscany had offered it to you already before."[66] During his time as a priest in Florence, Stensen was still comparatively heavily involved with natural research until 1677; for example, he provided Holger Jakobsen, who in the meantime in absentia had been appointed professor of history and geography at the University of Copenhagen, and the cousin of Caspar Bartholin Jr., Christoph Bartholin (1657–1714),[67] both of whom visited with him in Pisa from February 17 to March 6, 1677 in the course of their study travels, with ample opportunity to dissect fish in the grand-ducal winter residence. Stensen attended the pre sentations of their results they held before the crown prince each evening.[68] Three years after Stensen's death, Paullini betokened him as "highly esteemed friend and benefactor as long as he lived" and reported that Stensen kept in contact with him via letters as vicar apostolic in Hanover.[69]

Stensen's fostering of Lutherans, presumably also in hopes of their conversion, is in stark contrast to the claim by the Florentine librarian Antonio Magliabechi (1633–1714) that in January 1676 Stensen supposedly attempted to refuse the

---

Stenonis, Episcopum postea Titiopolitanum, & Vicar[ium] Apostolic[um], cujus curâ & impensis prodiit Florentiæ 1677. 4. (= quarto). Just as is noted in Marx, "Paullini", p. 84n119, no copy was to be found during the research for this study; the copy documented for the *Bodleian Library* of Oxford University in BBU 2 (1854), s.v. "Starcater, géant du Nord", col. 1713 along with BBU 1 (1854), p. IV could not be located despite a search on-site (*Rare Books Section, Departement of Special Collections & Western Manuscripts*) by curator Sarah Wheale in February 2012.

[65] Paullini studied medicine and Lutheran theology in several European cities; on Paullini: HGK, s.v. "Paullini, Christian Franz", p. 499; Dieter Lent, "Paullini, Christian Franz"; in: BBL, pp. 549–550; Marx, "Paullini", pp. 59–63. According to LkAE, Kirchenbuch Eisenach 1643–1648, p. 500[r] (baptism: 02/26/1643 st.v.) and ibid., Kirchenbuch Eisenach 1706–1719, p. 249[v] (funeral: 06/13/1712 st.n.), Paullini was baptized and buried a Lutheran.

[66] Letter excerpt (undated) in Paullini, *Cynographia* (Cap. III before Sectio I, here "Vota et censura excellentium virorum" [unpaginated], no. II): "Si hic Te haberem, de loco Pisis obtinendo inter Professores non desperarem, cum illum jam tum Tibi promiserit Sereniss[imus] Magnus He[*page break*]truriae Dux." Cf. E 112, p. 314, lines 31–32.

[67] On Christopher Bartholin: Hjalmar Sigvard A. Nygård, "Bartholin, Christopher"; in: DBL[3] 1 (1979), p. 474.

[68] Maar, *Holger Jacobæus' Rejsebog*, p. 137; Guerrini, "Biografia Rediana", p. 57. On Stensen's years as priest and teacher at the Tuscan Court see Sobiech, "Simplicity of Faith, Intuition and Giordano Bruno".

[69] Paullini, *Talpa*, p. 88: "Fautor & amicus, dum viveret, honoratissimus". In a letter to Heinrich Meibom jr. from Florence on 09/02/1670 (Bruun, "Fem nyfundne Niels Stensen-breve", p. 144, lines 15–16), Stensen speaks of "our old friendship" (vetus amicitia) which was to be renewed, and that he would welcome a regular exchange between them about their joint studies.

Reformed Gerhard Meier (1616–1695), rector of the Bremen Gymnasium Illustre, entry to the *Biblioteca Laurenziana* due to his confession.[70] One may assume that the somewhat cantankerous Magliabechi, with whom even the tactful Francesco Redi had his problems,[71] was simply acting out his disgruntlement – as in the case of his "rancor" around June 1667 caused by a witty remark by Stensen.[72] The details of this later case are unknown.

Dr. med. Johann Georg Rötenbeck (1647–at least 1710),[73] a Lutheran physician who corresponded with Malpighi about questions regarding the female ovary during his study trip through Italy, had, as Stensen wrote to Cosimo III from Rome on September 14, 1677,[74] converted a few weeks earlier in Rome shortly after Stensen had guided him through the Spiritual Exercises of St. Ignatius. Count Filippo del Pozzo (c. 1636–1719), who visited the sick in Roman hospitals as an expression of Christian charity, had introduced Rötenbeck to Stensen after taking note of the former's modesty and moral integrity.[75] Rötenbeck was ordained a priest by Stensen in 1679, and eventually worked as a missionary in Mecklenburg from 1683 as mentioned before.

Dr. med. Johannes Nicolaus Pechlin (1646–1706), son of a Lutheran predicant in Leiden, had studied under Drélincourt and Sylvius in that city before working as professor of medicine and botany in Kiel for a few years starting in 1673, then in 1682 became personal physician and in 1686 also librarian to the Duke of Holstein-Gottorp.[76] On December 25, 1675 st.v. he had conducted a public dissection in Kiel of a black African woman who had died there.[77] In 1688 he published – under the pseudonym "Janus Philadelphus" and providing a fictitious imprint – his *Consultatio desultoria de optima Christianorum secta, et vitiis Pontificiorum. Prodromus religionis medici* (Craft Consideration of the Best Christian Grouping, and on the Vices of the Papal Adherents. Precursor to a Religion of the Physician; Hamburg, 2nd ed.

---

[70] Nordström, "Magliabechi", pp. 19–20, 42. On Meier: Janse, "Reformed Theological Education", p. 39. On the behavior of the Italian librarians towards Protestants see Cavarzere, *Censura*, pp. 127–128; on Stensen's visits to the Florentine *Biblioteca Laurenziana* and to Roman libraries in the time prior to his conversion see Sobiech, *Herz, Gott, Kreuz*, pp. 38, 114. Für Stensen, the "conscience" (conscientia) of the individual was decisive in the question of conversion; he rejected compulsory measures by the state (ibid., pp. 335–343 ["Stensen und die Ökumene"], here 336–338).

[71] Adelmann, *Embryology*, vol. 1, p. 144.

[72] Concluded indirectly from E 26, p. 193, line 36–p. 194, line 6, here p. 193, line 37–p. 194, line 1: "risentimento", elicited, according to Stensen, by his "words of possibly too great familiarity" (parole [...], forse con troppa familiarità).

[73] On Rötenbeck: Scherz, Epistolae, vol. 1, pp. 49–50; Scherz, "Rötenbeck und Stensen"; Adelmann, *Embryology*, vol. 1, p. 405n4; Adelmann, *Correspondence*, vol. 2, no. 365, p. 764n1 (erroneous date of death "in Italy in 1677").

[74] E 137, p. 359, lines 17–20.

[75] E 290, p. 574, lines 21–30 & p. 575, lines 6–25. On Del Pozzo: E 49, p. 216n3.

[76] On Pechlin: DDPhS, s.v. "Pechlin, Johannes Nicolaas", cols. 1505–1507; Dietrich Korth, "Pechlin, Johann Nicolaus"; in: SHBL 7 (1985), pp. 164–166.

[77] See note 350, in Chap. 2 above.

1709) which, being a physician by profession and by practice, as he stated, he had written in his extremely rare hours of leisure.[78] Much more than by the apostasies of princes, courtiers and other worldly persons and their apologies, Pechlin writes, he was unsettled by apostasies of learned and intellectual men,[79] and adds: "None of these things, incidentally, moved or hurt me more than the apostasy of the recently, as I have heard, deceased Niels Stensen, because of which I once in fact began all this which is discussed [in this book]. For he was, if I am not mistaken, a virtuous man by nature, peaceful, whose judgment and wisdom in medical and mathematical things I was accustomed to setting great store by. Thus I thought it very much a wondrous thing that a man so mathematical and used to seeking for proof had converted to a grouping so depraved and based on absolutely no foundations."[80] After spotlighting the willingness of Mrs. Arnolfini to give her own life for Stensen's conversion if necessary,[81] the Florentine religious orders,[82] the position of the Pope, the outward ornamentation of churches and rites, regular prayers, the priestly lifestyle and stricter ecclesiastical discipline as significant factors for the conversion of Stensen, Pechlin goes on to state: "When he [= Stensen], through these impressive and eye-infatuating things, maybe also swayed by honors promised for the future, saw that what presents itself in medical and anatomical matters is mostly uncertain, and that that which appears certain remains on the surface and forms by far the smallest part of a true science, and after he had turned his mind away from evidence completely, he abandoned himself fully to gullibility, and through the imposing authority of the papal grouping, which appeared to be humility of the mind, he changed his profession simultaneously with his religion, whereby he no longer fulfilled the role of the physician, but that of the theologian, that of the priest, that of the titular bishop and the vicar apostolic. I expected from this man, however, like from nobody else, a truly herculean, most accomplished apology filled, as he otherwise cultivated, with mathematical proof; but lo and behold, an apology of his cause which has difficulties in all matters and in many parts is weaker than that [apology] of other things tends to be, and which rather, ordered for examination in Holland, Jena and Copenhagen, did not support his fame; as a result I thought him pitiable as, swept away by some lapse in his perception, he exchanged a better thing for a worse one. Nevertheless the man's holiness shone ceaselessly in his lifestyle, and in the disputations which he occasionally held, and in the letters some of which also graced me he put forth only love, patience and the customs of Christ to be replicated

---

[78] Pechlin, *De optima Christianorum secta*, "Monitio ad lectorem" (unpaginated), here first page.

[79] Ibid., p. 103.

[80] Ibid., pp. 103–104: "Nihil tamen omnium magis me commovit vel offendit, quam Nicolai Stenonii, pridem, ut audio, mortui, apostasia, cujus etiam causa hoc quidquid est negotii, olim suscepi. Erat enim, nisi me omnia fallunt, vir natura probus, pacatus, cujus ego judicio & pru[104] dentiae in rebus medicis atque mathematicis plurimum tribuere solebam. Adeo miraculo proximum duxi, ad sectam tam corruptam nullisque admodum fundamentis innixam transiisse hominem mathematicum & demonstrationes sectari solitum."

[81] See Sobiech, *Herz, Gott, Kreuz*, p. 41.

[82] In more detail ibid., pp. 39–41.

by imitation, and he disclosed, not secretly, but late, that he wanted to atone through the example of a life lived meagerly and ascetically, after having given away his goods to the poor, for the mistake he had made in the realm of the theoretical. And yet this same Stensen could with the same righteousness have been with us and through continued study of natural and medical things, as he seemed by nature to be made for, have served the honor of God the Greatest and Highest as well as the benefit of his neighbor."[83]

Stensen's writings from the field of controversial theology mentioned by Pechlin, whose appreciation and classification within the contemporary context are yet to be undertaken, were composed in a level-headed, natural-scientific style and without polemic. Unperturbed by the lack of understanding he was faced with from various Protestant circles, Stensen advocated "love for the brother" and in 1678, while at court in Hanover as vicar apostolic, he ordered "many rather hurtful" wordings to be removed from the manuscript of the *Philanthon vindicatus* by Dionysius von Werl OFMCap. in his function as book censor. With a view to the polemic still remaining after his restrained censoring, which was implicitly addressed to its Protestant readers, he hinted at his misgivings about the contents of the book by stating that he "could not prevent" its publication.[84] Even though Pechlin interpreted Stensen's humility in a sense supporting his own opinion, he nevertheless recognized it in its core value.

The physician Dr. med. Johann Heinrich Cohausen (1665–1750),[85] who served as physician at the court of the prince-bishops of Münster from 1700, asserts in his

---

[83] Pechlin, *De optima Christianorum secta*, pp. 104–106: "Hisce ille speciosis oculumque fasci-nantibus rebus forte etiam promissis in futurum honoribus, adductus, cum videret, quae in medicis atque anatomicis occurrunt, maximam partem incerta [105] esse, & quae certa videntur, in super-ficie haerere scientiaeque verae longe minimam facere partem, averso plane a demonstrationibus animo totum se credulitati dedit, & speciosa ille sectae pontificiae auctoritate, quae videbatur esse animi humilitas, cum religione pariter etiam professionis genus mutavit, non medici amplius, sed theologi, sed sacerdotis, sed Episcopi titularis & Vicarii Apostolici partes secutus. Exspectavi autem ab hoc viro, ut a nemine alio, herculeam plane & consumatissimam apologiam, & mathema-ticis, ut solebat alias, demonstrationibus plenam; sed ecce laborantem omnibus numeris causae defensionem & multis partibus imbecilliorem, quam esse illa aliorum solet, quaeque adeo in Hollandia, Jenae & Hauniae ad examen revocata famam non sustinuit; commiseratione ergo dig-num judicavi, qui sensuum quodam lapsu abreptus meliora deterioribus mutarit. Enituit tamen usque viri in vivendo sanctitas, & disputationibus, quas ultro citroque habuit, epistolisque, quarum una alteraque me quoque dignatus est, unice charitatem, patientiam, moresque Christi imitatione exprimendos, proposuit, ostenditque non obscure, quamvis sero, vitae parce & sobrie actae exem-plo, erogatisque in pauperes bonis, quem in theoreticis conceperat errorem, expiare voluisse. Potuisset tamen idem ille Stenonius eadem probitate apud nos esse, & propagato rerum naturalium medicarumque studio, ut ad quae na[106]tura factus videbatur, & honori Dei O[ptimi] M[aximi] & utilitati proximi servire." – Stensen's letters to Pechlin are presumably not preserved.

[84] E 148, p. 374, lines 29–30: "multa satis pungentia"; ibid., line 31: "dilectionem cum fratre"; Scherz, *Epistolae*, vol. 1, pp. 69–71, here 71: "non poteram prohibere". In the context of the decree on ecumenism *Unitatis redintegratio* (11/21/1964) of the Second Vatican Council see in detail Sobiech, *Herz, Gott, Kreuz*, p. 338.

[85] On Cohausen: Beauvois, *Cohausen*, pp. 26–27, 52–53; BEdtM 1 (2002), s.v. "Cohausen, Johann Heinrich", p. 108.

work *Clericus medicaster [...]* (The Cleric, a Bad Physician [...]; Frankfurt 1748), the most elaborate Catholic treatise of the time about medical activity by priests, that the office of the priest is incompatible with medical activity. At the end of the book, Cohausen mentions as examples of exceptions several medically trained and irreproachable priest-physicians who applied themselves to medical science.[86] He has this to say about Stensen: "I will now cultivate the glorious memory of the man who is famous in the republic of letters, and even around the entire world, the certainly very skillful physician in the first phase of his life, but in his second a priest and bishop worthy of the highest veneration. It is Niels Stensen or, as others call him, Stensen the famous Dane, whom the universities of Denmark, Holland, Tuscany and several others have seen as an extremely deft practitioner in medicine and anatomy; then the Roman Church venerated him, who had come to the righteous Faith by conversion, as priest, bishop of Titiopolis and vicar apostolic. How admirable his experience was in the art of dissecting is bespoken by the *Acta Danica* [= Bartholin's *Acta Hafniensia*] as well as by the books he published himself. But how much this man stood out in the pureness of his life, in the mortification of his body, in night vigils, fasting, in undissembled love, in the sacred teachings, in humility, in words and in every Christian perfection was known to everyone and everywhere from his public renown as long as he lived, as I can attest to myself from having been in Hildesheim and Hanover. This peerless man I therefore present as example to all priest-physicians. Never did he seek profit in the practice of medicine: He was not rich in property, not influential by wealth which the Prince of the Apostles prided himself in not having as well [cf. Acts 3:6a]. He valued nothing more than the augustness of priesthood; he gave everyone he associated with, the mundane persons as well as the clergymen, the nourishment of the sacred teachings, he gave salubrious advice, he gave fatherly comfort and the example of a holy life, when he, stepping into the footsteps of Christ and the apostles, abdicated completely from any endeavor for an influential position, from the love of worldly pleasures and from the resplendence of life."[87]

---

[86] Pompey, *Pastoralmedizin*, pp. 41–49 and a chapter overview of *Clericus medicaster* ibid., pp. 315–316. On the term "Medikaster" cf. Elkeles, "Medicus und Medikaster", p. 198.

[87] Cohausen, *Clericus medicaster* (Caput XI. De clericis veris ac probatis medicis discursus finalis), pp. 125–126: "Recolo nunc viri in republica litteraria imo orbe universo celebratissimi, prima quidem vitae statione expertissimi medici, altera vero sacerdotis & episcopi summe venerandi gloriosam memoriam. Est hic Nicolaus Stenonis, seu ut alii appellant, Stenonius nobilis Danus, quem universitates Daniae, Hollandiae, Hetruriae aliaeque plures viderunt in medicina & anatomia practicum dexterrimum, dein conversione ad fidem orthodoxam facta venerata est Romana Ecclesia sacerdotem, Episcopum Titiopolitanum, & Vicarium Apostolicum. Quam fuerit admiranda in arte prosecandi ejus experientia, & *Acta Danica*, & ab eo ipso in lucem dati libri loquuntur. Quantus autem vir in vitae integritate, in carnis mortificatione, in vigiliis, jejuniis, in charitate non ficta, in sancta doctrina, in humilitate, verbo, in omni Christiana perfectione extiterit, ex ejus dum adhuc viveret, fama publica omnibus & undique [126] innotuit, quod ipse testari possum, quando Hildesii & Hanoverae morabatur. Hunc igitur incomparabilem virum omnibus clericis medicis in exemplum statuo. Nunquam ex medicinae exercitio lucrum sectatus est: Non fuit dives opibus, non divitiis potens, quas & ipse apostolorum princeps se non habere gloriatur. Sublimitate sacerdotii nil judicavit majus; dedit omnibus, cum quibus est conversatus, tam secularibus quam

Pechlin's as well as Cohausen's descriptions, unanimous as they are in regard to Stensen' ethos and character, are like an illustration of what the Parisian medical practitioner, professor of anatomy at the Parisian *Collège de France* and *Jardin du roi* (from 1793: *Muséum national d'histoire naturelle*) and medical historian Dr. med. Antoine Portal (1742–1832),[88] emphasizes in the very detailed central article on Stensen in his bio-bibliographical work *Histoire de l'anatomie et de la chirurgie [...]* (Paris 1770), namely how "his knowledge, his enthusiasm for the sciences and the gentleness of his character" had allowed Stensen to make friends wherever he went. Portal goes on to say this: "Mr. Stensen was incidentally not of strong temperament, he possessed a tremendous sensitivity, which characterizes the refinement of his feelings. The death of Mr. Stensen aroused commotion within the learned world; the scientists had their eyes turned to him. His exceptional knowledge in anatomy and his not entirely expected conversion had to be epoch-making in the history of science."[89] More than 80 years after Stensen's death his conversion, his death and his personality were still exerting their influence.

3.4   In Public and in Politics

The "epoch-making" effect asserted by Portal can be understood quite concretely: Stensen's research results did not simply remain, as Cohausen put it, in the "republic of letters", but had a considerable influence on the "entire world" – first and foremost on Europe – , as has been shown with regard to the interest of the wider public: the women who approached André Graindorge in Caen in 1666 with their interest in Stensen's heart dissections, as well as the environs of the Pope in Rome, e.g. a copy of the Myology in Pope Clement IX's library.

The custodian of the Vatican Library Dr. iur. Stefano Gradi (*Croatian*: Stjepan Gradić) (1613–1683),[90] consultor to the Congregation of the Index from 1658, wrote the following to Prince Leopoldo de' Medici on June 18, 1667 during the conclave in Rome regarding Stensen's Myology, which he had also received: "Nothing could have seen the light of the world which would have satisfied my mind to a greater

---

ecclesiasticis, sacrae doctrinae pabulum, dedit salutare consilium, dedit paternam consolationem, & sanctae vitae exemplum, quando Christi & apostolorum vestigiis insistens omni opum studio, mundi deliciarum amore, vitae splendore se penitus abdicavit."

[88] On Portal: Pierre Chabbert, "Portal, Antoine"; in: DSB 11 (1975), pp. 99–100.

[89] Portal, *Histoire de l'anatomie*, pp. 159–183, here 159, 162: "Ses connoissances, son zele pour les Sciences, & la douceur de son caractere [...]. [162] M. Stenon n'étoit pas d'ailleurs d'un fort tempéramment, il étoit d'une sensibilité prodigieuse, ce qui marque la délicatesse de ses sentiments. La mort de M. Stenon fit du bruit dans l'univers savant, les Gens de Lettres avoient les yeux fixés sur lui. Son rare savoir en Anatomie & sa conversion peu attendue devoient faire époque dans l'histoire des Sciences".

[90] On Gradi: Tomaso Montanari, "Gradi (Gradič), Stefano"; in: DBI 58 (2002), pp. 361–363, here 362. Stensen was a guest of the *Bibliotheca Apostolica Vaticana*, see Sobiech, *Herz, Gott, Kreuz*, p. 114.

extent than the honor which your Highness has afforded me with the extremely appreciated gift of the anatomical work of Mr. Stensen, discussed and scientifically deliberated by me before it was published, and that now thereafter, with the same author [Stensen], I have the pleasure to get to know, to adore and to admire [it]."[91]

Frederik I, Duke of Saxe-Gotha-Altenburg ([* 1646] 1674–1691) reported in a note on May 12, 1669 in his hand-written journal of his educational journey to Italy, during which he came to know and appreciate Stensen, how on the occasion of an audience with the Florentine Grand Duke a "dähnischer Medicus oder Virtuosus" (Danish physician or virtuoso), Stensen, was admitted and gave a lecture about his geological research. Frederik was deeply impressed with Stensen and his talk, which resulted in the Gotha court and personal physician Dr. med. Daniel Ludwig (1625–1680) subsequently occupying himself with Stensen's theories as well.[92] Stensen, whose geological *Prodromus* had been published in April 1669,[93] thus indirectly contributed by way of his groundbreaking research – and his refutation of the theory of antiperistasis[94] – to the 'demythicization' of the nature of petrifaction, which most physicians considered a "joke of nature" (lusus naturae) mirroring supernal qualities under the surface of the Earth, and its unscientific use in medicine, e.g. against the plague and fevers.[95]

Two months later, in July 1669, the physician and natural scientist Dr. med. Johann Georg Volckamer Sr. (1616–1693),[96] founder of the Nuremberg *Societas Medica*, notified Stensen – who had apparently been in Southern Germany during the months of June and July – from Nuremberg that he was "wonderfully moved" by the benevolence that Stensen had shown him and his scientific colleagues during a visit including a public dissection.[97] Nothing more is known about this public dissection by Stensen.

As evidenced by Stensen's letter to the anatomist Heinrich Meibom Jr. on April 5, 1673 st.v., the latter was full of praise for Stensen in his lectures at the University of Helmstedt as well as mentioning Simon Paulli Jr. and Stensen in a "program" which he included in a letter to Paulli Jr. in Copenhagen.[98] This program points out a hitherto unknown and presumably coincidental parallel in the lives of Meibom and

---

[91] Mirto, "Lettere", pp. 380–381: "Non mi poteva giungere al mondo cosa che più contentasse il mio animo, quanto l'honore che mi fa V[ostra] A[ltezza] col suo preggiatissimo regalo dell'opera anatomica del Sig.r Stenonio, da me discorsa et disputata prima che uscisse alla luce, et hora dopo, col medesimo autore, che ho fortuna di conoscere, riverire et ammirare."

[92] Collet, *Welt in der Stube*, p. 182 with n. 549 (the *Accademia del Cimento* [sic!] had provided "den Rahmen dieser Vorführung", ie the setting for this demonstration); ibid., pp. 182–183. On Ludwig: Roob, "Ärzte", pp. 28–29.

[93] Scherz, *Geological Papers*, p. 25.

[94] See note 200, in Chap. 2 above.

[95] Collet, *Welt in der Stube*, pp. 183–187.

[96] Volckamer Sr. was a member of the *Academia Naturae Curiosorum* (est. 1652); on him: Renate Jürgensen/editorial, "Volckamer, Johann Georg"; in: Killy[2] 12 (2011), pp. 16–17.

[97] E 42, p. 209, lines 17–20, here 18: "mirifice[…] affectum".

[98] Bruun, "Fem nyfundne Niels Stensen-breve", p. 150, lines 14–15 & p. 153, lines 14–16, here 14–15: "programmate". Meibom's letter preceding the one by Stensen on 04/05/1673 st.v. as well

Stensen: at the beginning of 1673, Meibom had issued an invitation to the public dissection of a female corpse on Friday, February 28, 1673 st.v. at 10 a.m. – only 1 month after Stensen in his position as "Anatomicus regius" in Copenhagen – with the printed *Programma quo ad anatomen corporis foeminini in novo theatro primam omnium ordinum curiosos solemniter invitat* (Program with which [Professor Meibom] ceremoniously invites to the first dissection of a female body in the new [anatomical] theater the inquisitive of every standing).[99] Meibom had presumably sent the above-mentioned letter to Simon Paulli Jr. in Copenhagen several weeks after his own public dissection. Towards the end of the speech put down in the printed program, a brief foray through the history of anatomy, he discusses the anatomical theaters in other countries and mentions the Copenhagen anatomical theater, established by Paulli Jr. almost 30 years earlier (in 1645), which he cites as famous "due to the skill and experience of those who dissected there, [namely] Paulli, Bartholin, Stensen".[100] Stensen thanked Meibom for the "honorific mention", also in the name of Paulli Jr., which as Stensen emphasized he did with great pleasure, particularly in order to supplement that which was lacking in his own words through the name of the highly famous aged man alone.[101] What other statements in honor of Stensen were made by Meibom in his lectures can be guessed at if one bears in mind that in a letter to Stensen in Florence on March 15, 1670 he praised the latter's "excellence of talent, dexterity of the hand and other things" which, he added, he would rather admire and laud among friends and in public than to describe them in this letter.[102]

As late as 1676, while Stensen was a priest and tutor at the court of Cosimo III, there were efforts to entice him into assuming the vacant chair of anatomy at the Venetian University of Padua. Lorenzo Magalotti (1637–1712),[103] Florentine natural scientist and former secretary of the *Accademia del Cimento* who had praised the "unique modesty and the extremely agreeable character" of Stensen "with which he spices his teaching"[104] about 10 years earlier, made a note to himself on March 8,

---

as Meibom's letter sent to Paulli Jr. with the *Programma* are presumably not preserved, cf. ibid., p. 157n99 & pp. 161–162.

[99] Meibom, *Programma*, fol. A 1ʳ & A 4ᵛ; still unknown in Bruun, "Fem nyfundne Niels Stensen-breve", p. 162n115.

[100] Meibom, *Programma*, fol. A 3ᵛ–A 4ʳ, here A 4ʳ: "dexteritate & peritia eorum, qui ibi secuerunt Paulli, Bartholini, Stenonii".

[101] Bruun, "Fem nyfundne Niels Stensen-breve", p. 153, lines 15–20, here 15, 17: "honorificam […] mentionem".

[102] Ibid., p. 138, lines 18–20, here 18–19: "ingenij praestantia, manus solertia, et alia".

[103] On Magalotti: Cesare Preti/Luigi Matt, "Magalotti, Lorenzo"; in: DBI 67 (2006), pp. 300–304, here 301 on Stensen, claiming that Magalotti had visited Stensen in Denmark in 1674. That the two of them met during either of Magalotti's two sojourns in Copenhagen in 1674 is however unlikely according to Kardel/Maquet, *Biography and Original Papers*, p. 319n120.

[104] Letter to Alessandro Segni (1633–1697), Magalotti's predecessor as secretary of the *Accademia del Cimento*, on 08/24/1666 in Scherz, *Epistolae*, vol. 1, p. 25=Add. 4, p. 922, lines 22–23: "la singolar modestia e'l gentilissimo tratto col quale condisce la sua dottrina." On Segni: Mirto, "Segni", here p. 196 on Magalotti.

1676 in Vienna stating that on that evening the Ambassador of Venice – Francesco Michiel, ambassador from 1674 to 1677[105] – had asked him for a very personal opinion on Stensen's qualities and intimated that the University of Padua had an eye on Stensen with the intent of appointing him to the vacant chair of anatomy there. Magalotti's note continues to say that he headed off the ambassador's concern by telling him that the assignment had already been in preparation for 9 or 10 years at the grand-ducal court, but that Stensen was currently the hereditary prince's teacher.[106] Stensen's life subsequently took a different course.

Dr. med. Giovanni Maria Lancisi (1654–1720),[107] a member of several European scientific academies, taught at the Roman university *La Sapienza* from 1684 and from 1688 (with an interruption from 1689 to 1699) served as personal physician to three popes. Lancisi was at odds with himself as to whether to adhere to traditional medicine or lean towards modern experimental thinking and, among other undertakings, continued Stensen's heart research.[108] During a polemical controversy on the origin of scabies fought from August to October 1687 with the precocious 20-year-old physician Dr. med. Giovanni Cosimo Bonomo (1666–1696),[109] who had studied under Redi after receiving his doctorate in 1681 and presumably under Lancisi before then, and had discovered that scabies was of parasitic origin, Lancisi repeatedly prompted Bonomo to judge without passion in regard to this alleged discovery which of the two of them was advocating the truth, and urged him not to lose faith in the centuries-old traditions.[110] Lancisi, who cited "the sacred text in the Book of Deuteronomy" (the rules on clean and unclean food in Lev 11:4–8; Deut 14:4–8) and even a passage from the writings of Galilei in his own support,[111] also reproached Bonomo with the statement by the 26-year-old Stensen on the premature joy of anatomists from the latter's letter to Barbette: "Believe me, Mr. Bonomo, that that often happens which Niels Stensen wrote in a letter to Barbette when he told him that *like fear cheats the gullible, so does joy deceive the anatomists when some at first glance unfamiliar phenomenon which supports a preconceived opinion or provides opportunity for fresh consideration prompts them to shout Eureka*

---

[105] On Michiel: RDV 1 (1936), p. 548.

[106] Scherz, *Epistolae*, vol. 1, p. 25.

[107] On Lancisi: Stroppiana, "Lancisi", pp. 5, 12–13; Cesare Preti, "Lancisi, Giovanni Maria"; in: DBI 63 (2004), pp. 360–364. During the time when Stensen was working as "Anatomicus regius" in Copenhagen as well as thereafter, Lancisi – who had been awarded his doctor's degree on 09/12/1672 – was a student of the surgeon Giovanni Guglielmo Riva, whose vespertine academy meetings Stensen had attended in 1666 (see note 168, in Chap. 1 above), cf. ibid., p. 360.

[108] Stroppiana, "Lancisi", p. 10.

[109] On Bonomo: Giorgio Stabile, "Bonomo, Giovanni Cosimo"; in: DBI 12 (1970), pp. 338–341, here 338; Cesare Preti, "Lancisi, Giovanni Maria"; in: DBI 63 (2004), pp. 360–364, here 360–361.

[110] Faucci, *Polemica Bonomo-Lancisi*, p. 3 (letter on 12/10/1687), lines 8–9; ibid., p. 13 (letter on 09/20/1687), lines 34–35; ibid., p. 14 (the same letter), lines 3, 13, 35–36, 39–40; cf. Stroppiana, "Lancisi", p. 13.

[111] Faucci, *Polemica Bonomo-Lancisi*, pp. VII, 14 (letter on 09/201687), lines 19–21, here 20: "il sagro Testo nel Deuteronomio"; ibid., pp. VII, 21 (letter on 10/04/1687), lines 27–32.

repeatedly, while not allowing any time for more accurate examination. There are,
he continues, also other mental impulses which have made many men the authors of
outrageous treatises. Not that I think it necessary to cast doubt on your precise
observations, but that they do not represent all the possible ones, and that it is therefore necessary to provide further observations which restrict the assumed generality
of the own [observations]."[112] Redi, who had edited Bonomo's *Osservazioni intorno*
a'pellicelli del corpo umano [...] (Observations Regarding the Tiny Skin Worms of
the Human Body [...]; Florence 1687) and sent it to be printed but was likewise not
convinced by the latter's ideas, found Bonomo's style to be less than polite and his
passionate peremptoriness irreconcilable with academic decorum.[113]

One can only guess how Stensen, who had passed away a few months earlier,
would have evaluated Bonomo's microscopic epidemiological research – and in
particular, whether due to his own experience he would have recognized and appreciated Bonomo's treatise for its scientific value. Around 28 years later, at any rate,
Lancisi held a speech on the occasion of the inauguration of the *Accademia*
Lancisiana, an academy for youths studying medicine, in the rooms of the *Ospedale*
di Santo Spirito in Sassia on April 25, 1715 in the presence of Pope Clement XI ([*
1649] 1700–1721) and several cardinals. In this address, he made reference to various propositions of Galilei and Descartes in order to explain the significance and
boundaries of geometry for medicine[114] – a genuine topic of Stensen's, who had
spoken in Florence in 1669 of the "most solid demonstrations of the great Galilei".[115]

A typical product of the eighteenth century so rich in encyclopedias, thesauruses
and dictionaries is the edition of the *Dictionnaire historique-portatif [...]* (2 vols.,
Paris 1752) by Dr. theol. Jean-Baptiste Ladvocat (1709–1765) translated, corrected
and amended by the cleric regular Antonio Maria (De) Lugo CRS (c. 1715–1778),[116]
counted among the members of the *Accademia Pontificia Della Storia Romana*
(Pontifical Academy of Roman History; est. 1740) by Pope Benedict XIV. Published
under the title *Dizionario storico portatile [...]* (4 vols., Naples 1754–1755), this

---

[112] Ibid., p. 12 (letter on 09/20/1687), lines 35–44: "Mi creda sig.r Bonomo, che spesso succede ciò,
che scrisse Nicolò Stenone à Barbetta in una letera, dicendoli, che *ut credulos fallit terror, sic*
anatomicis imponit laetitia; dum visum quoddam prima specie insolitum, conceptae opinioni
favens vel novae meditationi ansam porrigens ad Ευρηκα ingeminandos illos propellit, nullo ad
accuratius examen praemittendum concesso loco. Sunt, segue egli a dire, *et alij animi motus qui*
multos monstrosoru[m] scriptoru[m] reddiderunt parentes. Non già, ch'io stimi doversi porre in
dubio le di lei acurate sperienze, ma che esse non siano tutte le possibili, e che vaglino per ciò darsi
altrui osservationi, che limitino la concepita generalità delle proprie." The passage cited by Lancisi
can be found in VitTrans., p. 212, lines 2–8.

[113] Bonomo, *Osservazioni* and on this Giorgio Stabile, "Bonomo, Giovanni Cosimo"; in: DBI 12
(1970), pp. 338–341, here 339–340.

[114] Lancisi, *De recta ratione*, pp. 14–15 (Galilei), 16–17 (Descartes). On the opening of the academy cf. Stroppiana, "Lancisi", p. 10; Cesare Preti, "Lancisi, Giovanni Maria"; in: DBI 63 (2004),
pp. 360–364, here 363. On Lancisi's speech in the context of corpuscular theories see Donato,
"Onere", pp. 85–86.

[115] Prodromus, p. 210, line 23 = Scherz, *Geological Papers*, p. 184, lines 31–32: "solidissimae
Magni Galilei demonstrationes"; on this, see also Scherz, *Geological Papers*, p. 229n105.

[116] On Lugo: AGCRS, Biografie C. R. S., n. 708, pp. 5, 7–8.

edition represented De Lugo's most-consulted work.[117] The entry on Stensen cites his Parisian *Discours* and Jacques-Bénigne Winslow, who reprinted it, with the words: "There exists by him [= Stensen] an excellent lecture on the anatomy of the brain and other learned writings. Mr. Winslow, his last nephew and famous anatomist, laudably maintained the reputation of this studied man."[118] Not much can be concluded about the degree of familiarity of a person from dictionary entries, however.

3.5 In the Age of Technology and Memory of the Church

The Jewish anatomist, surgeon and zoologist Ludvig Levin Jacobson (1783–1843) of Copenhagen, who was awarded an honorary degree from the University of Kiel in 1815 and the title of professor by King Frederik VI of Denmark ([* 1768] 1808–1839) in 1816 against the opposition of the University of Copenhagen, which was under the influence of the Lutheran state church, made this note for the presentation of his discovery of the so-called Jacobson's organ, a sensory organ found in many vertebrates, to the Danish *Videnskabernes Selskab* (Society of Sciences; est. 1742): "It has been over one and a half centuries now since our great Stensen, that assiduous and precise anatomist, enriched science with several important discoveries. Among them we already find a part of the organ that I wish to describe. The negligence with which all his discoveries have been treated affords the admirers of this science little honor: I am therefore doubly pleased that I, as a Dane, am able to continue an examination which had begun one and a half centuries ago by a Danish anatomist."[119] Jacobsen's criticism is also an indication that Stensen's work no longer seemed important even though his name was still being mentioned in medical publications.

Beginning with the 1870s, however, first signs of a Stensen-'Renaissance' in medicine became visible, with interest being taken in Stensen's life and work from a new perspective and more comprehensibly than before in an age of the booming

---

[117] On Ladvocat: Hugues Jean de Dianoux, "Ladvocat (Jean-Baptiste)"; in: DBF 19 (2001), col. 92.

[118] DSP 4 (1755), s.v. "Stenone (Nicolao)", p. 381, here r. col.: "Avvi un suo eccell[ente] discorso sopra l'Anatomia del cervello ed altre Op[ere] dotte. il Sig. Wenslow suo ultimo nipote, e cel[ebre] Anatomico, sostenne con gloria la riputaz[ione] di questo dotto Uomo."

[119] KU-BVFB, Hdskr. 2:II:40 a, pp. 5–7: "Over halvandet Aarhundrede, er det nu siden vor store Steno, denne ufortrödne og nöjagtige Anatom, berigede Vi[6]denskaben med flere vigtige Opdagelser. Blandt dem finde vi allerede en Deel af det Organ jeg vil beskrive […]. Den Skiödeslöshed, hvormed alle hans Opdagelser ere blevne behandlede, giöre Dyrkerne af denne Videnskab kun liden Ære: […] [7] […] Det er mig derfor dobbelt kiert, at jeg som dansk kan forsætte en Undersögelse, der allerede for halvandet Aarhundrede har været begyndt af en dansk Anatom." The ink of the manuscript is very faded. Cf. the transcription in Hollnagel-Jensen/ Andreasen, "Jacobson", p. 13; cf. also Melchiors, "Entdeckung". On Jacobson: Otto Carl Aagaard/ Ragnar Spärck, "Jacobson, Ludvig Levin"; in: DBL³ 7 (1981), pp. 207–208, here 207 r. col. on Stensen.

of natural science and technology in medicine, and thus in greater temporal distance to Stensen's century. Almost 100 years after Jacobsen's critical remark, Vilhelm E. Maar finally laid the foundation for medico-historical research into Stensen's publications in 1910 with a new edition which for the first time united all of the latter's writings from the field of natural science.

The twentieth century,[120] with the Stensen anniversaries in 1936, 1938, 1986 and 1988, the translation of Stensen's exhumed mortal remains from the crypt to the "Cappella Maria Santissima, detta 'la Ben Tornata'" (today's "Cappella Stenoniana") in the transept of the Florentine Basilica of San Lorenzo on the Feast of Christ the King on October 25, 1953, and, after initial efforts towards beatification in 1688/1700 and around 1800, the beatification process (1955–1988),[121] saw a new flourishing of the commemoration of the anatomist Niels Stensen. This was particularly promoted in the international medical community by the endeavors of the Redemptorist Fr. Gustav Scherz, born in Vienna and active in Copenhagen, and remained vibrant in two physicians' associations after Scherz' death.[122]

Dr. med. Dr. phil. Dr. iur. James J. Walsh (1865–1942), professor of neurology at the *School of Medicine* at *Fordham University* (New York) and author of, among others, books about the relationship of the Church with the natural sciences, e.g. the three-volume work *Catholic Churchmen in Science* (Philadelphia 1906/1909/1917), was convinced that one could learn to be less self-righteous from the study of medical history.[123] To Walsh, Stensen was "a man whose career of distinction in science was to prove that there was no opposition in ecclesiastical circles in Italy, during this century, to the development of natural science even in departments in respect to which the Church has, over and over again, been said to be particularly intolerant."[124]

---

[120] The treatment of Stensen in scientific and news literature against the background of the intellectual and ideological changes in the twentieth century, the two World Wars and the resulting disruptions in international exchange with regard to the question of medical ethics would be worthy of a separate study, eg with reference to the Stensen anniversaries in 1936, 1938, 1986 and 1988, and the translation of 1953.

[121] Sobiech, *Herz, Gott, Kreuz*, pp. 11–15. – See also the brief biography of Stensen published 10 days prior to his beatification (10/23/1988) in Schmitz, "Stensen".

[122] These were: 1) the "Niels-Stensen-Symposium", an association of Danish and North German dermatologists, est. 1975 by the Copenhagen professor of dermatology Dr. med. Niels Hjorth (1919–1990) for reestablishment of scientific contact. Hjorth served as president of the association together with Carl Schirren (according to information by telephone [07/04/2011] from Prof. em. Dr. Carl Schirren, it no longer exists; on Hjorth: Gunnar Lomholt, "Obituary: Niels Hjorth [1919–1990] "; in: IJD 30,12 [1991], p. 898; Carl Schirren, "Professor Dr. Niels Hjorth"; in: DermK 32,115 [March 1991], pp. 4–5), and 2) the "Niels-Stensen-Gemeinschaft e. V. (Ärzte, Apotheker, Pflegeberufe, Seelsorger im Gespräch)", est. 1982 by Prelate Dr. Stanis-Edmund Szydzik (1915–2001) and existing until 2000, see Estermann, *Szydzik*, pp. 91–92.

[123] Kirwin, "Walsh", p. 427.

[124] Walsh, *Churchmen in Science*, p. 138. – The Canadian-American multiple doctor and professor of anatomy and anthropology Sir Bertram C. A. Windle (1858–1929) (McCorkell, Windle, p. 54), who converted from Anglicanism to Catholicism in 1883, points out in his book *Twelve Catholic Men of Science* (London 1912), in which he draws on the works of Walsh in particular, that Stensen had been sent out as bishop by the Church with his groundbreaking research brought forth by him without inner reserve (Windle, "Stensen", p. 22).

Walsh's statement is corroborated by the fact that Rome and Italy offered far greater leeway in terms of freedom of thought and press than other European countries,[125] which Stensen likewise noticed during his lifetime.[126]

That the question of God arises in medicine – even though it has to be excluded methodically – was pointed out by the histologist Dr. med. Carla Freiin von Zawisch-Ossenitz (1888–1961), first female professor at the University of Graz and co-founder of the Austrian Catholic physicians' association "St. Lukas-Gilde" (Guild of St. Luke, est. 1932). Zawisch-Ossenitz had emigrated until 1946 after her ejection from academic service in 1938 following the occupation of Austria.[127] As dean of the Institute of Histology and Embryology at the University of Graz from 1947, she wrote to Fr. Scherz from Graz on July 10, 1953: "To once again see a truly brilliant person in the light of holiness can mean a lot in our time, especially for the scientist who almost always – consciously or unconsciously – in some way is grappling with and for God."[128]

A more personal testimony comes from Prof. Dr. med. Nevio Quattrin (1910–1993),[129] a Neapolitan hematologist who corresponded with Fr. Scherz. On September 30, 1959 Quattrin wrote to Scherz (Fig. 3.1): "Last Easter I had a brief audience with His Holiness [Pope John XXXIII ([* 1881] 1958–1963)][130] and also presented to him the publication on Stensen,[131] asking for him [Stensen] to be canonized as soon as possible. In spring I am invited to the University of Padua to talk about him on the occasion of the 300th anniversary of the discovery of the *Ductus parotideus*.[132] I greet you cordially in the Lord, your most devoted Nevio Quattrin."[133] In a personal testimony published in the *Stenoniana Catholica* in 1961, Quattrin explained for the first time how the human as well as the religious side of Stensen appealed to him; he was particularly fascinated and moved, however, by Stensen's moral purity[134] in body and soul, admired already by Redi and retained by Stensen despite his urbaneness, which to Quattrin was like a star lighting the way for physicians in a world characterized by dishonesty and corruption.[135] He affirmed

---

[125] Gross, *Rome*, p. 248; see in more detail Sella, *Italy*, pp. 144–160.

[126] See note 510, in Chap. 2 above.

[127] Horn/Dorffner, "Frauen", pp. 126–127; Kernbauer, "Zawisch-Ossenitz", pp. 266–268.

[128] KB, Tilg. 621 II 8 (Østrig).

[129] On Quattrin: AOV, [two vitae]; [editorial], "Morto Quattrin, padre dell'ematologia"; in: GVicenza, 03/17/1993 (with image), with thanks to Prof. Dr. Mariano Nardello (Vicenza).

[130] In Quattrin, *Stenone*, pp. 151–152, Quattrin reports on this audience.

[131] It is unclear which publication Quattrin refers to.

[132] Cf. on this lecture and on other conferences Quattrin, *Stenone*, p. 155.

[133] KB, Tilg. 621 II 12 (Italien til 1962 M–Z): "[…] Nella Pasqua scorsa ebbi una breve udienza con Sua Santità e Gli consegnai anche la pubblicazione su Stenone domandando che fosse fatto Santo al più presto. […] In primavera sono invitato all'Università di Parma per parlare su di Lui in occasione del 300.mo anniversario della scoperta del dotto parotideo. […] La saluto con ogni affetto nel Signore Suo dev[otissi]mo N[evio] Quattrin".

[134] See notes 43 & 44 above.

[135] Quattrin, *Stenone*, pp. 95, 97. The accompanying letter sent by Quattrin to Fr. Scherz from Naples on 11/02/1961 together with the manuscript (not preserved) intended for printing in the *Stenoniana Catholica* can be found in KB, Tilg. 621 II 12 (Italien til 1962 M–Z).

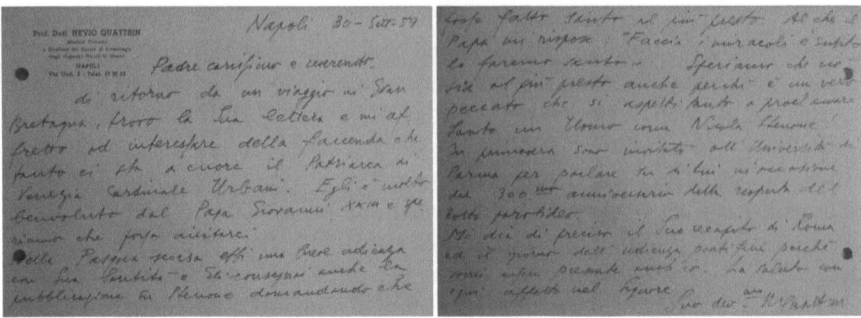

Fig. 3.1 Card sent by the Neapolitan hematologist Nevio Quattrin (Naples, September 30, 1959) to Fr. Gustav Scherz CSsR in Copenhagen
KB, Tilg. 621 II 12 (Italien til 1962 M–Z)

this opinion in his biography of Stensen published in 1987.[136] That Quattrin's thoughts are still influential in Italy is substantiated by a contribution in the Italian *Giornale della Previdenza dei Medici e degli Odontoiatri* (Journal of Prevention for Physicians and Dentists) from 2008 in which Stensen is identified as "one of the few truly 'blessed' of our profession".[137]

Of great importance for the medico-historical research on Stensen is the professor of anatomy Dr. med. Adolf Faller (1913–1989),[138] director of the Institute of Anatomy at the Faculty of Mathematics and Natural Sciences at the University of Fribourg (Switzerland). According to Günter Rager, professor of anatomy and embryology and director of the Institute of Anatomy and Special Embryology, Faller conducted work in the dissecting room, which as "proof of the reverence which the Christian shows for the human corpse" has been adorned with a large crucifix by Marcel Feuillat (1896–1962) since the founding of the institute, "with the highest discipline and almost religious respect".[139] Next to the crucifix is an engraved marble plaque affixed by Rager on the occasion of the 50-year anniversary of anatomy in Fribourg in 1988 and the beatification of Stensen in the same year, which replaces an inscription on the wall of the same words presumably placed

---

[136] Quattrin, *Stenone*, pp. 154–157, here 155 with reference to Quattrin, "Presenza".

[137] Sterpellone, "Stenone", p. 45: "Tra i pochi veri 'beati' della nostra categoria."

[138] On Faller: Institut für Anatomie und spezielle Embryologie der Universität Freiburg (Schweiz), *Nachtrag*, p. 38 (vita in tabular form); Günter Rager, "Adolf Faller (1913–1989)"; in: Acta anat. 137 (1990), p. 279; extended in Rager, "Adolf Faller: Anatom und Medizinhistoriker 1913–1989"; in: Gesnerus 47 (1990), pp. 118–121, here 119 on Stensen; Rager, "Gründung", pp. 20–22; Ruffieux, *Personnes*, s.v. "Faller, Adolf", p. 944; Pierre Sprumont, "A propos du décès d'un de nos membres d'honneur: Adolf Faller, un idéaliste obstiné"; in: BSFSN 78 (1989), pp. 51–55, here 53 on Faller's works on Stensen. A collection of Faller's publications along with a bibliographical overview (up to 1982/1983) as well as the rest of his library are located in: UniFR-DepMed.

[139] Rager, "Institut", pp. 859–861 with citations on pp. 861, 859. On the crucifix cf. Faller, *Abenteuer*, p. 17.

there by Faller before June 21, 1978.[140] The words are a quotation from Stensen's inaugural address in Copenhagen concerning the goal of anatomy (Fig. 1.5). Rager also ends his overview of the institute's history published in celebration of its anniversary with the same quote.[141] Thus both the crucifix and the quote allude strikingly to anatomy and the anatomist as "pointer in the hand of God" (radius in manu Dei).

The Danish nuclear physicist Dr. phil. Niels Bohr (1885–1962),[142] professor of theoretical physics at the University of Copenhagen from 1916, emphasized in his paper *Physical Science and the Problem of Life*, published in 1957 and based on his *Steno-Forelæsning* (Stensen Lecture) held at the *Dansk Medicinsk Selskab* (Danish Medical Society; est. 1919) in Copenhagen in February 1949,[143] that Stensen's Parisian *Discours*, in which he had highlighted the insufficiency of contemporary knowledge, was proof of the "great observational power and open-mindedness" characteristic of Stensen's entire body of scientific work.[144] Despite the fundamentally different epistemology in physics as compared with Stensen's time, his "striving for beauty and harmony" remained a corrective for it.[145] That Bohr was substantially involved in the initiation of the monument to Stensen in Denmark's capital (Fig. 1.2), which was eventually built in 1963, is evidenced by a letter sent to him on October 4, 1953 by Gustav Scherz from Copenhagen: "I grasp the opportunity to thank and congratulate you for your initiative regarding the monument to Stensen in Copenhagen."[146] On September 28, 1958, Bohr assured Fr. Scherz that the latter knew Bohr's "sincere admiration for Niels Stensen".[147] Stanley L. Jaki OSB (1924–2009), priest and physicist, likewise emphasized with reference to Bohr's appraisal the forward-looking significance of Stensen's statements in the *Discours*.[148]

By contrast, the Regensburg theologian and religious scholar Norbert Schiffers (1927–1988) asserted "that it is not possible for a natural scientist of today to take a point of view outside his science as unbiasedly as a Niels Stensen in order to observe from there the harmony of his own scientific fabric", as today's scientist had to

---

[140] According to an e-mail (11/06/2012) by Prof. em. Günter Rager; 06/21/1978 as terminus ante quem due to the mention in Faller, *Abenteuer*, p. 17.

[141] Rager, "Gründung", p. 31; the fresco of Stensen, created at no cost by Oscar Cattani (1887–1960), is located in the lecture hall of the institute, cf. Giovanni, "Projets", p. 873 (color photograph).

[142] On Bohr: Jens Rud Nielson [to be corrected into: Nielsen], "Memories of Niels Bohr"; in: PhT 16,10 (1963), pp. 22–30.

[143] On Bohr's manuscripts cf. Favrholdt, *Complementarity beyond Physics*, esp. p. [586], no. 29 & p. [590], no. 49.

[144] Bohr, "Problem of Life", p. [117].

[145] Ibid., p. [123].

[146] NBA, Niels Bohr General Correspondence, 1910–1962, Folder 27: "Jeg benytter Lejligheden til at takke og lykønske Dem til Deres Initiativ vedrørende Stensens Monument i København."

[147] Ibid.: "oprigtige beundring for Niels Stensen".

[148] Jaki, *Brain*, pp. 120–121. On Jaki: Paul Haffner, "Jaki, Stanley L."; in: NCE[2] Supplement 2010, vol. 2, pp. 576–578.

beware an ideologization of his science.[149] Due to Stensen's "longing for the ascertainment of harmony in the diversity of life", which "could only satisfy itself through the ultimately all-governing inclusion of belief in a God who held together this multiplicity", the "critical and differentiating examination of our means of communicating about reality" fell short, wherefore Niels Bohr was revering a great compatriot "with whom he shared the longing for harmony, but not the path to cognition of that harmony".[150] The question of whether Schiffers hits the mark in regard to Bohr's position in philosophy of science shall be left to research into the history of physics. That Bohr was simply 'revering' a great compatriot is at any rate highly questionable; what is clear, however, is that Schiffers' interpretation does not do Stensen justice for two reasons:

1. because Stensen *methodically* separated scientific method from Faith, exemplified by his discovery of and research on the ovaries, and
2. in terms of philosophy, the beginning of which – according to Plato – is amazement; and this is precisely what Stensen had expressed in his publications prior to his conversion.

Dietmar Mieth should therefore be endorsed in counting Stensen, the "bishop and expert in anatomy", among those theologians who include e.g. Albertus Magnus, and who in an alliance of natural science and the Christian Faith – as pointed out, according to Mieth, in the encyclical *Fides et ratio* (September 14, 1998) by Pope John Paul II ([* 1920] 1978–2005) – have contributed to the elimination of prejudices.[151]

In the English part of his sermon on October 23, 1988 during the beatification ceremony in Rome, Pope John Paul II emphasized Stensen's "acute powers of observation and his calm objectivity",[152] and in the following German part addressed "the present scientists and researchers, wherever and in whichever area they are active".[153] James Tait Goodrich of the New York *Albert Einstein College of Medicine, Yeshiva University* muses as follows in regard to Stensen's beatification in his commentary published in the periodical *Neurosurgery* in 2010: "With all this wonderful research, amazingly sanguine ideas, and a natural skill at scientific investigation one wonders why he ended up an itinerant traveling titular bishop in poverty trying to bring Catholicism to the heathens of Europe. Even more amazing is that a recent pope considered his efforts so important he has recently become beatified and is now recognized as our patron saint of scientists."[154]

---

[149] Schiffers, *Fragen der Physik*, p. 128; similar to Schiffers later in Thielicke, *Anthropology*, p. 181. On Schiffers: RGG⁴ index vol. (2007), s.v. "Schiffers, Norbert", col. 351.

[150] Schiffers, *Fragen der Physik*, pp. 167, 127 (with reference to Stensen's belief in providence).

[151] Mieth, "Wissenschaft", pp. 405–406, here 406; cf. Mieth, "Science", p. 61: "bishop *and* anatomist".

[152] AAS 81 (1989), pp. 290–296, here 292.

[153] Ibid., p. 294. On the Christian calling of the scientist from the point of view of Pope John Paul II see Austriaco, *Biomedicine and Beatitude*, pp. 208–210.

[154] Goodrich, [*Comment*].

An article published in the periodical *Aesthetic Plastic Surgery*, co-authored by Dr. med. Bernard E. Simon (1912–1999), a plastic surgeon at *Mount Sinai Hospital* in New York who provided his surgical services free of charge to 25 maimed Japanese women with disfigured faces resulting from the nuclear bombing of Hiroshima,[155] states that despite the "[architecturally] isolated and poorly decorated chapel of the large church", i.e. the "Cappella Stenoniana" in the Basilica of San Lorenzo in Florence,[156] there was "a pilgrimage of faithful academics who visit his remains, and above all young medical students who turn to him to obtain grace and fervor for their studies, leaving behind small notes and flowers with their requests to be fulfilled."[157] This is still true today (Fig. 3.2). Marco Marzollo wrote in 1967 that, in the event of Stensen's beatification, he would be "one of the first to feel compelled to carry a bouquet of white lilies to his tomb in San Lorenzo."[158] The altar stipes of the Lady Chapel inside the Crypt of Bishops at the Osnabrück Cathedral of St. Peter houses a foot relic of Stensen gifted to the bishop of Osnabrück, Titular Archbishop Dr. theol. Wilhelm Berning ([* 1877] 1914–1955) in 1955 by Msgr. Giuseppe Capretti (1899–1973), "Priore mitrato di San Lorenzo" (Mitred Prior of San Lorenzo) since 1948.[159] Accompanying the gift was Capretti's wish that the relic be made accessible for veneration by the faithful following the hoped for beatification of Stensen; this wish was eventually granted.[160]

Newer English-language overview articles in various medical periodicals betoken Stensen as a "saint" – the correct designation would be "blessed" – and as a "bishop".[161] This also shows a great appreciation for his life and achievements.

That eponyms – part of the system of rewards in science – ,[162] and hence the eponym *Ductus Stenonis*, have been disestablished in anatomy is regrettable since it causes the human reference to be lost, as emphasized by the Hamburg endocrinologist

---

[155] On Simon: William H. Honan, "B. E. Simon, 'Hiroshima Maidens' Surgeon, 87"; in: NYT-BS 30,8 (August 1999), p. 1230 (NY Times from 08/05/1999). The overview on the use of the nuclear bomb in Ludwig E. Feinendegen, "Hiroshima/Nagasaki"; in: LBioeth. 2 (1998), pp. 218–220 lacks ethical reflection.

[156] On this, see Bierbaum/Faller/Traeger, *Stensen*, p. 163.

[157] Grappolini/Signorini/Simon, "Stenone", p. 96.

[158] Marzollo, "Musculo", pp. 3–4 ("Prefazione" from 11/30/1967), here 4: "uno dei primi che si sentirà spinto a portare un fascio di candidi gigli sulla Sua tomba in San Lorenzo."

[159] I.e. with episcopal insignia. On Capretti: BSLF, In morte […], p. [8].

[160] The letter by Capretti to Berning from 07/17/1955 accompanying the relic can be found in BAOS, 01-89-10-02. In the eulogy during the funeral for Capretti on 02/26/1973 in the Basilica of San Lorenzo, his great dedication to Stensen's beatification is emphasized (BSLF, In morte […], p. [5]). On Berning: Klemens-August Recker/Wolfgang Seegrün, "Berning, Hermann Wilhelm"; in: BDL II, pp. 422–427.

[161] See note 51, in Chap. 1 above; cf. the reference to Stensen's beatification in the French overview essay Van Besien/Van Besien, "Stenon", pp. 78, 81.

[162] Merton, "Priorities", p. 459.

Fig. 3.2 Niels Stensen's marble sarcophagus (early Christian, fourth century) in the "Cappella Stenoniana", a side chapel to the north of the nave of the Basilica of San Lorenzo in Florence
It is always covered with many sheets of paper (ca. 12,000 from 1990 to 2006 alone) with which pilgrims from throughout the world pray for Stensen's intercession with God for different requests, cf. Sobiech, Capella Stenoniana

Gerhard Bettendorf (1926–2009).[163] According to Bettendorf, scientific progress is not a linear development; rather, it is always people who produce advancements.[164]

In the same vein, namely pertaining to the human aspect of all science, are the words of the internist and endocrinologist Hendrik Lehnert, who ended his speech "Internal Medicine Between Zeitgeist and Changing Times", held on May 1, 2011 at the 117th Congress of the *Deutsche Gesellschaft für Innere Medizin* (German Society of Internal Medicine; Wiesbaden, April 30–May 4, 2011), with the statement that "our forward striving" on the way to a human society of knowledge, characterized by the fact "that while with the help of science we can accumulate ever more knowledge, we can nevertheless *not* know for certain what we will know in the future", had been described "hardly ever better" than in Stensen's well-known motto – "Beautiful is that which is seen, more beautiful that which is known, most beautiful by far that which is not known".[165]

---

[163] EndokrG, p. VII. On Bettendorf: Meinert Breckwoldt/Wilhelm Braendle/Herbert Kuhl, "Nachruf auf Herrn Prof. Dr. med. Gerhard Bettendorf * 04.05.1926 † 20.04.2009"; in: EndokrI 33 (2009), p. 67.

[164] EndokrG, p. 633.

[165] Lehnert, *Innere Medizin*, pp. 23–24 with citations on p. 24. Lehnert quotes the third element of the motto thus: "[…] bei weitem am schönsten jedoch das, was wir nicht wissen" ([…] by far the most beautiful that which we do not know).

Chapter 4
Niels Stensen's Lasting Significance

In what way can the Danish anatomist Niels Stensen contribute to encouraging medical research as well as professional medical practice in their ethos of serving and supporting human life? This question shall be examined in closing with the help of selected aspects of Stensen's own ethos and bioethics.

The anatomist Stensen was characterized in his research by high levels of autonomy and judgment that avoided prejudice, as well as of self-reflection and awareness for philosophy of science, especially concerning the error-proneness of scientific observations – those of others as much as his own. Stensen's excellent capability for reflection is expressed not least by the fact that the continuing succession of his own discoveries surprised him and retained in his eyes some remnants of inexplicability. Out of humility he did not query the reasons for this, as he knew they would remain hidden to him and he also saw the danger of ascribing to himself something which he owed to another: God.[1]

As scientist and as bishop, Stensen saw the benefit of science conducted lege artis and held, as he mentioned in the *Discours*, an overly high opinion of the "learned men" (hommes de lettres) in general, which is mirrored in his respect for their scientific accomplishments. In regard to the consequences of medical science, Stensen was optimistic since for him, as he explained as "Anatomicus regius" in Copenhagen in 1673, the acquisition of knowledge in medicine, if the latter was conducted "with relatively low expectations and a relatively complete suppression of prejudice",[2] promised a greater chance of healing. He was nevertheless aware that despite the best efforts, various disruptive factors originating from the person of the researcher himself, among them his own advancement, had to be reckoned with in the course of scientific examinations; this again evidences Stensen's extraordinary capability for reflection.

---

[1] Prodromus, p. 184, line 37–p. 185, line 11 = Scherz, *Geological Papers*, p. 138, lines 15–25.

[2] Prooemium, p. 255, lines 5–7, here 6–7: "minori cum praesumptione & pleniori exstirpatione praejudiciorum".

© Springer International Publishing Switzerland 2016
F. Sobiech, *Ethos, Bioethics, and Sexual Ethics in Work and Reception
of the Anatomist Niels Stensen (1638-1686)*, Philosophy and Medicine 117,
DOI 10.1007/978-3-319-32912-3_4

Although Stensen was dependent on the goodwill of sponsors after the end of his studies in 1663/1664 due to his parents' modest financial endowment, and even found himself in a precarious situation concerning his research until spring 1666, he remained comparatively autarkic in the organization of his activities and the selection of his research topics even after his engagement at the Medici court. Until he was 37, at which age he was ordained to the priesthood, he did not lead an easy or long-term materially secure life, but a highly self-determined life filled entirely by tasks he chose for himself. This was owed in part to the fact that his great scientific as well as interpersonal aptitude allowed him to 'open many doors' during his travels. The Grand Duke of Florence, for example, granted him complete liberty in the choice of his topics of research. As Stensen was apparently still used to conducting his own research in the evening, and thus going to bed late, from his study days, he stayed in bed for some time after waking up in the morning, occasionally slept until 9 a.m. and even then did not arise immediately – something he reflected on critically as a bishop at age 45 in his Hamburg confession notes.[3] This relative freedom in his choice of topics and time management was likely pivotal for his decision in 1675 to be ordained a secular priest as opposed to joining a religious order – although he had considered various ones – , a fact he likewise contemplated in the aforementioned notes.[4] Concerns about strict religious discipline and possibly about the inevitable novitiate had apparently played a role here.[5] That Stensen often slept while travelling at night during his visitation of the Münster diocese as suffragan bishop – in order to gain more time for his daytime pastoral activities – was perhaps a remnant of his earlier, more independent time management.[6]

It is admirable, Remacle Rome emphasizes, how Stensen did not lose track of his research despite "a certain disorderliness in his work"[7] owing to his frequent changes in location and the challenges posed to him in these various places, instead founding the field of geology as the history of the Earth and thus, in this area, the science of evolution. This "innovation"[8] in geology only began to become accepted during the first half of the nineteenth century, however. Rome sums up as follows: "A question that he addresses is rarely thoroughly examined; he leaves it in order to cast his eyes on another which has struck him during his dissection. The discoveries which are rather close to being ingenious are found noted down in a few lines, sometimes in a single line, among others of lesser importance. It is not astonishing that certain ones among them would remain unnoticed, that other scientists would

---

[3] Op. 4, p. 443, lines 30–32; see also Sobiech, *Herz, Gott, Kreuz*, pp. 86n338 & pp. 106–107. – Descartes, who was sickly as a child like Stensen, slept ten hours a day at age 35 – which he recommended – and used the time after waking up for meditation; his lay medical self-observation from the age of 40, however, aimed at ensuring as long a life as possible (Shapin, "Descartes the Doctor", pp. 139–140).

[4] In detail in Op. 4, p. 437, lines 9–10.

[5] This can be surmised from E 292, p. 580, lines 2–6.

[6] Add. 43, p. 985, lines 6–8; in more detail in Sobiech, *Herz, Gott, Kreuz*, p. 86.

[7] Rome, "Sténon", p. 537: "un certain désordre dans son travail".

[8] Prodromus, p. 220, line 37 = Scherz, *Geological Papers*, p. 204, line 9: "novitate".

make them again after him, and that they would bona fide believe to be the first to have made them."[9] Stensen's work sometimes suffered delays due to his agile and rambling intellect: similarly to the events in Amsterdam, where Blasius' mistakes were his source of new discoveries, Stensen apologized to Thomas Bartholin from Leiden on January 9, 1662 for having been sidetracked by his own dissections during which each casual observation resulted in another.[10] Robert K. Merton cites Stensen as proof that multiple variegated discoveries are typical for outstanding talents. Even Stensen's contemporaries noticed his extraordinary knack for discovery and research which barely afforded him the time to properly and comprehensively appreciate his individual findings.[11] The Amsterdam practicing physician Dr. med. Justus Schrader (1646–after 1720) writes in his *Observationes et historiae [...]* (Observations and Histories [...]; Amsterdam 1674) that Stensen had mentioned his discovery that the testes of women were "analogous" to the ovary "quasi in passing" in his treatise.[12] Stensen's anatomical writings are in a way timeless in their succinct brevity and linguistic pithiness.

Using the new methods of his time – albeit in often unconventional approaches – Stensen conducted basic research as an anatomist. On the one hand, this was due to the fact that it conformed to the natural inclinations he had come to feel in the course of his studies – fueled by the arising necessity to defend his scientific discoveries as well as by the publication of Descartes' *De homine*. On the other hand, he did so also because he was convinced that only precise knowledge of the anatomy and physiology of the body would allow the discovery of appropriate remedies for diseases. Simultaneously, Stensen warned against the overestimation of human cognition and hence, in modern terminology, against 'scientific faith' resp. scientism. That he attempted to keep his observations free of unproven and metaphysical assertions, and withheld initial judgment in uncertain questions, even in regard to theology, is a sign of his brilliancy.

With regard to his profession, Stensen had decided – as the result of an internal transformative process lasting several years – to give up his anatomical and geological research in order to dedicate himself exclusively to his pastoral activity. This

---

[9] Rome, "Sténon", p. 537: "une question qu'il aborde est rarement traitée à fond, il l'abandonne pour en envisager une autre qui s'est présentée à lui au cours de sa dissection. Des trouvailles, qui sont bien proches d'être géniales, se trouvent consignées en quelques lignes, en une ligne parfois, au milieu d'autres moins importantes. Il n'est pas étonnant que certaines d'entre elles aient passé inaperçues, que d'autres savants les aient refaites après lui, et qu'en toute bonne foi, ils aient cru être les premiers à les avoir faites."

[10] SudOr., pp. 8–10 = E 5, p. 147, lines 7–9.

[11] Merton, "Matthew Effect", p. 451.

[12] Schrader, *Observationes et historiae*, Praefatio (unpaginated), here fol. * 12ᵛ: "obiter quasi [...] analogos". On Schrader and his work: DDPhS, s.v. "Schrader, Justus", cols. 1773–1774, here 1774. In Schrader, *Observationes et historiae*, Praefatio (unpaginated), here fol. * 12ʳ⁺ᵛ, Schrader mentions the unpublished (and presumably lost) conclusion by the professor of anatomy at the Dordrecht Athenaeum Dr. med. Willem Langly (1614–1668), drawn in 1657 by way of comparative anatomy, that the *testes muliebres* contained and discharged eggs. On Langly: DDPhS, s.v. "Langly (Langli), Willem", cols. 1141–1142.

presumably originated from a feeling that he had invested enough into the natural sciences, maybe even too much, as it had also already caused delay on his path to conversion.[13] A profound change had taken place in Stensen's life; his spirituality and ethos remained unaltered at their core, however, as evidenced by his coat of arms (Figs. 1.1, 2.9). That he compared his activity as pastor with that of a physician, eg in *Parochorum hoc age*, shows that despite his criticism of the far-reaching deficiencies of the medicine of his time, he nevertheless possessed great appreciation for the medical profession.

Stensen stands for truthfulness in research, publishing, defending against hassling and applying for positions, eg concerning his possible appointment to a chair at the University of Copenhagen in 1664. In this case he was helped by the faith he held from around 1664 onward, as expressed by his prayer documented in the first part of the "Schweriner Gebetbuch" (Schwerin Prayer Book), in the providence of a personal God which he attempted to turn everything over to by refraining from embracing the influence of third parties so as to be certain that not he himself had procured a position, but that it was divine willing.[14]

He was also prepared to suffer injustice concerning priority of his discoveries so long as no lies were told and he did not damage anyone who had supported him – in the case of the plagiarism by Blasius, his professors Sylvius and Van Horne. With no intention of returning in kind the curses, defamations and spitefulness of Blasius, Hoboken and Deusing, Stensen defended himself by pointing out to these men that they were acting wrongly. Though he only began around 1664 to avoid anything which might smack of "human, not gospel-compliant" wisdom,[15] his ethics characterized by Christian values already shine through in his behavior towards his early detractors (cf. Matt 5:39–40 par; John 18:22–23). His spiritually deepened humility following his conversion caused him to disregard his own person even more in favor of De Graaf. As a scientist and as a priest, Stensen's greatest concern always was not to say or, worse still, to write in overzealousness or anger a single word that he would have to take back later. To him, defending an incorrect scientific position unemotionally and level-headedly due to ignorance was better than doing so by deriding and ridiculing the opposite side.[16]

Stensen's character was one of restraint and depth, with pronounced modesty despite his accomplishments. His humor, visible in the painting showing him as "Anatomicus regius" in Copenhagen (Fig. 2.6), was dry, enigmatic and reflected great experience of life. The *Chaos* manuscript of his last study year in Copenhagen contains anecdotes taken from various books and his conversations with Ole Borch.[17] The French Jansenist spiritual writer and hagiographer Laurent Blondel (1671–1740),[18] who bases his description of Stensen's life on an authoritative

---

[13] See in detail in Sobiech, *Herz, Gott, Kreuz*, pp. 68–75.

[14] In the context of his religious development see in more detail ibid., pp. 35–36, 328–329.

[15] See note 35, in Chap. 3 above.

[16] MuscGland., p. 169, lines 13–14 with reference to the anatomy of the heart.

[17] Cf. Ziggelaar, *Chaos-Manuscript*, p. 481.

[18] On Blondel: Antony McKenna, "Blondel, Laurent, dit «le frère Laurent»"; in: DPR, p. 181.

source, presumably a manuscript by Stensen's sister Anne Kitzerow née Stensdatter (c. 1635–1703) which was later lost,[19] writes that Stensen had "a magnanimous and very courteous heart" as well as "a pleasant, amiable and charming humor"; that he was "naturally eloquent" and "spoke little, and always at the right time".[20] Despite his urbaneness, Stensen was inwardly independent of worldly prestige to a high degree – although he reproached himself for the opposite in his Hamburg confession notes.

His knowledge of human nature already as a student is evidenced eg in his letter to Thomas Bartholin from Leiden on April 30, 1663 in which he indicates Hoboken's and Blasius' literary activity, targeted at him but ultimately self-damaging, as proof that "precisely those possess the greatest audacity whose experience is rather humble".[21] Because Stensen was sensitive and highly self-critical, he often deliberated own deficiencies which he perceived as painful in his later years. This affected his spiritual life and his work as a bishop noticeably. Great analytical intelligence, when paired with a sincere character aware of the constraints of time, is not in danger of flagging and sedately coming to terms with the circumstances; it can, however, count on resistance from the more aligned. Hence Stensen found himself opposed by structures and a clique of aristocracy in the Prince-Bishopric of Münster which fostered the ordination of unworthy men, and could do little about it. He attempted to vicariously atone for this, and eventually requested to be relieved by Rome of his office as suffragan bishop of Münster when his conscience could no longer tolerate the situation.

The Leuven endocrinologist Jean Lederer (1910–2003) claims that as a result of his commitment to the poor, Stensen could also be viewed as a progenitor of sociology.[22] While Lederer fails to offer a justification for his assertion, Stensen's attitude is nevertheless untypical for his time and thus noteworthy: he did not merely give alms – a practice for which the Church was sometimes criticized in modern times because, as the critics said, it did not contribute to eliminating the causes of poverty – but recognized the sinful structures underlying poverty.[23] During his extensive geological research trip undertaken from 1668 to 1670 at the request of Grand Duke Ferdinando II, which brought him to Lower Hungary among other

---

[19] Sobiech, *Herz, Gott, Kreuz,* p. 112. On Kitzerow: Add. 1, pp. 905–906 (no. 12); Kardel/Maquet, *Biography and Original Papers,* p. 18.

[20] Blondel, *Vies de saints* [a], cols. 733–747 (Le XXV. jour de Novembre. Le venerable Nicolas Stenon evêque de Titiopolis, & vicaire apostolique dans le pays Septentrionaux), here 734 = Blondel, *Vies de saints* [b], cols. 1575–1590 (Le XXV. jour de Novembre. Le venerable Nicolas Stenon evêque de Titiopolis, & vicaire apostolique dans le pays septentrionaux), here 1576: "un coeur génereux & tout à fait obligeant [in Blondel, *Vies de saints* [b]: un coeur généreux, & tout-à-fait obligeant]: une humeur douce, affable & insinuante […] naturelement eloquent [in Blondel, *Vies de saints* [b]: naturellement éloquent] […] parloit peu, & toujours à propos".

[21] NovMusc., p. 160, lines 21–27, here 23–24 = E 13, p. 178, lines 31–37, here 34–35: "illorum demum maximam esse temeritatem, quorum minima est experientia".

[22] Lederer, "Sténon", p. 114. On Lederer: Christian Beckers/Martin Buysschaert, "In memoriam Jean Lederer"; in: LouvMéd. 122 (2003), pp. 225–227.

[23] Concluded from E 216, p. 466, line 35.

places, Stensen witnessed hazardous labor conditions, eg in mines. In his role as tutor, he acquainted crown prince Ferdinando III about these, speaking about the few "lazy rich men" for whom the poor gave up their lives in excessive struggle.[24] Stensen's function as legal advisor to Grand Duke Cosimo III should also be mentioned, during which he likewise expressed his concern[25] for the poor and was reimbursed by Cosimo for all debts incurred from his care for destitute and penniless theology students and converts.

Stensen's research on human procreation and embryology are highly relevant in terms of ethics. The Hamburg dermatologist and andrologist Carl Schirren emphasizes that it would be "completely unthinkable" to Stensen to "conduct research on human embryos".[26] This is not contradicted by Stensen's examination of the dysplastic fetus which had apparently died after birth. According to Günter Rager, embryology confirms the view that the human in the womb is a human being in the full meaning of the word from the time of conception: "The embryo possesses from the time of conception a human-specific and individual-specific genome. It does not develop into a human being, but instead unfurls during its development the human possibilities already enclosed within it."[27] In terms of legal policy and factual attitude towards induced abortions – which he unmistakably identified as "crimes"[28] – Stensen, who advised Cosimo III in secular criminal cases as well,[29] with a view to the embryological results and also due to his own empathy, which untypically for his time extended even to the suffering of animals, would never have subscribed to a perspective ignoring the lot of the embryo or fetus.

Furthermore, Günter Rager states that the uterus, ie the nidation of the fertilized egg cell, is "not constitutive" for the development of the embryo "as the latter already carries the full potentiality for its final shape within itself."[30] Although embryology was still being established in Stensen's time, his departure from the Aristotelian notion of successive ensoulment following his discovery of the ovary and his investigations into the path of the male sperm, his labeling of the child in the womb as a full-fledged human and his ever-exercised caution show that he would also reject any graduation into levels of humanness.

As far as the *end* of life is concerned, it can be assumed that Stensen would be deeply skeptical towards the alleged brain death criterion, if it were given to him for judgment: for in his Parisian *Discours* he countered the anatomists who self-assuredly always believed to know the purpose of certain organs by pointing out their own ignorance.[31]

---

[24] With sources Sobiech, *Herz, Gott, Kreuz*, p. 85n327; Sobiech, "Stensen und der Bergbau", pp. 293–294, 298–299.

[25] By agreement in E 111, p. 313, lines 31–32.

[26] Schirren, *Stensen*, p. 41.

[27] Rager, *Person*, pp. 197–207, here 203.

[28] See note 548, in Chap 2 above.

[29] See note 529, in Chap 2 above.

[30] Rager, *Person*, pp. 183–249, here 246 & p. 190.

[31] In more detail in Discours, p. 24, line 37–p. 25, line 3.

The Pastoral Constitution *Gaudium et spes* of the Second Vatican Council emphasizes that life is to be protected with the greatest care from the moment of conception.[32] The *Berliner Ordinarienkonferenz* (Berlin Ordinaries' Conference) under the presidency of the Bishop of Berlin Alfred Cardinal Bengsch ([* 1921] 1961–1979),[33] in its remarks – resolved on September 9, 1968 in Magdeburg in the former GDR and betokened by the Conference as "Advice for pastoral reflection" – on the encyclical *Humanae vitae*, enunciated something that

1. corresponds to Stensen's caution as researcher and
2. appears like a commentary on Stensen's path of life, correlating to his essential stance as a scientist ordained to the priesthood and to the episcopal office, not least in regard to the methodological separation of medicine and theology and the resulting allocation of tasks:

> We speak not as scientists; it is not primarily our task to debate questions about the concept of nature or medical problems. Rather, we must come to a judgment based on our Faith and help others to reach it as best we can. But we may rest assured that very few people on Earth can truly speak on the basis of their own clear research results (and not even all of these [can speak] competently about ethical norms); all others follow an authority, be it that of a scientist or public opinion or simply the majority. And the fallibility of these 'Authorities', in fact their lack of competence in questions of morality, is evident.[34]

The relationship between faith and science touched upon here, as well as the likewise mentioned danger of narcissism in regard to one's own opinions, are expressed particularly succinctly in a retrospect by Stensen – which partially passes into prayer – from the beginnings of his time as suffragan bishop in Münster: "God has granted you the possibility to make the acquaintance of many, in leading positions as well as people of all provenance, He has granted you the ability to discover much in matters of nature which is necessary for the rectification of many errors of philosophers and physicians, He has granted you the ability to gain insight into many controversies of faith, He has granted you the practice of several languages. If in all of this you seek only yourself, namely your pleasure, your benefit, your honor, then you seek the ephemeral, the vanity of vanities; if however you work therein at the glorification of God, then you pursue the will of God and the cause of eternal happiness, you are free of inane concerns and avoid future worries. He [= God] has given you a body that is healthy and suitable for many functions, He has taught you much through the words of humans, and through writings, He has shown you many works [of nature] that were visible to none before you. He has persuaded people of all ranks to be well-disposed to you in many places around the world, but He also converted your soul and gave it the desire for an eternity in the enjoyment of Yourself

---

[32] AAS 58 (1966), pp. 1025–1120, here 107, no. 51.

[33] On Bengsch: Josef Pilvousek, "Bengsch, Alfred"; in: LThK³ 2 (1994), col. 229.

[34] Hinweise zur pastoralen Besinnung nach der Enzyklika Humanae Vitae [Notes on pastoral reflection according to the encyclical Humanae Vitae]; in: DDBK 1 (1998), pp. 472–482, here 472–473. For details on this pastoral treatise see Schmitz, "Bengsch", esp. pp. 54–55, 60–61; Deckers, *Lehmann*, p. 154 with n. 52 on p. 157.

[= God]."[35] Only the orientation towards the otherworldly goal of man, which allows one to serve mankind without fear of man, truly arbitrates between science and faith and prevents egoism and pride as well as incorrect decisions resulting inevitably from these two bad attitudes.

Hartmut Kreß rightfully emphasizes that the empirical natural sciences and modern medicine – with Stensen being a representative of both – have contributed to the valuation of prenatal existence, meaning that human life begins with insemination resp. conception.[36] The additional question arises, however, to what degree science is sometimes overlaid – even if only in an implied manner – by zeitgeist-conditioned notions, irreconcilable with the dignity of man and subsumed under the term "scientific ideologies" (idéologies scientifiques) according to Georges Canguilhem, in the biosciences and their ethics. Science not only requires revision due to the constant accretion of knowledge, but similarly needs ethical reflection, which in turn depends on a foundation expressed in Stensen's case in his ethos characterized by the connection of Christian humility and love, as well as in his spirituality of the human body as "interpreter" of the love of God.

That scientific ethics becomes quite problematic without such a foundation is evidenced by Van Rensselaer Potter who, while identifying himself as a "pragmatic mechanist",[37] adheres to the concept "that man is a machine"[38] – more precisely a "cybernetic machine".[39] Marianna Gensabella Furnari criticizes Potter's reductionist image of the human being from a philosophical point of view, stating that man's freedom in fact comprises silent introversion and self-questioning.[40] Potter's statements about the 'free' will of man, which remain within the realm of machine theory,[41] and his concept based upon them must therefore be rejected early on from a philosophical perspective. Stensen, on the other hand, had a heuristic approach to the Cartesian machine concept, apparent in the fact that he described as admirable the process – invisible to the human senses – of the relationship between the soul and the body resp. between the mind, the most important executive organ of the

---

[35] Op. 9, p. 489, line 27–p. 490, lines 2, 6–12, here p. 489, line 27–p. 490, lines 2, 6–10: "Dedit tibi Deus, ut innotesceres multis, et magnatibus et omnis generis hominibus, dedit, ut in naturalibus multa videres ad philosophorum et medicorum multos errores tollendos necessaria, dedit, ut in controversiis fidei multa agnosceres, dedit aliquot linguarum exercitium. Si in his omnibus te tantum quaeris, scilicet tuam voluptatem, tua commoda, tuum honorem, quaeris peritura, vanitatem vanitatum; si autem Dei per illa manifestationem operaris, voluntatem Dei agis, rem aeterni gaudii agis, a vanis sollicitudinibus liber es, futuras sollicitudines vitas. […] Dedit corpus sanum et multis functionibus aptum, dedit mentem multis facultatibus dotatam, docuit plurima per hominum voces, tum scripta[,] monstravit multa opera nulli alii ante te visa. Omnium hominum ordines [= Omnium ordinum homines] movit ad favendum tibi in plurimis locis terrae, sed et animam convertit, aeternitatis desiderium in tui fruitione indidit."

[36] Kreß, *Medizinische Ethik*, pp. 40–41, 162–163.

[37] Potter, *Bioethics*, p. 11.

[38] Ibid., p. 10.

[39] Ibid., p. 36.

[40] On this, see Furnari, "Scientist Demanding Wisdom", p. 38.

[41] On this, see esp. Potter, *Bioethics*, pp. 62–63.

soul, and the body.[42] Besides the rational fathoming of nature, which he conducted in methodically innovative ways, Stensen also saw significance in the unfathomable and mysterious. In one sermon he asks the rhetorical question: "Who assists you that your soul stays united with the body, who established this law, who upholds it?"[43] A one-sided notion of reality in anthropology is incompatible with Stensen's image of man.

Potter developed his concept "humility with responsibility" because a "moratorium on new knowledge"[44] seemed illusionary, and reasoned: "We have to proceed *as if* we believed that the solution to man's major problems includes nothing that isn't 'available to the minds of men,' with just the added ingredient of humility ('fear of the Lord' [Ps 110(111):10]) that admits the possibility that natural forces may elude our attempts to build the kind of Utopia we can imagine. Whether a belief in a Deity is required or not is less important to me than the question of whether we proceed with humility or with arrogance, whether we respect the forces of Nature or whether we assume that science can do anything, whether we look at our heritage or whether we ignore it. It seems to me we have no choice other than to try to deal with dangerous knowledge by seeking more knowledge [= reference to Chapter 5]. We have already decided to tamper with the system [= nature and the biology of man], now we can do no less than to proceed with humility, respect the forces of Nature, and with respect to our ethical heritage, 'prove all things; hold fast that which is good. [1 Thess 5:21]' The situation is urgent."[45] One might initially deliberate in regard to these statements by Potter whether he may have had in mind some form of natural law (cf. Rom 2:14–16) with his principally welcome appeal for "humility with responsibility" – not least with a view to his bible citations; he presumably did not intend such a thing, however, as the positions he advances fundamentally antagonize it.[46]

Potter concretizes what he means by "respect to our ethical heritage" thus: "We can no longer afford to continue to be obscure in discussing what aspects of religion are worthy of mankind and what aspects are in danger of hastening our destruction."[47] Stensen would have agreed, for to him from a non-revelation-bound standpoint, which he himself supported for several years, as well as from his methodology of controversial theology, that society was happiest in which everybody strived to perfect themselves according to the laws of the best religion.[48] Since Potter ultimately considers every religion to be identical to superstition and thus dangerous for the

---

[42] E 65, p. 249, lines 12–17=Adelmann, *Correspondence*, vol. 2, no. 271, pp. 597–598.

[43] Sermo 23, p. 271, lines 36–37: "quis te adjuvat, ut anima tua corpori unita maneat, quis hanc legem dedit, quis conservat?"

[44] Potter, *Bioethics*, p. 11.

[45] Ibid., pp. 10–12, here 11–12.

[46] Ibid., p. 72 merely mentions "some conscientious physician".

[47] Ibid., p. 85.

[48] E 163, p. 396, line 38–p. 397, line 8; see also in more detail in Sobiech, *Herz, Gott, Kreuz*, pp. 52–54.

survival of humanity, his citation of Ps 110(111):10,[49] which forms the beginning of Stensen's so-called Schwerin Prayers,[50] is divested of its meaning.[51]

On the one hand, Potter exemplifies a standpoint that intends to empower the natural sciences to "a kind of ethical revolution" akin to the scientific-technical revolution – whose precursor was the Scientific Revolution of the early modern period –, although he is somewhat skeptical that its results have an influence on the actions of the world's population, and deduces: "It will have to be evangelized, and *the chances that this will occur are desperately slim.*"[52] On the other hand he argues that we should "pin our faith not on science alone, but on a search for wisdom, a wisdom that will recognize man's spiritual needs as well as his physical needs, a wisdom that will conquer by force of persuasion."[53]

Owing to his creation spirituality, Stensen would have understood and appreciated Potter's request for responsible handling of natural resources, though he would not have agreed with the latter's solutions in the manner they are proposed. Stensen would at first support Potter in two basic positions:

1. That the path to ever more knowledge is inexorable and that there can be no "moratorium" in this regard as God by no means despises, according to Stensen, the diligence of the researcher[54]; rather, "great things can be hoped for" from science conducted lege artis.[55] Incidentally, Potter's appeal to strive for more knowledge is reminiscent of the first two elements of Stensen's motto "Beautiful is that which is seen, more beautiful that which is known, most beautiful by far that which is not known". When Matthew Cobb claims Stensen's research to be one of the causes for a materialistic conception of the actual origins of all life,[56] it must be kept in mind that knowledge of the interrelations in nature, even if it is accompanied by a 'demystification' as eg in the case of the discovery of the muscular structure of the heart, is in itself always non-judgmental. But knowledge can be abused, a fact which Stensen frequently and emphatically pointed out. Stensen was one of the forefathers of modern-day bioresearch, though the sheer range of responsibility, which has grown in recent years due to expeditious developments, was in its infancy in his time. For example, follicle-producing stem cells in the ovaries were discovered in mice in a 2004 study, and in humans in 2012.[57]

---

[49] Potter, *Bioethics*, pp. 9, 184.

[50] Sobiech, *Herz, Gott, Kreuz*, pp. 108–109.

[51] Cf. Furnari, "Scientist Demanding Wisdom", pp. 39–40.

[52] Potter, *Bioethics*, p. 50.

[53] Ibid., p. 52.

[54] See note 184, in Chap. 2 above.

[55] See note 201, in Chap. 2 above; in contrast to the medicine of his time, see note 199, in Chap. 2 above.

[56] See note 201, in Chap. 1 above.

[57] Confirmation by other researchers is still pending, cf. White/Woods/Takai/Ishihara/Seki/Tilly, "Oocyte formation", pp. 413, 419–420 (Discussion).

2. Stensen would agree with Potter that "more knowledge" *can* be a point of departure for a better set of ethics, eg in terms of the anatomy and physiology of the sexual organs and embryology with regard to the concept of human sexuality and the moment of ensoulment of the embryo.[58] That this represents only a possibility and by no means an inevitable consequence is indicated in Stensen's warning against the "vanity of vanities" elevating "your pleasure, your benefit, your honor" – under exclusion of other people – to the measure of all things.

The problem thus lies in what Potter, who approved of "the scientific method in seeking wisdom",[59] meant by "wisdom": for wisdom, according to Potter "the knowledge of how to use knowledge"[60] – ie the balancing of knowledge with other knowledge for the sake of the social good –, can be found in the same way as other knowledge. Here Potter and Stensen, who in contrast to Potter never absolutized the model of man as a machine, part ways. For Stensen, the scientist's humility was not a mechanistic postulate, but connected to the cognition that nature is a creation of God. Potter saw no ulterior goal for man in God; all that counted for him was the this-worldly survival of humanity. Hence when his own death was approaching, he viewed its experiencing as one "of completion yet mystery" and recognized "the necessity for personal death as a part of the further evolution of nature".[61] That Stensen rejected a mechanistic position ultimately leading to atheism is evidenced by his definition of the function of the sensory organs: to provide the mind with those requirements of things for examination sufficient to obtain a knowledge of things which enables man to achieve his goal.[62] From Stensen's perspective, Van Rensselaer Potter can thus be counted among the former's "friends in the Netherlands"[63] who wanted to make philosophy the judge of the insights granted by grace.

The Cartesian Licent. phil. Dr. med. Arnout Geulincx (1624–1669), who was befallen by several calamities which influenced his writings and who taught at the University of Leiden from the end of 1658 – with moral philosophy being one of his later subjects –,[64] posited as the new cardinal virtues, initially in 1664 and eventually

---

[58] Paolo Zacchia, for example, justified the canonical penal law relating to induced abortion – which still existed at his time, ie existed again in graduated form as a result of the Constitution *Sedes Apostolica* (1591) – in part with the philosophical and natural scientific state of knowledge of his time (Spitzer, *Zacchia*, p. 25). The life of the French geneticist and "Servus Dei" (servant of God) Prof. Dr. med. Jérôme Lejeune (1926–1994), whose Roman process was officially opened on 02/21/2013, is significant for the knowledge of trisomy 21 (Down syndrome), cf. http://www.amislejeune.org.

[59] Potter, *Bioethics*, p. 50.

[60] Ibid., p. 9 with reference to p. 1.

[61] In the words of Peter J. Whitehouse, "The Rebirth of Bioethics: A Tribute to Van Rensselaer Potter"; in: GlobBioeth. 14 (2001), pp. 37–45, here 39.

[62] See note 173, in Chap. 2 above.

[63] See note 169, in Chap. 1 above.

[64] Geulincx taught moral philosophy already before 1665, and officially since 1667; on him: Han van Ruler, "Geulincx, Arnold"; in: DDPh. 1 (2003), pp. 322–331, here 324; DDPhS, s.v. "Geulincx, Arnold", cols. 665–666.

in *De virtute et primis ejus proprietatibus, quae vulgo virtutes cardinales vocan-tur[,] tractatus ethicus primus* (On Virtue and Its Foremost Characteristics, Commonly Known as Cardinal Virtues, First Ethical Treatise; Leiden 1665), "dili-gentia" (diligence), "obedientia" (obedience), "justitia" (justice), and "humilitas" (humility).[65] In his book, Geulincx also expressed his admiration for what had been discovered with the help of the microscope.[66] He attempted to base ethics, which to him formed the centerpiece of philosophy, solely on the "natural light", albeit with no intent of provoking a conflict with Revelation.[67] It is not known whether Stensen was aware of Geulincx' writings; at any rate, Stensen strived for something similar as an aid to faith, as evidenced by his letter to Malpighi on November 24, 1671 in which he writes that he hoped to make visible with his anatomical research what "the natural light" of reason was and was not capable of in its cognition of nature in order to lead his "friends in the Netherlands" back to true Christian humility.[68]

The request by Nida-Rümelin mentioned at the end of Chap. 1 to complement the handed-down "ethos of epistemological rationality" of the early modern era and the European Enlightenment with the dimension of "responsible scientific practice"[69] has received significant impulses from Stensen's call for a humble and pastoral attitude by the physician and an ethos of the scientist which inspects human inade-quateness, is oriented towards anatomical results and respects the life of every human being, even the unborn, due to its ensoulment at the moment of insemina-tion. Bernard N. Nathanson had similarly spoken of the "eighteenth-century cer-tainty", which he himself had indulged in according to his own words and which he recognized to have been disturbed by the notion of a delivering and merciful God.[70] This is comparable to the path that the anatomist Stensen had travelled in his reli-gious quest from 1662/1663 up to his conversion in 1667.

*

*   *

In closing, a further reflection on the ethos of the anatomist and physician may be offered. The Latin original version of the main title of this book, "Radius in manu Dei" (Pointer in the Hand of God), can be interpreted in two different ways. Because the Latin noun "radius" can also mean "ray", eg of the sun, a further dimension of meaning relating to anatomy and Stensen's spirituality is attained with the help of

---

[65] Geulincx, *De virtute*, fol. ** 1^r–** 5^v (Lector benevole), here ** 1^{r+v}.

[66] Ibid., fol. ** 4^{r+v}.

[67] Battail, "Geulincx", p. 392.

[68] E 65, p. 249, lines 26–37, here 27 = Adelmann, *Correspondence*, vol. 2, no. 271, p. 598: "il lume naturale"; cf. E 65, p. 249, line 34 = Adelmann, *Correspondence*, vol. 2, no. 271, p. 598: "il lume della natura".

[69] See note 202, in Chap. 1 above.

[70] Nathanson, *Hand of God*, pp. 193–194 with citation on p. 194. – For comparison with the "cer-tainty" (certitudo) highlighted by Stensen as a central theme with regard to his conversion and sprituality cf. Sobiech, *Herz, Gott, Kreuz*, p. 391, s.v. "Sicherheit".

Gottfred Eickhoff's sculpture "Steno" unveiled in Copenhagen in 1963 (Fig. 1.2). In the photograph taken of the sculpture in the month of November, there is visible a sheaf of sunlight (Fig. 1.4) which,[71] originating from God as part of Creation and a natural gift, falls directly onto the head of the female corpse with Stensen's hand resting on the side of its chest. This casts an apposite light on the research as well as the public activity of the anatomist: with God's help, the rays of the sun, the anatomist as an element of Creation is enabled to interact in a researching and healing manner with Creation as an instrument of God. Such spiritual interpretations of the natural light of the sun were familiar to Stensen as well.[72] If one bears in mind that Stensen's concept of the 'circulation of love' between God and human being also applies to the researching anatomist and medical practitioner, then the expression "radius in manu Dei" (pointer/ray in the hand of God) incorporates a call to the anatomist and physician to become aware of his referentiality to the Creator through the function of the human body as "interpreter" of the love of God, and to see by merit of this referentiality to the triune[73] God his responsibility towards every human being.

---

[71] It was captured unintentionally when the photo was taken (11/10/2011) and subsequently gave rise to these thoughts. Cf. note 534, in Chap. 2 above.

[72] See note 443, in Chap. 2 above.

[73] On Stensen's Trinitarian 'short formula' of faith see Sobiech, *Herz, Gott, Kreuz*, pp. 217–230.

Bibliography

Where more than one work by an author or editor is included, the references are sorted chronologically. For Latin titles, the early modern notation including punctuation is principally maintained, ie ligatures (eg æ) are resolved (except &), u and v are normalized according to their phonetic value, and periods after cardinal numbers and accents are not reproduced. E caudata and abbreviations are resolved, with the latter having their abbreviation sign removed and the complement placed in parentheses. Capitalization is left unchanged only for proper nouns in the broadest sense.

Emphasis in titles, academic titles of authors, dedications, and information regarding printer or publisher are generally not included.

The abbreviated forms used in footnotes for encyclopedias, periodicals and certain works by Niels Stensen reproduced in later editions can be found under *Abbreviations, A*.

In citations from primary sources, emphasis corresponds to that of the source material. In citations from early modern prints, page breaks are also identified. In all other regards, citation of Latin sources follows the practice for references mentioned above. For sources in Italian, u and v are normalized according to their phonetic value. For individual Latin terms taken out of their source context, case and person are either matched to the syntax of the English translation into which the respective term is inserted, or are cited in the basic grammatical form derived from the original passage.

Archive Materials

Included here are the prints listed in *Bibliography, C* for which only one copy could be verified, as well as the signatures of the prints used for the illustrations in this book (cf. *Bibliography, B*).

© Springer International Publishing Switzerland 2016   199
F. Sobiech, *Ethos, Bioethics, and Sexual Ethics in Work and Reception of the Anatomist Niels Stensen (1638-1686)*, Philosophy and Medicine 117, DOI 10.1007/978-3-319-32912-3

The *Bodleian Library* at Oxford University (see above footnote 64, in Chap. 3) and the *Hauptstaatsarchiv Dresden* (see above footnote 579, in Chap. 2) were also consulted.

The university archives of the Westfälische Wilhelms-Universität Münster and the *Service Archives-Documentation* in Neuilly-sur-Seine were used for some of the works listed in *Bibliography, C.* This is indicated by ** (for WWU Münster) and *** (for *Service Archives-Documentation*) in the respective bibliography entries.

AGCRS	**Archivio Generalizio – sezione storica, Chierici Regolari Somaschi, Rome** Biografie C. R. S., n. 708: Lugo (De) Antonio Maria
AOV	**Accademia Olimpica, Vicenza** [two biographies (mach. ms. recto), each entitled "Quattrin Nevio", period ending in 1982, 2 folios]
APUG	**Archivio Storico della Pontificia Università Gregoriana, Rome** 576, fol. 44^r–45^v: Lettere autografe di Niccolò Stenone di Copenaghen celebre anatomico e poi vescovo di Zitopoli e vicario apostolico dirette al Gran Duca di Toscana
ARSI	**Archivum Romanum Societatis Iesu, Rome** Hist. Soc. 49: Defuncti 1670–1700 Rh. Inf. 15: Rhen[i] Inf[erioris] Indiam petentes 1616–1740 Rh. Inf. 23: Rh[eni] I[nferioris] Cat[alogi] Tr[iennales] 1678 Rh. Inf. 46: Rhenus. Necrolog[ia] 1620–1625. Rhen[i] Inf[erioris] Necrolog[ia] 1626–1670 Rh. Inf. 57^{II}: Hist[oriae] et annuae 1687–1692
ASF	**Archivio di Stato di Firenze, Florence** Mediceo del principato 5510, fol. 156^r
BAM	**Bistumsarchiv Münster** Coll. Kleruskartei
BAOS	**Bistumsarchiv Osnabrück** 01-89-10-02: Deutsches Niels-Stensen-Komitee 1950–1959; Translation der Gebeine (1953) und Pilgerreise (1953)
BL	**British Library, London** General Reference Collection 548.e.32.(2.): Stensen, *Apologiae prodromus*
BNCF	**Biblioteca Nazionale Centrale di Firenze, Florence** Magl. XXV, 42: Bisdosso o vero diario di Francesco Bonazini. Tomo primo Ms. Gal. 252 (Div. 4^a, t. 142), fol. 109^r–110^v (Vincenzio Viviani, da Firenze, 26 gennaio 1666, ab Inc[arnatione = March 25] [=1667], a Bruto Annibale della Molara, a Pisa. Minuta autografa, cf. CG-BNCF III 2.2, p. 419, no. 55)
BSLF	**Insigne Basilica di San Lorenzo, Florence** In morte di Mons. Giuseppe Capretti Priore Mitrato della Insigne Basilica di S. Lorenzo. Commemorazione tenuta dal

canonico monsignore Ottorino Agresti nella Basilica di San
Lorenzo il 26 febbraio 1973 [unpaginated print, 6 pages with 2
pages of addenda, here p. (7): "Mons. Giuseppe Capretti nei
Ricordi del Curato/Sac. Emilio Grosso – Curato di S. Lorenzo"
and p. (8): "Monsignor Giuseppe Capretti" (biographical data)]

KB   **Det Kongelige Bibliotek, Copenhagen**
              34:3,–79 2°: Bartholin, *Decanus Facultatis Medicae* (Latinsk
              Universitetsprogram udstedt af det medicinske Fakultet ved
              Niels Stensens tiltrædelsesforelæsning. 28. Jan. 1673)
              Ny kgl. Samling 4019 4°: Nicolai Stenonis Elementorum myo-
              logiae specimen […] (Renskrift med egenhændige rettelser
              m.m., 1667) [manuscript of Stensen, *Elementorum myologiae
              specimen*]
              Ny kgl. Samling 4963 4° (Gustav Scherz' samlinger om Niels
              Stensen), VIII: [Société Internationale de Chirurgie], *Guide*
              Tilg. 621: Gustav Scherz' brevarkiv [unpaginated]

KU-BVFB   **Det Biovidenskabelige Fakultetsbibliotek, University of
              Copenhagen**
              Hdskr. 2:II:40 a: No 807 og 874. Anatomisk Beskrivelse af et
              hidintil ukiendt Afsondringsorgan i Næsen ved Ludevig
              Jacobson […] [48 pages]

KU-MM   **Det Sundhedsvidenskabelige Fakultet, Institut for
              Folkesundhedsvidenskab, Medicinsk Museion, University
              of Copenhagen**
              Billede no. I-000263

LBH   **Gottfried Wilhelm Leibniz Bibliothek – Niedersächsische
              Landesbibliothek, Hanover**
              LBr. 943, fol. 168
              M-A 12 (piece 5 in [Anatomica varia 1600–1709]): Jessenius
              von Jessen, *Invitatio*
              Ms. XLII 1902, fol. 16^v [with wax seal]

LkAE   **Landeskirchenarchiv Eisenach**
              Kirchenbuch Eisenach 1643–1648
              Kirchenbuch Eisenach 1706–1719

NBA   **Niels Bohr Archive, Copenhagen**
              Niels Bohr General Correspondence, 1910–1962, Folder 27:
              Scherz, Gustav, 1953–1959. Regarding: Niels Stensen

SUB Göttingen **Niedersächsische Staats- und Universitätsbibliothek
              Göttingen, Georg-August-Universität Göttingen**
              8 THER 711: Blasius, *Medicina generalis*

UB Greifswald **Universitätsbibliothek, Ernst-Moritz-Arndt-Universität
              Greifswald**
              520/Va 375 adn21: Helwig, *B[enevolo] l[ectori] s[alutem]*

UniFR-DepMed **Departement of Medicine, Fachbibliothek Anatomie,
              Université de Fribourg Suisse/Universität Freiburg i. Ü.,
              Fribourg (Switzerland)**

[Collection of publications up to 1982/1983 by Prof. Dr. Adolf Faller, 10 vols. with tables of contents, mach. Mss.]: A. Faller – works from 1941 to 1949 (nos. 1–50); 1950–1954 (nos. 51–106); 1954–1958 (nos. 107–179); 1959–1962 (nos. 180–233); 1963–1965 (nos. 234–283); 1966–1968 (nos. 284–324); 1969–1970 (nos. 325–351); 1971–1972 (nos. 352–376); 1973–1977 (nos. 377–439); 1978–1982 (nos. 440–484) [with 1 bibliographical unit for 1983; others for 1983 in: Institut für Anatomie und spezielle Embryologie der Universität Freiburg (Schweiz), *Nachtrag*, pp. 40, 46, 48]

UOL   **University of Oklahoma Libraries, History of Science Collections, Bizzell Memorial Library, Norman, OK, USA**
[No signature]: Copy of Stensen, *Elementorum myologiae specimen* from the private library of Pope Clement IX

Photographs

The photographs (Figs. 1, 1.1, 1.2, 1.3, 1.4, 1.5, 2.1, 2.2, 2.3, 2.4, 2.5, 2.6, 2.7, 2.8, 2.9, 3.1, 3.2) are located on pages xi, 4, 12, 13, 14, 17, 43, 51, 76, 83, 92, 94, 121, 122, 151, 178, 182.

Photo credits:

Fig. 1.2: Frank Sobiech (2011);	Fig. 2.4: KB (2012);
Fig. 1.3*: Frank Sobiech (2011);	Fig. 2.5: KB (2013);
Fig. 1, 3.2: Frank Sobiech (2014);	Fig. 2.6: KU-MM (2003);
Fig. 1.4: Sebastian Olden-Jørgensen (2011);	Fig. 2.7, 1.1: LBH (2012);
Fig. 1.5: Robert Kretz (2012);	Fig. 2.8: UOL (2012);
Fig. 2.1: SUB Göttingen (2013);	Fig. 2.9: APUG (2014);
Fig. 2.2: BL (2015);	Fig. 3.1*: Frank Sobiech (2011).
Fig. 2.3: UB Greifswald (2013);	* With permission by KB.

Primary Sources and Secondary Literature

All citations in footnotes are given in shortened form consisting of the last name of the author and a shortened form of the main title of the source.

On the full citation format for editions of the works of Niels Stensen marked with a preceding asterisk (*), see *Abbreviations, A.*

Citations from websites not representing essays, articles from encyclopedias, reviews and obituaries are referenced only in the footnotes accompanying the main text body. The last date of access for all websites is March 4, 2016.

Acierno, Louis J. "Etienne-Louis Fallot: Is It His Tetralogy?" ClinCard. 22 (1999), pp. 321–322 = in *Profiles in Cardiology: A Collection of Profiles Featuring Individuals Who Have Made Significant Contributions to the Study of Cardiovascular Disease*, edited by J. Willis Hurst, C. Richard Conti and W. Bruce Fye, ch. 1, pp. 163–164. Mahwah, NJ: Foundation for Advances in Medicine and Science, 2003.****

**** *Pages from the book publication are cited in parentheses.*

Adamietz, Joachim, ed. *Juvenal: Satiren. Lateinisch-deutsch.* Darmstadt: Wissenschaftliche Buchgesellschaft, 1993.

Adelmann, Howard B. *Marcello Malpighi and the Evolution of Embryology.* Ithaca, NY: Cornell University Press, 1966.

Adelmann, Howard B., ed. *The Correspondence of Marcello Malpighi.* Ithaca, NY: Cornell University Press, 1975.

Adelmann, Howard B. "A Supplement to the Correspondence of Marcello Malpighi." *JSH 33,1* (1978), pp. 53–73.

Alexander VII, ed. *Index librorum prohibitorum Alexandri VII, Pontificis Maximi iussu editus. Iuxta exemplar excusum.* Rome, 1667.

Altieri Biagi, Maria Luisa, and Bruno Basile, ed. *Scienziati del seicento.* La letteratura italiana; Storia e testi 34,2. Milan: Riccardo Ricciardi, 1980.

Altoviti, Filippo Neri, ed. *L'obbligo de['] parochi dimostrato con evidenza. Da zelante, e dotto Prelato* [= Niels Stensen]. *Opera utilissima. Tradotta dal Latino d'ordine. Per istruzione, e profitto de['] Parochi della sua Diocesi.* Florence, 1684.

Andrault, Raphaële. *Niels Stensen (Nicolas Sténon): Discours sur l'anatomie du cerveau (1669).* Textes de philosophie 2. Paris: Éditions Classiques Garnier, 2009.

[Anonymous]. *Miracle arrivé dans la ville de Genève en ceste année 1609. D'une femme qui a faict un veau, à cause du mespris de la puissance de Dieu, & de Madame saincte Marguerite.* Paris (Iouxte la copie imprimée à Tonon, prés ladite ville de Genève), 1609.

Armogathe, Jean-Robert. "Cartesian Physics and the Eucharist in the Documents of the Holy Office and the Roman Index (1671–6)." Translated by Patrick Moran. In *Receptions of Descartes: Cartesianism and Anti-Cartesianism in Early Modern Europe*, edited by Tad M. Schmaltz, pp. 149–170. London: Routledge, 2005.

Ateneo di Salò. *Omaggio a Marco Marzollo medico e scrittore.* Raffa di Puegnano: Ateneo di Salò, 2003.

Attavanti, Giuseppe Ottavio, ed. *L'obbligo de' parochi dimostrato con evidenza. Da zelante, e dotto Prelato* [= Niels Stensen]. *Opera utilissima. Fatta di nuovo dare alle Stampe*, 2nd ed. Florence, [1685].

Aucante, Vincent. *La philosophie médicale de Descartes.* Diss. phil., Paris, 1998. Paris: Presses universitaires de France, 2006.

Austriaco, Nicanor Pier Giorgio. *Biomedicine and Beatitude: An Introduction to Catholic Bioethics.* Washington, DC: Catholic University of America Press, 2011.

Bach, Thomas, Olaf Breidbach and Dietrich von Engelhardt, ed. *Lorenz Oken: Gesammelte Werke.* Vol. 1, *Frühe Schriften zur Naturphilosophie.* Weimar: Böhlau, 2007.

Bacon, Francis. *The Two Bookes [...] of the Proficience and Advancement of Learning, Divine and Humane*. London, 1605.

Bacon, Francis. *De sapientia veterum liber*. London, 1609.

Bacon, Francis. *De dignitate & augmentis scientiarum, libri IX [...]. Editio nova,cum indice rerum ac verborum locupletissimo*. Leiden, 1645.

Baldewein, Christian Adolph. *Phosphorus hermeticus, sive magnes luminaris*. Frankfurt, 1675.

Baldini, Ugo, and Leen Spruit, ed. *Catholic Church and Modern Science: Documents from the Archives of the Roman Congregations of the Holy Office and the Index. Sixteenth-Century Documents*. 3 vols. and 1 index vol. Fontes Archivi Sancti Officii Romani: Series Documentorum Archivi Congregationis pro Doctrina Fidei 5. Vatican City: Libreria Editrice Vaticana, 2009.

Bambacari, Cesare Nicolao. *Descrizione Delle Azioni, e Virtù dell'illustrissima signora Lavinia Felice Cenami Arnolfini*. Lucca, 1715.

Barber, Bernard. "Resistance by Scientists to Scientific Discovery." In *The Sociology of Science*, 5th ed, edited by Bernard Barber and Walter Hirsch, pp. 539–556. New York: The Free Press, 1968.

Barnes, S. Barry, and Robert G. A. Dolby. "The Scientific Ethos: A Deviant Viewpoint." *Archives Européene de Sociologie* 11 (1971), pp. 3–25.

Barth, Hans-Martin. *Atheismus und Orthodoxie: Analysen und Modelle christlicher Apologetik im 17. Jahrhundert*. Habil. theol., Erlangen-Nürnberg, 1969/1970. Forschungen zur systematischen und ökumenischen Theologie 26. Göttingen: Vandenhoeck & Ruprecht, 1971.

Bartholin, Caspar Sr. *De studio medico inchoando, continuando, & absolvendo consilium breve atque extemporaneum*. Copenhagen, 1628.

Bartholin, Thomas. *De cometa, consilium medicum, cum monstrorum nuper in Dania natorum historia*. Copenhagen, 1665.

Bartholin, Thomas. *Epistolarum medicinalium centuria III. historiis medicis aliisque ad rem medicam spectantibus plena*. Copenhagen, 1667[a].

Bartholin, Thomas. *Epistolarum medicinalium centuria IV. variis observationibus curiosis & utilibus referta*. Copenhagen, 1667[b].

Bartholin, Thomas, ed. *Catalogus & valor medicamentorum simplicium & compositorum in officinis Hafniensibus prostantium. Apotecker Taxt Paa alt hvis i Kiøbenhaffn hos de fire privilegerede Apotheckere til Kiøbs findis. Apothecker Taxa Aller dehrer Medicamenten und Waaren / welche man bey den vier Privilegirten Apotheckern in Kopenhagen zu kauffe findet*. Copenhagen, 1672.****

**** *Frontispiece*: Catalogus medicamentorum officinalium cum taxa pharmaceutica Hafniensi 1671.

Bartholin, Thomas. *Decanus Facultatis Medicae in Reg[ia] Acad[emia] Hafniensi [...] Anatomes studiosis s[alutem]: [...]; there follows the text of the program] Hafniae d[ie] 28. Januarii anno 1673 [...]*. [Copenhagen, 1673].

Bartholin, Thomas. *Cista Medica Hafniensis eller Det medicinske fakultets brevkiste som er fuld af forskelligartede råd, kure, sjældne tilfælde, biografier over københavnske læger og andre ting som vedrører anatomien, botaniken samt kemien. Dernæst følger En kortfattet beskrivelse af Anatomihuset i København forfattet af*

Thomas Bartholin. Edited by Niels W. Bruun and Hans-Otto Loldrup. Copenhagen: Dansk Farmaceutforening, 1982.

Bartholin, Thomas. *Anatomihuset i København kortfattet beskrevet.* Edited by Niels W. Bruun. Copenhagen: Loldrup, 2007.

Battail, Jean-François. "Arnold Geulincx." In *Die Philosophie des 17. Jahrhunderts.* Vol. 2, *Frankreich und Niederlande,* edited by Jean-Pierre Schobinger, pp. 375–397, 462–465. Grundriss der Geschichte der Philosophie, new edition 1983–. Basel: Schwabe, 1993.

Baumann, Hans-Ullrich. *Über Mehrfachentdeckungen: Eine Darstellung des allgemeinen Problems mit Beispielen aus drei Teilgebieten der theoretischen Medizin.* Diss. med., Münster, 1972. Münstersche Beiträge zur Geschichte und Theorie der Medizin 7. Münster: Universität Münster, 1972.

Beauvois, A[thanase Mary Alphonse]***. *Un praticien allemand au XVIIIᵉ siècle. Jean- Henri Cohausen (1665–1750). Docteur en Médecine et en Philosophie. Premier Médecin des Princes évêques de Munster (1700–1719). Archiâtre des Bailliges d'Horstmar-Ahaus. Doyen des Praticiens du diocése de Munster.* Diss. med., Paris, 1899/1900. Paris: Maloine, 1900.

Becher, Johann Joachim. *Methodus didactica seu clavis et praxis super novum suum organon philologicum, Das ist: Gründlicher Beweis / daß die Weg und Mittel / welche die Schulen bißhero ins gemein gebraucht / die Jugend zu Erlernung der Sprachen / insonderheit der Lateinischen / zu führen / nicht gewiß / noch sicher seyen / sondern den Regulen und Natur der rechten Lehr / und Lern-Kunst schnurstracks entgegen lauffen / derentwegen nicht allein langwilig sondern auch gemeiniglich unfruchtbar / und vergeblich ablauffen: Samt Anleitung zu einem besseren,* 2nd ed. Frankfurt, 1674.

Bek-Thomsen, Jakob. "Diplomatic Skills: Tracing the Functions of Nicolaus Steno's Networks". Lecture at the conference "Science, Learning and Censorship", London, March 27, 2011. Under: http://backdoorbroadcasting.net/wp-content/documents/JakobBekThomsen/BekThomsen_2011_DiplomaticSkills.pdf (paginated, 13 pages).

Bellini, Lorenzo, and Gerard Leonard Blasius. *Exercitatio anatomica de structura & usu renum, cui renum monstrosorum exempla, ex medicorum celebrium scriptis addidit Gerardus Blasius.* Amsterdam, 1665.

Benter, Philipp, Wolfgang Benter, and Thomas Benter. "Niels Stensen – Arzt, Naturforscher und Priester"="Niels Stensen – physician, naturalist and priest". *DMW 138* (2013), pp. 2678–2681.

Bergdolt, Klaus. *Das Gewissen der Medizin: Ärztliche Moral von der Antike bis heute.* Munich: C.H. Beck, 2004.

Bergmann, Anna. *Der entseelte Patient: Die moderne Medizin und der Tod.* Berlin: Aufbau-Verlag, 2004.

Bernardi, Walter. *Il paggio e l'anatomista: Scienza, sangue e sesso alla corte del Granduca di Toscana.* Florence: Le Lettere, 2008.

Bernier, Jean. *Essais de medecine Où il est traité de l'histoire de la medecine et des medecins. Du devoir des Medecins à l'égard des malades, & de celui des malades à l'égard des Medecins. De l'utilité des remedes, & de l'abus qu'on en peut faire*. Paris, 1689.

Bernoulli, René. "Raimundus Sebundus: Prinzipien seiner Anthropologie." In *Philosophische Tradition im Dialog mit der Gegenwart: Festschrift für Hansjörg A. Salmony*, edited by Andreas Cesana and Olga Rubitschon, pp. 9–28. Basel: Birkhäuser, 1985.

Bertha, Ludwig. *Medicus Christianus detegens sanguineis lacrymis deplorandam ferrei hujus saeculi caecitatem praesentium et imminentium plagarum originem. Praescribens remedia, tam ex Sacra Scriptura quam SS.* [= Sanctis] *Patribus desumpta ad omnem Christianae reipublicae statum sanandum. Opus quadripartitum, universis propriam, vel alienam salutem quaerentibus utilissimum, maxime tamen concionatoribus, confessariis, caeterisque, de quorum manibus justus judex requiret sanguinis sui pretium fidelium animas. Accessit index pro singulis dominicis, & festis totius anni in usum praedicantium coram quocumque statu.* Antwerp, 1665.

Bierbaum, Max. *Niels Stensen: Von der Anatomie zur Theologie. 1638–1686.* Münster: Aschendorff, 1959.

Bierbaum, Max, and Adolf Faller. *Niels Stensen: Anatom, Geologe und Bischof 1638–1686*, 2nd ed. Münster: Aschendorff, 1979.

Bierbaum, Max, Adolf Faller, and Josef Traeger. *Niels Stensen. Anatom, Geologe und Bischof 1638–1686*, 3rd ed. Münster: Aschendorff, 1989.

Blasius, Gerard Leonard. *Medicina generalis, nova accurataque methodo fundamenta exhibens*. Amsterdam, 1661.

Blasius, Gerard Leonard. *Institutionum medicarum compendium, disputationibus XII in illustri Amstel[odamensi] Athenaeo publice ventilatis, absolutum.* Amsterdam, 1667.

Blondel, Laurent. *Les vies des saints pour chaque jour de l'anée* [sic!], *tirées des auteurs originaux: Avec une priere à la fin de chaque vie & un Martyrologe*. Paris, 1722[a].

Blondel, Laurent. *Les vies des saints pour chaque jour de l'année, tirées des auteurs originaux: Avec une priere à la fin de chaque vie, & un martyrologe*. Paris, 1722[**** b].

**** *Revised edition*.

Bohr, Niels. "Physical Science and the Problem of Life." In *Niels Bohr: Collected Works*. Vol. 10, *Complementarity beyond Physics (1928–1962)*, edited by David Favrholdt, pp. [113]–[123]. Amsterdam: North-Holland, 1999.

Bonomo, Giovanni Cosimo. *Osservazioni intorno a' pellicelli del corpo umano [...], e da lui con altre Oßervazioni scritte in una lettera all'illustriss[imo] sig. Francesco Redi*. Florence, 1687.

Bonse, Hans-Michael. "Die Erstbeschreibung der Fallotschen Tetralogie durch Niels Stensen im Lichte moderner kinderkardiologischer Hypothesen der formalen Genese." Diss. med. dent., Münster, 1986. Mach. ms.

Borck, Cornelius, Volker Hess, and Henning Schmidgen, ed. Introduction to *Maß und Eigensinn: Studien im Anschluß an Georges Canguilhem*, pp. 7–41. Munich: Wilhelm Fink, 2005.

Borel, Pierre. *Historiarum, et observationum medicophysicarum centuriae IV. In quibus non solum multa utilia, sed & rara, stupenda ac inaudita continentur. Accesserunt d[omini] Isaaci Cattieri, doctoris Monspeliensis, & medici regij, observationes medicinales rarae, dom[ino] Borello communicatae: et Renati Cartesii vita eodem P[etro] Borello authore. Quae omnia nunc primum in lucem prodeunt.* Paris, 1656.

Boselli, Francesco. *Amaltheum medico-historicum, tres in apparatus digestum, doctrinae varietate, cum omnibus historicis & politicis, tum promovendis inprimis & promotis medicis non minus utile quam jucundum. Accessit geminum corollarium: encomii in Academia Patavina medicinae professorum ab anno 1631, & elogii publica in bibliotheca ibidem heroum depictorum.* Padua, 1668.

Boudewijns, Michiel. *Ventilabrum medico-theologicum quo omnes casus, tum medicos, cum aegros, aliosque concernentes eventilantur, et quod SS. PP.* [= Sanctis Patribus] *conformius, Scholasticis probabilius, & in conscientia tutius est, secernitur: Opus cum theologis & confessarijs, tum maxime medicis perquam necessarium.* Antwerp, 1666.

Boulliau, Ismael. *De natura lucis.* Paris, 1638.

Bouuaert, Claeys. "Contribution à l'histoire de la députation envoyée à Rome en 1677–1679 par la faculté de théologie de Louvain: La participation de Martin Steyaert." In *Miscellanea historica in honorem Alberti de Meyer Universitatis Catholicae in oppido Lovaniensi iam annos XXV professoris*, pp. 1130–1145. Publicaties op het gebied der Geschiedenis en der Philologie 3,23. Leuven: Universiteitsbibliotheek, 1946.

Brandl, Leopold. *Die Sexualethik des heiligen Albertus Magnus: Eine moralgeschichtliche Untersuchung.* Studien zur Geschichte der katholischen Moraltheologie 2. Regensburg: F. Pustet, 1955.

Brockliss, Laurence, and Colin Jones. *The Medical World of Early Modern France.* Oxford: Clarendon Press, 1997.

[Browne, Thomas]. *Religio Medici: The Sixth Edition, Corrected and Amended. With Annotations Never Before Published, Upon All the Obscure Passages Therein. Also Observations by Sir Kenelm Digby, Now Newly Added.* London, 1669.

Bruch, Richard. "Schutz des vorpersonalen menschlichen Lebens im Mutterleib in moraltheologischer Sicht." *ThPQ* 120 (1972), pp. 112–131.

Bruni, Roberto L. "Editori e tipografi a Firenze nel Seicento." *StSec.* 45 (2004), pp. 325–419.

Bruun, Niels W. "Fem nyfundne Niels Stensen-breve." *FoF* 47 (2008), pp. 115–165.

Buder, Christian Gottlieb, ed. *Nützliche Sammlung Verschiedener Meistens ungedruckter Schrifften, Berichte, Urkunden, Briefe, Bedencken Welche Zu Erläuterung Der Natur und Völcker besonders Teutschen Staats- und Lehn-Rechten auch Kirchen-Politischen und gelehrten Historien dienen können. Mit Einigen Anmerckungen erläutert.* Frankfurt, 1735.

Busenbaum, Hermann. *Medulla theologiae moralis, facili ac perspicua methodo resolvens casus conscientiae, ex variis probatisque authoribus, concinnata [...]. Poenitentibus aeque ac confessariis perquam utilis. Editio novissima. Recognita ab uno e Societate, & a multis mendis repurgata, quae in praecedentibus irrepserant.* Lyon, 1671.

Cangiamila, Francesco Emanuele. *Embriologia sacra, ovvero dell'uffizio de' sacerdoti, medici, e superiori, circa l'eterna salute de' bambini racchiusi nell'utero, libri quattro.* Milano, 1751.

Canguilhem, Georges. "Descartes et la technique." In *Travaux du IXᵉ Congrès International de Philosophie: Congrès Descartes. II. Études Cartésiennes. IIᵐᵉ Partie*[:] *III. La méthode et les mathématiques. IV. La physique. V. La morale et la pratique. VI. Histoire de la pensée de Descartes*, edited by Raymond Bayer, pp. 77–85. Actualités scientifiques et industrielles 531. Paris: Hermann et Cⁱᵉ, 1937.

Canguilhem, Georges. *Idéologie et rationalité dans l'histoire des sciences de la vie: Nouvelles études d'histoire et de philosophie des sciences*, 2nd ed. Paris: J. Vrin, 1988.

Casella, Luciano, and Enrico Coturri, ed. *Niccolò Stenone: Opere scientifiche.* Vol. 1. Florence: La Nuova Europa, 1986.

Cavarzere, Marco. *La prassi della censura nell'Italia del Seicento: Tra repressione e mediazione.* Temi e testi 92. Rome: Storia e Letteratura, 2011.

Ceyssens, Lucien. *Le cardinal François Albizzi (1593–1684): Un cas important dans l'histoire du jansénisme.* Spicilegium Pontificii Athenaei Antoniani 19. Rome: Pontificium Athenaeum Antonianum, 1977.

Chiodi, Maurizio. *Modelli teorici in bioetica.* Quaderni FAD 1. Milan: F. Angeli, 2005.

Christian V, ed. *Forordning om Medicis oc Apotecker [et]c. Constitutiones regiae de medicis & pharmacopoeis &c. Königliche Verordnung angehend die Medicos und Apotheker [et]c.* Copenhagen, 1672.

Cigogna, Emmanuele Antonio. *Delle inscrizioni veneziane raccolte ed illustrate.* Vol. 3. Venice, 1830.

Ciuti, Francesco. *L'Ospedale di Santa Maria Nuova di Firenze: Assistenza, sanità, medicina nella Toscana moderna (secoli 16.–18.).* Diss. phil., Pisa, 2011 (*unprinted*).

Cobb, Matthew. *The Egg & Sperm Race: The Seventeenth-Century Scientists Who Unravelled the Secrets of Sex, Life and Growth.* London: Free Press, 2006.

Cohausen, Johann Heinrich. *Clericus medicaster, in quo sacrarum litterarum auctoritate, sanctorum Patrum sententia, sacrorum canonum decretis, recta ratione atque experientia demonstratur: sacerdotem inprimis curatum praxeos medicae exercitium non decere.* Frankfurt, 1748.

Collet, Dominik. *Die Welt in der Stube: Begegnungen mit Außereuropa in Kunstkammern der Frühen Neuzeit.* Veröffentlichungen des Max-Planck-Instituts für Geschichte 232. Göttingen: Vandenhoeck & Ruprecht, 2007.

Committee for Selecting and Procuring the Collective Scientific Exhibit, Donated by the Danish Government to the Museum of Science and Industry at Chicago, ed. *Denmark and some Danish Scientists of Note.* Copenhagen: Egmont H. Petersen (printer), 1933.

Committee of Dutch Scientists = Commissie van Nederlandsche geleerden, ed. *The Collected Letters of Antoni van Leeuwenhoek = Alle de brieven van Antoni van Leeuwenhoek*. Vol. 2 [1673–1676]. Amsterdam: Swets & Zeitlinger, 1941.

Connery, John R. *Abortion: The Development of the Roman Catholic Perspective*. Chicago: Loyola University Press, 1977.

Cook, Harold J. "The New Philosophy in the Low Countries." In *The Scientific Revolution in National* Context, edited by Roy Porter and Mikuláš Teich, p. 115–149. Cambridge: Cambridge University Press, 1992.

Corte, Bartolomeo. *Notizie istoriche intorno a' medici scrittori milanesi, e a' principali ritrovamenti fatti in Medicina dagl'Italiani [...]*. Milano, 1718.

Cunningham, Andrew. *The Anatomist Anatomis'd: An Experimental Discipline in Enlightenment Europe*. Farnham: Ashgate, 2010.

Danneberg, Lutz. *Die Anatomie des Text-Körpers und Natur-Körpers: Das Lesen im liber naturalis und supernaturalis*. Säkularisierung in den Wissenschaften seit der Frühen Neuzeit 3. Berlin: De Gruyter, 2003.

[Dansk medicinsk-historisk Selskab]. *Historical Symposium on Nicolaus Steno (1638–1686) on* [corr. ****: and] *the Brain Research in the Seventeenth Century* [**** An International Symposium København, August 18, 19, 20, 1965]. Organised by Dansk medicinsk-historisk Selskab in connection with the International Congress of Neurosurgery Copenhagen, August 23– 27, 1965, sponsored by the International Academy of the History of Medicine[;] the International Brain Research Organization (IBRO) with co-operation from Medicinsk-historisk Museum and Niels Steensen's Hospital, Copenhagen. [Copenhagen, 1965].

**** *Front cover*.

Daur, Klaus-Detlef, ed. *Sancti Aurelii Augustini epistulae CI–CXXXIX*. Aurelii Augustini Opera; Pars III, 3 / Corpus Christianorum; Series Latina 31 B. Turnhout: Brepols, 2009.

Dauwe, Jef. "De Leuvense Boekdrukkers van Sassen alias Sassenus." *AL 2* (1973), pp. 235–273.

De Angelis, Simone. *Anthropologien: Genese und Konfiguration einer "Wissenschaft vom Menschen" in der Frühen Neuzeit*. Habil. phil., Bern, 2008. Historia Hermeneutica; Series Studia 6. Berlin: De Gruyter, 2010.

De Cabriada, Juan. *Carta filosofica, medico-chymica. En que se demuestra, que de los tiempos, y experiencias se han aprendido los Mejores Remedios contra las Enfermedades. Por la nova-antigua medicina*.**** Madrid, 1686 (corr. 1687).

**** *Frontispiece*: De los tiempos, y experiencias el mejor remedio al mal. Por la nova-antigua medicina. Carta phillosophica [sic!] medica chymica.

Deckers, Daniel. *Der Kardinal: Karl Lehmann – Eine Biographie*. Munich: Pattloch, 2002.

De Deckker, Patrick. "Dom Remacle Rome (1893–1974)." In *Proceedings of the 5th International Symposium on Evolution of Post-Paleozoic Ostracoda (Hamburg, 18–25 August 1974)*, edited by Gerhard Hartmann, pp. 7–9. Abhandlungen und Verhandlungen des Naturwissenschaftlichen Vereins in Hamburg, n.s. 18/19 [Supplement]. Hamburg: Paul Parey, 1976.

De Fiores, Stefano. "Maria in der Geschichte von Theologie und Frömmigkeit." In *Handbuch der Marienkunde*. Vol. 1, *Theologische Grundlegung, Geistliches Leben*. 2nd ed., edited by Wolfgang Beinert and Heinrich Petri, pp. 99–266. Regensburg: F. Pustet, 1996.

De Graaf, Reinier. *Opera omnia*. Leiden, 1677.

Delisle, Candice. "The Letter: Private Text or Public Place? The Mattioli-Gesner Controversy about the Aconitum Primum." *Gesnerus 61* (2004), pp. 161–176.

De Micheli Serra, Alfredo, and Raúl Izaguirre Ávila. "A Saint in the History of Cardiology." *ACM 84* (2014), pp. 47–50.

De Nave, Francine, and Marcus De Schepper, ed. *De Geneeskunde in de Zuidelijke Nederlanden (1475–1660): Tentoonstelling Museum Plantin-Moretus, 1 September – 25 November 1990*. Antwerp: Museum Plantin-Moretus, 1990.

De Renzi, Silvia. "Medical Competence, Anatomy and the Polity in Seventeenth-Century Rome." In *Spaces, Objects and Identities in Early Modern Italian Medicine*, edited by Sandra Cavallo and David Gentilcore, pp. 79–95. Malden, MA: Blackwell, 2008.

Descartes, René. *Discours de la méthode pour bien conduire la raison, & chercher la verité dans les sciences. Plus La dioptrique. Les meteores. Et La geometrie. Qui sont des essais de cete Méthode*. Leiden, 1637.

Descartes, René. *Specimina philosophiae: seu dissertatio de methodo recte regendae rationis, & veritatis in scientiis investigandae: dioptrice, et meteora. Ex Gallico translata, & ab auctore perlecta, variisque in locis emendata*. Amsterdam, 1644.

Descartes, René. *Opera philosophica*. 2nd ed. Amsterdam, 1650.

Descartes, René. *De homine. Figuris et Latinitate donatus a Florentio Schuyl*. Leiden, 1662.

Deusing, Anton. *Appendix ad dissertationem de hepatis officio seu vindiciae hepatis redivivi, leni correctione tangentes sequiorem interpretationem clarissimi viri d[omini] J[oannis] van Horne, med[icinae] doct[oris] anatomiae & chirurgiae in acad[emia] Lugd[unensi] Bat[avorum] professoris*. Groningen, 1661.

Dew, Nicholas. *The Pursuit of Oriental Learning in Louis XIV's France*. Diss. phil., Oxford, 1999 (*unprinted*).

Dew, Nicholas. *Orientalism in Louis XIV's France*. Oxford: Oxford University Press, 2009.

Dilthey, Wilhelm. "Die Autonomie des Denkens im 17. Jahrhundert." In *Stoizismus in der europäischen Philosophie, Literatur, Kunst und Politik: Eine Kulturgeschichte von der Antike bis zur Moderne*, edited by Barbara Neymeyr, Jochen Schmidt and Bernhard Zimmermann, vol. 2, pp. 977–995. Berlin: De Gruyter, 2008.

Donato, Maria Pia. "L'onere della prova. Il Sant'Uffizio, l'atomismo e i medici romani." *Nuncius 18,1* (2003), pp. 69–87.

Dörnemann, Michael. *Krankheit und Heilung in der Theologie der frühen Kirchenväter*. Diss. theol., Bochum, 2003. Studien und Texte zu Antike und Christentum 20. Tübingen: Mohr Siebeck, 2003.

Drélincourt, Charles. *Praeludium anatomicum, quod Lugdunensium in Amphitheatro ad primam anatomes suae ἐγχείρησιν adhibuit. Editio altera.* Leiden, 1672.

Duffin, Jacalyn. "Questioning Medicine in Seventeenth-Century Rome: The Consultations of Paolo Zacchia." *CBMH 28* (2011), pp. 149–170.

Dupont, Jean-Claude. "Un autre embryon? Quelques relectures classiques de l'embryologie antique." In *L'embryon: formation et animation. Antiquité grecque et latine. Tradition hébraïque, chrétienne et islamique*, edited by Luc Brisson, Marie-Hélène Congourdeau and Jean-Luc Solère, pp. 255–269. Histoire des doctrines de l'antiquité classique 38. Paris: J. Vrin, 2008.

Elkeles, Barbara. "Aussagen zu ärztlichen Leitwerten, Pflichten und Verhaltensweisen in berufsvorbereitender Literatur der frühen Neuzeit." Diss. med., Hanover, 1979. Mach. ms.

Elkeles, Barbara. "Medicus und Medikaster. Zum Konflikt zwischen akademischer und 'empirischer' Medizin im 17. und frühen 18. Jahrhundert." *MhJ 22* (1987), pp. 197–211.

Engel, George L. "Wie lange noch muß sich die Wissenschaft der Medizin auf eine Weltanschauung aus dem 17. Jahrhundert stützen?" In *Uexküll: Psychosomatische Medizin*, 5th ed., edited by Rolf H. Adler et al., pp. 3–11. Munich: Urban und Schwarzenberg, 1996.

Enke, Ulrike, ed. *Samuel Thomas Soemmerring: Schriften zur Embryologie und Teratologie.* Samuel Thomas Soemmerring; Werke 11. Basel: Schwabe, 2000.

Enke, Ulrike. "Von der Schönheit der Embryonen: Samuel Thomas Soemmerrings Werk Icones embryonum humanorum (1799)." In *Geschichte des Ungeborenen. Zur Erfahrungs- und Wissenschaftsgeschichte der Schwangerschaft, 17.–20. Jahrhundert*, edited by Barbara Duden, Jürgen Schluhmbohn and Patrice Veit, pp. 205–235. Veröffentlichungen des Max-Planck-Instituts für Geschichte 170. Göttingen: Vandenhoeck & Ruprecht, 2002.

Epple, Alois. *KZ Türkheim: Das Dachauer Außenlager Kaufering VI*, Bielefeld: Lorbeer-Verlag, 2009.

Ernesti, Jörg. *Ferdinand von Fürstenberg (1626–1683): Geistiges Profil eines barocken Fürstbischofs.* Habil. theol., Mainz, 2003. Studien und Quellen zur westfälischen Geschichte 51. Paderborn: Bonifatius, 2004.

Ernst, Michael. *Ärztliches Handeln und ethische Fragen am Lebensende im Ventilabrum medico-theologicum von Michael Boudewyns (1666).* Diss. med., Würzburg, 2011. Medizinhistorische Schriften 7. Cologne: WiKu-Verlag, 2012.

Estermann, Felicitas. *Die Menschen so zu lieben …: Leben und Wirken des Seelsorgers Stanis-Edmund Szydzik.* Bonn: Siering, 2005.

Étaix, Raymond, Charles Morel, and Bruno Judic, ed. *Grégoire le Grand: Homélies sur l'évangile.* Bk. 1, *Homélies I–XX. Texte latin, introduction, traduction et notes.* Sources chrétiennes 485. Paris: Cerf, 2005.

Fabroni, Angelo. *Lettere inedite di uomini illustri. Tomo secondo.* Florence, 1775.

Fabroni, Angelo. *Vitae Italorum doctrina excellentium qui saeculis XVII. et XVIII. floruerunt.* Vol. 11. Pisa, 1785.

Fabroni, Angelo. *Historiae Academiae Pisanae volumen III*. Pisa, 1795. Reprint, Bologna, 1971.

Faller, Adolf. "Eintrittskarten für das Anatomische Theater." *Acta anat. 2* (1946/1947), pp. 401–402.

Faller, Adolf. *Nicolaus Stenonis: Anatom, Geologe und Bischof. Das Abenteuer eines reich bewegten Lebens. Abschiedsvorlesung von Prof. Dr. med. Adolf Faller. Mittwoch, den 21. Juni 1978, 18.15 h.* Freiburger Universitätsreden, n.s. 32. Fribourg: Universitätsverlag, 1978.

Faller, Adolf. "Die 'Tetralogie von Fallot': Zur geschichtlichen Entwicklung von Diagnose und Therapie eines kongenitalen Herzsyndroms von Niels Stensen bis zur modernen Herzchirurgie." *Gesnerus 39* (1982), pp. 321–346.

Faller, Adolf. "Welchen Platz nimmt Stensens anatomische Forschung in Lorenz Heisters Chirurgie und Anatomie ein?" *Gesnerus 40* (1983), pp. 55–65.

Fangerau, Heiner. "Ethik der medizinischen Forschung." In *Geschichte, Theorie und Ethik der Medizin: Eine Einführung*, edited by Stefan Schulz, Klaus Steigleder, Heiner Fangerau and Norbert W. Paul, pp. 283–300. Frankfurt: Suhrkamp, 2006.

Farinacci, Prospero. *Tractatus de haeresi. In quo per quaestiones, regulas, ampliationes, limitationes, quidquid iure civili & canonico; quidquid sacris conciliis, summorumque pontificum constitutionibus sancitum, & communiter in ea materia receptum; quidquid denique in praxi servandum, brevi methodo illustratur. Cum argumentis, summariis, & indice locupletissimo. Editio novissima*. Lyon, 1650.

Faucci, Ugo. *La polemica Bonomo-Lancisi sull' 'origine acarica della scabbia': Pubblicazione del codice della Biblioteca Lancisiana di Roma, fatta celebrandosi in Livorno, il giorno 20 Giugno 1937 – XV, il 250° anniversario della scoperta di Giovan Cosimo Bonomo e di Diacinto Cestoni* [...]. Società Italiana di dermatologia e sifilografia; Onoranze a Giovan Cosimo Bonomo ed a Diacinto Cestoni / Estratto dal "Bollettino consorziale" Anno XXIII – 1937 – XV. Livorno: Belforte, 1937.

Favrholdt, David, ed. *Niels Bohr: Collected Works*. Vol. 10, *Complementarity beyond Physics (1928–1962)*. Amsterdam: North-Holland, 1999.

Filippini, Nadia Maria. "Die 'erste Geburt': Eine neue Vorstellung vom Fötus und vom Mutterleib (Italien, 18. Jahrhundert)." In *Geschichte des Ungeborenen: Zur Erfahrungs- und Wissenschaftsgeschichte der Schwangerschaft, 17.–20. Jahrhundert*, edited by Barbara Duden, Jürgen Schluhmbohn and Patrice Veit, pp. 99–127. Veröffentlichungen des Max-Planck-Instituts für Geschichte 170. Göttingen: Vandenhoeck & Ruprecht, 2002.

Findlen, Paula. "Controlling the Experiment: Rhetoric, Court Patronage and the Experimental Method of Francesco Redi." *HS 31* (1993), pp. 35–64.

Findlen, Paula. *Possessing Nature: Museums, Collecting, and Scientific Culture in Early Modern Italy*. Studies on the History of Society and Culture 20. Berkeley, CA: University of California Press, 1994.

Finger, Stanley, and Marco Piccolino. *The Shocking History of Electric Fishes: From Ancient Epochs to the Birth of Modern Neurophysiology*, Oxford: Oxford University Press, 2011.

Fiorentini, Girolamo. *Disputatio de ministrando baptismo humanis foetibus abortivorum. In hac secunda impressione ab eodem authore S[acrae] Indicis Congregationis iussu recognita, & declarata.* Lucca, 1665.

Fischer-Defoy, Werner. "Die Studienreise des nachmaligen Jenenser Professors Krause (1666–1670)." *DMW 36* (1910), pp. 324–326, 371–372.

Flourens, Marie Jean Pierre. "Sulla curabilità degli ascessi del cervello. – Nota comunicata all'Accademia delle scienze nella seduta del 17 nov. 1862." *AUn. 183 = 4. Ser. 47* (1863), pp. 171–175.

Frankl, Viktor E. *... trotzdem Ja zum Leben sagen: Ein Psychologe erlebt das Konzentrationslager*, 29th ed. Munich: Deutscher Taschenbuch-Verlag, 2008.

Friedensburg, Walter. *Geschichte der Universität Wittenberg.* Halle an der Saale: Max Niemeyer, 1917.

Friedrich, Susanne. *Drehscheibe Regensburg: Das Informations- und Kommunikationssystem des Immerwährenden Reichstags um 1700.* Institut für Europäische Kulturgeschichte der Universität Augsburg; Colloquia Augustana 23. Berlin: Akademie-Verlag 2007.

Furnari, Marianna Gensabella. "The Scientist Demanding Wisdom: The 'Bridge to the Future' by Van Rensselaer Potter." *PBM 45* (2002), pp. 31–42.

Fye, W. Bruce. "Niels Stensen." *ClinCard. 19* (1996), pp. 440–441 = in *Profiles in Cardiology: A Collection of Profiles Featuring Individuals Who Have Made Significant Contributions to the Study of Cardiovascular Disease*, edited by J. Willis Hurst, C. Richard Conti and W. Bruce Fye, ch. 1., pp. 27–28. Mahwah, NJ: Foundation for Advances in Medicine and Science, 2003.

Gadebusch Bondio, Mariacarla. "Anatomie der Hand und anatomisches Handwerk vor und nach Andreas Vesal." In *Die Hand: Elemente einer Medizin- und Kulturgeschichte*, pp. 79–116. Kultur: Forschung und Wissenschaft 14. Berlin: Lit, 2010.

Gaither, Kecia, Andrea Ardite, and Sarita Dhuper. "In Utero Diagnosis of Agenesis of the Ductus Arteriosus in a Twin Pregnancy: An Unusual Case Presentation." *ISRN-OG* 12/02/2010 (published in 2011), http://www.isrn.com/journals/obgyn/2011/258431 (PDF file etc., paginated, 3 pages).

Gasparri, Pietro, ed. *Codicis iuris canonici fontes.* Vol. 3, *Romani pontifices A. 1867–1917. N. 545–713.* Rome: Typis Polyglottis Vaticanis, 1933.

Gasparri, Pietro, ed. *Codicis iuris canonici fontes.* Vol. 1, *Concilia generalia – Romani pontifices usque ad annum 1745. N. 1–364.* Rome: Typis Polyglottis Vaticanis, 1947.

Gasparri, Pietro, ed. *Codicis iuris canonici fontes.* Vol. 4, *Curia Romana S. C. S. Off. – S. C. Ep. et Reg. N. 714–2055.* Rome: Typis Polyglottis Vaticanis, 1951.

Gassendi, Pierre. *Animadversiones in decimum librum Diogenis Laertii, qui est de vita, moribus, placitisque Epicuri [...].* Lyon, 1649.

Gélis, Jacques. *L'arbre et le fruit: La naissance dans l'Occident moderne (XVIᵉ–XIXᵉ siècle).* Paris: Fayard, 1984.

Geoffroy de Grandmaison, Charles-Alexandre. *La congrégation (1801–1830)*, 2nd ed. Paris: E. Plon, Nourrit et Cⁱᵉ, 1890.

Geulincx, Arnout. *De virtute et primis ejus proprietatibus, quae vulgo virtutes cardinales vocantur[,] tractatus ethicus primus.* Leiden, 1665.

Giese, Ernst, and Benno von Hagen. *Geschichte der Medizinischen Fakultät der Friedrich-Schiller-Universität Jena.* With a preface by Prof. Dr. med. Hellmuth Kleinsorge. Jena: Fischer, 1958.

Giglioni, Guido. "The Machines of the Body and the Operations of the Soul in Marcello Malpighi's Anatomy." In *Marcello Malpighi: Anatomist and Physician,* edited by Domenico Bertoloni Meli, pp. 149–174. Biblioteca di Nuncius; Studi e testi 27. Florence: L.S. Olschki, 1997.

Gios, Pierantonio, ed. *Lettere di Gregorio Barbarigo a Cosimo III de' Medici (1680– 1697).* San Gregorio Barbarigo; Fonti e ricerche 5. Padua: Istituto per la Storia Ecclesiastica Padovana, 2003.

Giovanni, Edgardo. "Les projets et tentatives de création d'une faculté complète de médecine à Fribourg." In *Histoire de l'Université de Fribourg Suisse 1889–1989: institutions, enseignement, recherches = Geschichte der Universität Freiburg Schweiz 1889–1989: Institutionen, Lehre und Forschungsbereiche.* Vol. 2, *Les Facultés = Die Fakultäten,* edited by Roland Ruffieux et al., pp. 866–875, 915. Fribourg: Academic Press Fribourg, 1991.

Goltz, Dietlinde. "Der leere Uterus: Zum Einfluß von Harveys De generatione animalium auf die Lehren von der Konzeption." *MhJ 21* (1986), pp. 242–268.

Goodrich, James Tait. [**** *Comment*]. *NSurg. 67* (2010), p. 9.
**** *The comment refers to Perrini/Lanzino/Parenti, Stensen.*

Gördes, Elisabeth. *Heilkundige in Münster i. W. im 16. und 17. Jahrhundert.* Diss. phil., Münster, **1916/1918. Beiträge für die Geschichte Niedersachsens und Westfalens 46. Hildesheim: A. Lax, 1917.

Gotfredsen, Edvard. "The Life and Work of Nicolaus Stenonis." In *Atti del XIV° Congresso internazionale di storia della medicina. Roma – Salerno 13–20 settembre 1954,* edited by Adalberto Pazzini, vol. 1, pp. 150–153. Rome, [1960].

Grappolini, Simone, Massimo Signorini, and Bernard Everett Simon. "Niccolò Stenone: A Life between Science and Faith." *APS 22* (1998), pp. 90–96.

Grell, Ole Peter. "Caspar Bartholin and the Education of the Pious Physician." In *Medicine and the Reformation,* edited by Ole Peter Grell and Andrew Cunningham, pp. 78–100. London: Routledge, 1993.

Grell, Ole Peter. "Between Anatomy and Religion: The Conversions to Catholicism of the Two Danish Anatomists Nicolaus Steno and Jacob Winsløw." In *Medicine and Religion in Enlightenment Europe,* edited by Ole Peter Grell and Andrew Cunningham, pp. 205–221 Aldershot: Ashgate, 2007.

Grell, Ole Peter. "'Like the Bees, Who Neither Suck nor Generate Their Honey from One Flower': The Significance of the Peregrinatio Academica for Danish Medical Students in the Late Sixteenth and Early Seventeenth Centuries." In *Centres of Medical Excellence? Medical Travel and Education in Europe, 1500–1789,* edited by Ole Peter Grell, Andrew Cunningham and Jon Arrizabalaga, pp. 171–189. Farnham: Ashgate, 2010.

Greyerz, Kaspar von. "Religion und Natur in der Frühen Neuzeit: Aspekte einer vielschichtigen Beziehung." In *"Die Natur ist überall bey uns": Mensch und Natur in der Frühen Neuzeit*, edited by Sophie Ruppel and Aline Steinbrecher, pp. 41–58. Zürich: Chronos, 2009.

Groh, Dieter. *Schöpfung im Widerspruch: Deutungen der Natur und des Menschen von der Genesis bis zur Reformation*. Frankfurt: Suhrkamp, 2003.

Gross, Hanns. *Rome in the Age of Enlightenment: The Post-Tridentine Syndrome and the Ancien Regime*. Cambridge: Cambridge University Press, 1990.

Grün, Hugo. "Die Medizinische Fakultät der Hohen Schule Herborn." *NassA 70* (1959), pp. 55–144.

Guenellon, Pieter. *Epistolica dissertatio, de genuina medicinam instituendi ratione*. Amsterdam, 1680.

Guerrini, Luigi. "Due lettere inedite di Tommaso Frosini a Francesco Redi sul De Motu Animalium di Giovanni Alfonso Borelli." *Nuncius 13,1* (1998), pp. 193–208.

Guerrini, Luigi. "Contributo critico alla biografia Rediana: Con un studio su Stefano Lorenzini e le sue 'Osservazioni intorno alle torpedini'." In *Francesco Redi: Un protagonista della scienza moderna. Documenti, esperimenti, immagini*, edited by Walter Bernardi and Luigi Guerrini, pp. 47–69. Biblioteca di Nuncius; Studi e testi 33. Florence: L.S. Olschki, 1999.

Gysel, Carlos. "Le conflit, à propos de la parotide, entre Nicolas Sténon (1638–1686) et Gérard Blasius (1625–1682)." *AOS no. 196* (1996), pp. 529–548.

Hack, Achim Thomas. *Gregor der Große und die Krankheit*. Päpste und Papsttum 41. Stuttgart: Anton Hiersemann, 2012.

Hack, Tobias. *Der Streit um die Beseelung des Menschen: Eine historisch-systematische Studie*. Diss. theol., Freiburg, 2010. Studien zur theologischen Ethik 131. Fribourg: Academic Press Fribourg, 2011.

Hack-Molitor, Gisela. *On Tiptoe in Heaven: Mystik und Reform im Werk von Sir Thomas Browne (1605–1682)*. Diss. phil., Heidelberg, 1998. Anglistische Forschungen 295. Heidelberg: C. Winter, 2001.

Haller, Albrecht von. *Bibliotheca anatomica. Qua scripta ad anatomen et physiologiam facientia a rerum initiis recensentur. Tomus I. Ad annum 1700*. Zürich, 1774.

Harvey, William. *Exercitationes de generatione animalium. Quibus accedunt quaedam de partu[,] de membranis ac humoribus uteri[,] & de conceptione*. Amsterdam, 1651.

Heida, Ulrike. *Niels Stensen in den Beziehungen zu medizinischen Fachkollegen seiner Zeit [on the front cover: Niels Stensen und seine Fachkollegen]*. Diss. med., Hamburg, 1986. Grosse-Scripta 12. Berlin: Grosse, 1986.

Helm, Jürgen. "Religion and Medicine: Anatomical Education at Wittenberg and Ingolstadt." In *Religious Confessions and the Sciences in the Sixteenth Century*, edited by Jürgen Helm and Annette Winkelmann, pp. 51–68. Studies in European Judaism 1. Leiden: Brill, 2001.

Helwig, Christoph Sr. *B[enevolo] l[ectori] s[alutem]. Quod felix & faustum & Academiae huic salutare faciat Deus T[er] O[ptimus] M[aximus] amplissima facultas medica Gryphiswaldensis in solennis Anatomes publica panegyri curiosorum oculis sistet cadaver infelicis feminae, [...]*. Greifswald, 1677.

Herberger, Patricia, and Michael Stolleis. *Hermann Conring 1606–1681: Ein Gelehrter der Universität Helmstedt. Ausstellung der Herzog August Bibliothek Wolfenbüttel im Juleum Helmstedt 12. Dezember 1981 bis 31. März 1982, im Alten Rathaus zu Norden Frühsommer 1982, im Museum für das Fürstentum Lüneburg Herbst 1982*. Ausstellungskataloge der Herzog August Bibliothek 33. Wolfenbüttel: Herzog August Bibliothek, 1981.

Hoboken, Nicolaas. *Anatomia secundinae humanae, quindecim figuris ad vivum propria autoris manu delineatis, illustrata. [...] Cum annexo s[ive] spicilegio epistolarum, rem potissimum generatoriam referentium*. Utrecht, 1669.

Høeg, Eiler. *Licent. med. Johann Valentin Wille (Johannes Valentinus Willius): Læge hos Abrahamstrups (Jægerspris) Ejer Over-Jægermester Vincentz Joachim Hahn i Aarene 1674–1676 samt kgl. dansk Feltmedikus*. Copenhagen: Levin & Munksgaard, 1934.

Hollnagel-Jensen, Oluf Cecilius, and Erik Andreasen: "Om Ludvig Jacobson og et glemt haandskrift om det Jacobsonske Organ." *DVÅ 11* (1944), pp. 5–25.

Holomanova, Anna, Anna Ivanova, and Ingrid Brucknerova. "Niels Stensen – Prestigious Scholar of the 17th Century." *BLLi. 103,2* (2002), pp. 90–93.

Holzberg, Niklas, ed. *Publius Ovidius Naso: Liebeskunst. Ars amatoria. Heilmittel gegen die Liebe. Remedia amoris. Lateinisch-deutsch*, 4th ed. Düsseldorf: Artemis & Winkler, 1999.

Holzberg, Niklas, ed. *Publius Ovidius Naso: Briefe aus der Verbannung. Tristia – Epistulae ex Ponto. Lateinisch und deutsch*, 5th ed. Translation by Wilhelm Willige. Mannheim: Artemis & Winkler, 2011.

Horn, Sonia, and Gabriele Dorffner. "'… männliches Geschlecht ist für die Zulassung zur Habilitation nicht vorgesehen'. Die ersten an der medizinischen Fakultät der Universität Wien habilitierten Frauen." In *Töchter des Hippokrates: 100 Jahre akademische Ärztinnen in Österreich*, edited by Birgit Bolognese-Leuchtenmüller and Sonia Horn, pp. 117–138. Vienna: ÖÄK-Verlag, 2000.

Hübener, Wolfgang. "Der Praxisbegriff der aristotelischen Tradition und der Praktizismus der Prämoderne." In *Theoria cum praxi: Zum Verhältnis von Theorie und Praxis im 17. und 18. Jahrhundert. Akten des III. internationalen Leibnizkongresses, Hannover, 12. bis 17. November 1977*. Vol. 1, *Theorie und Praxis, Politik, Rechts- und Staatsphilosophie*, pp. 41–59. Studia Leibnitiana; Supplementa 19. Wiesbaden: F. Steiner, 1980.

Hübener, Wolfgang. *Zum Geist der Prämoderne*. Würzburg: Königshausen & Neumann, 1985.

Huisman, Tim. *The Finger of God: Anatomical Practice in 17th-Century Leiden*. Diss. phil., Leiden, 2008. Leiden: Primavera Pers, 2009.

Hurd-Mead, Kate Campbell. *A History of Women in Medicine: From the Earliest Times to the Beginning of the Nineteenth Century*. Haddam, CT: The Haddam Press, 1938. Reprint, New York: AMS Press, 1977.

[Innocent XI, ed.]. *Decret de N[otre] S[aint] P[ère] le Pape Innocent XI. contre plusieurs propositions de morale. Suivant les Exemplaires de Rome, De l'Imprimerie de la Reverendissime Chambre Apostolique*. N.p., 1679.

Institut für Anatomie und spezielle Embryologie der Universität Freiburg (Schweiz), ed. *Prof. Dr. med. A. Faller, 1913–1983 zum 70. Geburtstag: Nachtrag 1973–83 zu der vom Institut für Anatomie und spezielle Embryologie der Universität Freiburg (Schweiz) herausgegebenen Broschüre zum 60. Geburtstag.* Date and place of publication not specified.

Iofrida, Manlio. "Niels Stensen." In *Die Philosophie des 17. Jahrhunderts.* Vol. 1, *Allgemeine Themen: Iberische Halbinsel. Italien,* edited by Jean-Pierre Schobinger, pp. 890–897, 962–963. Grundriss der Geschichte der Philosophie; Ueberweg, 17. Jahrhundert 1/2. Basel: Schwabe, 1998.

Israel, Jonathan I. *Radical Enlightenment: Philosophy and the Making of Modernity 1650–1750.* Oxford: Oxford University Press, 2001.

Jaki, Stanley L. *Brain, Mind and Computers,* 2nd ed. South Bend, IN: Gateway Editions, 1978.

Jamison, Andrew. *National Components of Scientific Knowledge: A Contribution to the Social Theory of Science.* Diss. phil., Göteborg, 1983. Lund: Research Policy Institute, 1982.

Janse, Wim. "Reformed Theological Education at the Bremen Gymnasium Illustre." In *Bildung und Konfession: Theologenausbildung im Zeitalter der Konfessionalisierung,* edited by Herman J. Selderhuis and Markus Wriedt, pp. 31–49. Spätmittelalter und Reformation, Neue Reihe 27. Tübingen: Mohr Siebeck, 2006.

Jessenius von Jessen, Johannes. *Pro anatome sua actio, & ad spectandum invitatio.* Wittenberg, 1600.

Jones, David Albert. *The Soul of the Embryo: An Enquiry into the Status of the Human Embryo in the Christian Tradition.* London: Continuum, 2004.

Judic, Bruno, ed. *Grégoire le Grand: Règle pastorale.* Bk. 2. Sources chrétiennes 382. Paris: Cerf, 1992.

Kaiser, Wolfram, and Arina Völker. "Die Medizin im 17. Jahrhundert." In *Naturwissenschaftliche Revolution im 17. Jahrhundert,* edited by Günter Wendel, pp. 231–248. Beiträge zur Wissenschaftsgeschichte 7. Berlin: Deutscher Verlag der Wissenschaften, 1989.

Kardel, Troels. "Steno on Hydrocephalus: Introduction to Niels Stensen's Letter 'On a Calf with Hydrocephalus', with a Short Biography." *JHNS* 2,2 (1993), pp. 171–178.

Kardel, Troels. *Steno: Life, Science, Philosophy. With Niels Stensen's Prooemium or Preface to a Demonstration in the Copenhagen Anatomical Theater in the Year 1673, and Holger Jacobæus: Niels Stensen's Anatomical Demonstration no. XVI, and Other Texts Translated from Latin.* Acta historica scientiarum naturalium et medicinalium 42. Copenhagen: Danish National Library of Science and Medicine, 1994.

Kardel, Troels, ed. *Steno on Muscles Containing Stensen's Myology in Historical Perspective: Niels Stensen's New Structure of the Muscles and Heart [1663] and Specimen of Elements of Myology [1667].* With facsimile of first editions annotated by Harriet Hansen and August Ziggelaar SJ. Transactions of the American Philosophical Society 84,1. Philadelphia: American Philosophical Society, 1994.

Kardel, Troels, and Paul Maquet, ed. *Nicolaus Steno: Biography and Original Papers of a 17th Century Scientist.* Berlin: Springer, 2013.

Kenny, Neil. *The Uses of Curiosity in Early Modern France and Germany.* Oxford: Oxford University Press, 2004.

Kernbauer, Alois. "Carla Zawisch-Ossenitz: Eine biografische Skizze der ersten Professorin der Grazer Universität." In *Frauenstudium und Frauenkarrieren an der Universität Graz*, edited by Alois Kernbauer and Karin Schmidlechner-Lienhart, pp. 265–270. Publikationen aus dem Archiv der Universität Graz 33. Graz: Akademische Druck- und Verlagsanstalt, 1996.

Kervran, Roger. *Laennec: His Life and Times.* Oxford: Pergamon Press, 1960.

King, Lester S. *The Road to Medical Enlightenment (1650–1695).* London: Macdonald, 1970.

Kircher, Athanasius. *Magnes sive de arte magnetica opus tripartitum, quo praeterquam quod universa magnetis natura, eiusque in omnibus artibus & scientijs usus nova methodo explicetur, e viribus quoque & prodigiosis effectibus magneticarum, aliarumq[ue] abditarum naturae motionum in elementis, lapidibus, plantis & animalibus elucescentium, multa hucusque incognita naturae arcana per physica, medica, chymica & mathematica omnis generis experimenta recluduntur. Editio secunda post Romanam multo correctior.* Cologne, 1643.

Kirwin, Harry W. "James J. Walsh: Medical Historian and Pathfinder." *CHR 45* (1960), pp. 409–435.

Klomps, Heinrich. *Ehemoral und Jansenismus: Ein Beitrag zur Überwindung des sexualethischen Rigorismus.* Habil. theol., Bonn, 1963. Cologne: J.P. Bachem, 1964.

Koch, Carl Hendrik. "Dänemark." In *Die Philosophie des 17. Jahrhunderts.* Vol. 4, *Das Heilige Römische Reich Deutscher Nation: Nord- und Ostmitteleuropa*, edited by Helmut Holzhey and Wilhelm Schmidt-Biggemann, pp. 1246–1258. Grundriss der Geschichte der Philosophie; Ueberweg: 17. Jahrhundert 4/2. Basel: Schwabe, 2001.

Koch, Hans-Theodor. "Die Wittenberger medizinische Fakultät (1502–1652): Ein biobibliographischer Überblick." In *Medizin und Sozialwesen in Mitteldeutschland zur Reformationszeit*, edited by Stefan Oehmig, pp. 289–348. Schriften der Stiftung Luthergedenkstätten in Sachsen-Anhalt 6. Leipzig: Evangelische Verlagsanstalt, 2007.

Koch, Jeanette E. "Niccolò Stenone (Niels Steensen), 1638–1686. Bibliografia." *MNIR 38=n.s. 3* (1976), pp. 135–157.

Kochuthara, Shaji George. *The Concept of Sexual Pleasure in the Catholic Moral Tradition.* Diss. theol., Rome, 2006. Tesi Gregoriana; Serie Teologia 152. Rome: Editrice Pontificia Università Gregoriana, 2007.

Kohl, Wilhelm, ed. *Das Bistum Münster.* Vol. 7, *Die Diözese*, bk. 4. Germania sacra; Historisch-statistische Beschreibung der Kirche des Alten Reiches; n.s. 37,4. Berlin: De Gruyter, 2004.

Kohl, Wilhelm, ed. *Das Bistum Münster.* Vol. 8, *Das (freiweltliche) Damenstift Nottuln.* Germania sacra; Historisch-statistische Beschreibung der Kirche des Alten Reiches; n.s. 44. Berlin: De Gruyter, 2005.

Kohlsaat, Heike. "Die Untersuchungen von Niels Stensen unter besonderer Berücksichtigung seiner Erkenntnisse über die Haut und ihre Anhangsorgane." Diss. med. dent., Hamburg, 1967. Mach. ms.

[Kölliker, Albert von]. *Zur Geschichte der medicinischen Facultät an der Universität Würzburg*. Speech at the celebration commemorating the endowment of the Julius-Maximilians-Universität, held by Dr. Albert von Kölliker, Professor of Anatomy and Rector, on January 2, 1871. Würzburg, n.d.

Korff, Rüdiger. *Das Berufsethos in der Chirurgie Lorenz Heisters (1683–1758)*. Diss. med., Zürich, 1975. Zürcher Medizingeschichtliche Abhandlungen; n.s. 107. Zürich: Juris-Verlag, 1975.

Kragh, Helge. "From Medieval Scholarship to New Science: Circa 1000–1730." In *Science in Denmark: A Thousand-Year History*, edited by Helge Kragh, Peter C. Kjærgaard, Henry Nielsen, Kristian Hvidtfelt Nielsen, pp. 17–125, 551–553. Aarhus: Aarhus University Press, 2008.

Kraus, Max-Joseph. "Niels Stensen in Leiden." Diss. med., Budapest, 1999. Under: http://www.xammax.medmatic.com/texte/stensenportal.htm (PDF file and others, paginated, 94 pages).

Kreß,   Hartmut.   *Medizinische   Ethik: Gesundheitsschutz   – Selbstbestimmungsrechte – heutige Wertkonflikte*, 2nd ed. Ethik – Grundlagen und Handlungsfelder 2. Stuttgart: Kohlhammer, 2009.

Lackmann, Heinrich, ed. *Katholische Reform im Niederstift Münster: Die Akten der Generalvikare Johannes Hartmann und Petrus Nicolartius über ihre Visitationen im Niederstift Münster in den Jahren 1613 bis 1631/1632*. Westfalia sacra 14. Münster: Aschendorff, 2005.

Lackmann, Heinrich, Tobias Schrörs, Reimund Haas, and Reinhard Jüstel, ed. *Katholische Reform im Fürstbistum Münster unter Ferdinand von Bayern: Die Protokolle von Weihbischof Arresdorf und Generalvikar Hartmann über ihre Visitationen im Oberstift Münster in den Jahren 1613 bis 1616*. Westfalia sacra 16. Münster: Aschendorff, 2012.

Lancisi, Giovanni Maria. *Dissertatio de recta medicorum studiorum ratione instituenda, habita ad novae Academiae alumnos, & medicinae tyrones in Archinosocomio S[ancti] Spiritus in Saxia*. Rome, 1715.

Lanza, Antonio. "La questione del momento in cui l'anima razionale è infusa nel corpo." *BollFil. 4* (1938), pp. 211–266, 335–367, and *5* (1939), pp. 36–97, 206–273.

Lanza, Antonio. *La questione del momento in cui l'anima razionale è infusa nel corpo*. Biblioteca del bollettino filosofico 2. Rome: Pontificio Ateneo Lateranense, 1939.****

**** *Extended book version of Lanza, "Momento"*.

* Larsen, Knud, and Gustav Scherz, ed. *Nicolai Stenonis opera theologica cum prooemiis ac notis Germanice scriptis*, 2nd ed. Vol. 1. Copenhagen: Nyt Nordisk Forlag, 1944.

* Larsen, Knud, and Gustav Scherz, ed. *Nicolai Stenonis opera theologica cum prooemiis ac notis Germanice scriptis*. Vol. 2. Copenhagen: Nyt Nordisk Forlag, 1947.

Lazzarelli, Giovanni Francesco. *La Cicceide legitima*. Studi Pichiani 11. Florence: Olschki, 2007.

Lederer, Jean. "Le bienheureux Nicolas Sténon ou l'intégration de la science et de la foi." *RQS 165,1* (1994), pp. 11–36, and *165,2* (1994), pp. 89–118.

Lehmann, Karl. *Das Eintreten für das Lebensrecht des ungeborenen Kindes als christlicher und humaner Auftrag*. Opening speech during the autumn plenary meeting of the German Bishops' Conference, September 23, 1991. Der Vorsitzende der Deutschen Bischofskonferenz 16. Bonn: Sekretariat der Deutschen Bischofskonferenz, 1991.

Lehmann, Karl. "Das Eintreten für das Lebensrecht [I] des ungeborenen Kindes als christlicher und humaner Auftrag." In *Herausforderung Schwangerschaftsabbruch: Fakten – Argumente – Perspektiven*, edited by Johannes Reite and Rolf Keller, pp. 34–60. Freiburg: Herder, 1992.

Lehmann, Karl. "Das Eintreten für das Lebensrecht [II] des ungeborenen Kindes als christlicher und humaner Auftrag." *IKZ 21* (1992), pp. 103–122.

Lehmann, Karl. *Das Recht, ein Mensch zu sein: Zur Grundfrage der gegenwärtigen bioethischen Probleme*. Opening speech during the autumn plenary meeting of the German Bishops' Conference in Fulda, September 24, 2001. Der Vorsitzende der Deutschen Bischofskonferenz 22. Bonn: Sekretariat der Deutschen Bischofskonferenz, 2001.

Lehnert, Hendrik. *Innere Medizin zwischen Zeitgeist und Zeitenwende*. Speech by the Chaiman of the Deutsche Gesellschaft für Innere Medizin, Prof. Dr. Hendrik Lehnert (Lübeck) on May 1, 2011 in Wiesbaden. Under: http://www.dgim.de/portals/pdf/kongresse/DGIM%20PraesidentenredeDruck-Version%20final%202011-05-23.pdf (paginated, 24 pages).

Leibbrand, Werner. *Der göttliche Stab des Äskulap: Vom geistigen Wesen des Arztes*, 3rd ed. Salzburg: Müller, [1952].

Leibniz-Archiv der Niedersächsischen Landesbibliothek Hannover, ed. *Gottfried Wilhelm Leibniz: Mathematischer, naturwissenschaftlicher und technischer Briefwechsel*. Vol. 2, *1676–1679*. Berlin: Akademie-Verlag, 1987.

Leibniz-Forschungsstelle der Universität Münster, ed. *Gottfried Wilhelm Leibniz: Philosophischer Briefwechsel 1663–1685*, 2nd ed. Berlin: Akademie-Verlag, 2006.

Leone, Salvino. "The Ancient Roots of a Recent Debate." In *Identity and Statute of Human Embryo: Proceedings of Third Assembly of the Pontifical Academy for Life (Vatican City, February 14–16, 1997)*, edited by Juan de Dios Vial Correa and Elio Sgreccia, pp. 28–47. Vatican City: Libreria Editrice Vaticana, 1998.

Leone, Salvino. *La prospettiva teologica in bioetica*. Facoltà Teologica di Sicilia; Istituto Siciliano di Bioetica; Collectio moralis 5. Acireale: Istituto Siciliano di Bioetica, 2002.

Lesky, Erna. "Die Entdeckung der Funktion des Säugetierovars durch Nicolaus Stensen." In *Steno and Brain Research in the Seventeenth Century: Proceedings of the International Historical Symposium on Nicolaus Steno and Brain Research in the Seventeenth Century Held in Copenhagen 18–20 August 1965*, edited by Gustav Scherz, pp. 235–251. Analecta Medico-Historica Academiae Internationalis Historiae Medicinae 3. Oxford: Pergamon, 1968.

Lesky, Erna. "Männliches und weibliches Prinzip im Zeugungsgeschehen: Von Aischylos bis Stensen." *FdA 2* (1971), pp. 2–12.

Lessius, Leonhard. *Hygiasticon seu vera ratio valetudinis bonae et vitae una cum sensuum, judicii, & memoriae integritate ad extremam senectutem conservandae [...] Subjungitur tractat[us] Ludovici Cornari Veneti, eodem pertinens, ex Italico in Lat[inum] serm[onem] ab ipso Lessio translatus.* Molsheim, 1670.

Liceti, Fortunio. *De monstris: Ex recensione Gerardi Blasii, [...] qui monstra quaedam nova & rariora ex recentiorum scriptis addidit. Editio novissima. Iconibus illustrata.* Padua, 1658.

Lindeboom, Gerrit A. *Descartes and Medicine.* Nieuwe Nederlandse Bijdragen tot de Geschiedenis der Geneeskunde 1. Amsterdam: Rodopi, 1979.

Lindeboom, Gerrit A. "Leeuwenhoek and the problem of sexual reproduction." In *Antoni van Leeuwenhoek 1632–1723: Studies on the Life and Work of the Delft Scientist Commemorating the 350th Anniversary of his Birthday*, edited by Lodewijk C. Palm and Harry A. M. Snelders, pp. 129–152. Nieuwe Nederlandse Bijdragen tot de Geschiedenis der Geneeskunde en der Natuurwetenschappen 8. Amsterdam: Rodopi, 1982.

Lobato, Abelardo. "Presentazione." In *Etica dell'atto medico*, pp. 5–10. Philosophia 7. Bologna: Edizioni Studio Domenicano, 1991.

Lombardi, Daniela. *Povertà maschile, povertà femminile: L'ospedale dei Mendicanti nella Firenze dei Medici.* Bologna: Il Mulino, 1988.

López Piñero, José María. *La medicina y las ciencias biológicas en la historia valenciana.* Valencia: Ajuntament de València, 2004.

Lorenzini, Stefano. *Osservazioni intorno alle torpedini [...].* Florence, 1678.

Lux, David S. *Patronage and Royal Science in Seventeenth-Century France: The Académie de Physique in Caen.* Ithaca, NY: Cornell University Press, 1989.

* Maar, Vilhelm, ed. *Nicolai Stenonis opera philosophica.* Vol. 1. Copenhagen: V. Tryde, 1910[a].

* Maar, Vilhelm, ed. *Nicolai Stenonis opera philosophica.* Vol. 2. Copenhagen: V. Tryde, 1910[b].

Maar, Vilhelm, ed. *Holger Jacobæus' Rejsebog (1671–1692).* Copenhagen: Gyldendal, 1910[c].

Maar, Vilhelm. "Et Blad af Domus Anatomicas Historie." In *Thomas Bartholin: 1616 – 20. Oktober – 1916. Mindeskrift paa 300 Aarsdagen for Hans Fødsel*, pp. 14–19. Copenhagen: J.L. Lybecker, 1916.

Maclean, Ian. "The Medical Republic of Letters Before the Thirty Years War." *IHR 18* (2008), pp. 15–30.

Maehle, Andreas-Holger. *Kritik und Verteidigung des Tierversuchs: Die Anfänge der Diskussion im 17. und 18. Jahrhundert.* Habil. med., Göttingen, 1990. Stuttgart: F. Steiner, 1992.

Maehle, Andreas-Holger. "Werte und Normen: Ethik in der Medizingeschichte." In *Medizingeschichte: Aufgaben, Probleme, Perspektiven*, edited by Norbert Paul and Thomas Schlich, with the assistance of Stefanie Kuhne, pp. 335–354. Frankfurt: Campus, 1998.

Mägdefrau, Werner. "Der Aufstieg und die erste Blütezeit der Universität Jena in der zweiten Hälfte des 17. Jahrhunderts." Essay. In *Geschichte der Universität Jena 1548/1558–1958: Festgabe zum vierhundertjährigen Universitätsjubiläum*. Vol. 1, *Darstellung*, edited by Max Steinmetz, pp. 111–165. Jena: G. Fischer, 1958.

Mägdefrau, Werner. "Der Aufstieg und die erste Blütezeit der Universität Jena in der zweiten Hälfte des 17. Jahrhunderts." Notes. In *Geschichte der Universität Jena 1548/1558–1958: Festgabe zum vierhundertjährigen Universitätsjubiläum*. Vol. 2, *Quellenedition zur 400-Jahr-Feier 1958. Archivübersichten, Quellen- und Literaturberichte, Anmerkungen, Abbildungskatalog, Literaturverzeichnis, Personen- und Ortsregister, Abkürzungsverzeichnis*, edited by Max Steinmetz, pp. 475–485 Jena: G. Fischer, 1962.

Malpighi, Marcello. *De pulmonibus observationes anatomicae*. Bologna, 1661.

Malpighi, Marcello. *Dissertatio epistolica de bombyce*. London, 1669.

Malpighi, Marcello. *Opera posthuma: In quibus excellentissimi authoris vita continetur, ac pleraque quae ab ipso prius scripta aut inventa sunt confirmantur, & ab adversariorum objectionibus vindicantur. Supplementa necessaria, & praefationem addidit, innumerisque in locis emendavit Petrus Regis [...]. Editio ultima figuris aeneis illustrata, priori longe praeferenda*. Amsterdam, 1698.

Manni, Domenico Maria. *Vita del letteratissimo monsig. Niccolò Stenone di Danimarca vescovo di Titopoli* [sic!] *e vicario apostolico*. Florence, 1775.

Marini-Bettòlo, Giovanni Battista. "The Niels Stensen Memorial Plaque: Background Notes." In *Blessed Niels Stensen and His Memorial Plaque in the Pontifical Academy of Sciences*, edited by Pontificia Academia Scientiarum, pp. 33–34. Vatican City: Ex Aedibus Academicis in Civitate Vaticana, 1989.

Marques Filho, José, and Márcio Fabri dos Anjos. "Van Rensselaer Potter e a Religião na Bioética" *Bioethikos 5* (2011), pp. 427–433. Under: http://www.saocamilo-sp.br/pdf/bioethikos/89/A9.pdf.

Martin, Jacques. *Heraldry in the Vatican = L'araldica in Vaticano = Heraldik im Vatikan*, edited by Peter Bander van Duren. Gerrards Cross: Van Duren, 1987.

Martin, Josef, ed. *Sancti Aurelii Augustini De doctrina christiana: De vera religione*. Aurelii Augustini Opera; Pars IV, 1 / Corpus Christianorum; Series Latina 32. Turnhout: Brepols, 1962.

Marx, Karl Friedrich Heinrich. "Zur Beurtheilung des Arztes Christian Franz Paullini." *AKGWG 18* (1873), pp. 53–92.

Marzollo, Marco. "Anatomia e fisiologia del musculo nell'opera di Nicolò [*sic!*] Stenone." Collana monografica del "Bollettino della Società medico-chirurgica Bresciana"; supplemento al IV fascicolo 1968. Brescia: Ospedale civile "La Memoria", 1969.

Mauch, Mercedes. *Senecas Frauenbild in den philosophischen Schriften*. Diss. phil., Freiburg, 1996. Studien zur klassischen Philologie 106. Frankfurt: P. Lang, 1997.

Mazzolini, Renato G. "Kiel 1675: La dissezione pubblica di una donna africana." In *Per una storia critica della scienza*, edited by Marco Beretta, Felice Mondella and Maria Teresa Monti, pp. 371–393. Quaderni di Acme 26. Milan: Cisalpino, 1996.

McCorkell, Edmund J. "Bertram Coghill Alan Windle, F.R.S., F.S.A., K.S.G., M.D., LL.D., Ph.D., Sc.D." *CCHA-R 25* (1958), pp. 53–58.

Meibom, Heinrich Jr. *De vasis palpebrarum novis epistola.* Helmstedt, 1666.

Meibom, Heinrich Jr. *Programma [...] quo ad anatomen corporis foeminini in novo theatro primam omnium ordinum curiosos solemniter invitat.* Helmstedt, 1673.

Meisen, Valdemar. "Niccolò Stenone: Celebre anatomista danese." In *Atti dell'VIII° Congresso internazionale di storia della medicina: Roma – dal 22 al 27 settembre 1930,* edited by Pietro Capparoni, pp. 179–189. Pisa: V. Lischi, 1931.

Meisen, Valdemar. "Nicolaus Steno (Niels Steensen)." In *Prominent Danish Scientists through the Ages: With Facsimiles from Their Works,* pp. 36–43. Copenhagen: Levin & Munksgaard, 1932.

Melchior, Johannes C. et al., ed. *Københavns Universitet 1479–1979: Udgivet af Københavns Universitet ved 500 års jubilæet.* Vol. 7, *Det lægevidenskabelige Fakultet.* Copenhagen: Gads Forlag, 1979.

Melchiors, August C. C. "Stensen und L. L. Jacobsons Entdeckung." *StenCath.* 2 (1956), p. 39.

Meli, Domenico Bertoloni. *Mechanism, Experiment, Disease: Marcello Malpighi and Seventeenth-Century Anatomy,* Baltimore: Johns Hopkins University Press, 2011.

Merton, Robert K., George G. Reader, and Patricia L. Kendall, ed. *The Student-Physician: Introductory Studies in the Sociology of Medical Education.* 2nd ed. Cambridge, MA: Harvard University Press, 1969.

Merton, Robert K. "Notes on Problem-Finding in Sociology." In *Sociology Today: Problems and Prospects,* edited by Robert K. Merton and Leonard Broom, pp. ix–xxxiv. New York: Basic Books, 1959.

Merton, Robert K. *Social Theory and Social Structure,* 3rd ed. New York: Free Press, 1968.

Merton, Robert K. "Priorities in Scientific Discovery: A Chapter in the Sociology of Science." In *The Sociology of Science,* 5th ed., edited by Bernard Barber and Walter Hirsch, pp. 447–485. New York: Free Press, 1968.

Merton, Robert K. "The Matthew Effect in Science." In *The Sociology of Science: Theoretical and Empirical Investigations,* 2nd ed. with an introduction by Norman W. Storer, pp. 439–459. Chicago: University of Chicago Press, 1974.

Messbarger, Rebecca. *The Lady Anatomist: The Life and Work of Anna Morandi Manzolini.* Chicago: University of Chicago Press, 2010.

Metzler, Josef. "Die Kongregation in der zweiten Hälfte des 17. Jahrhunderts." In *Sacrae congregationis de propaganda fide memoria rerum.* Vol. 1, *1622–1700,* pp. 244–305. Rome: Herder, 1971.

Meyer, Wilfried, and Wolfgang Schwarz. "Eine neuentdeckte 'Vita' Niels Stensens aus der Zeit um 1700." *DHVG 56* (1988), pp. 63–70.

Miert, Dirk van. *Humanism in an Age of Science: The Amsterdam Athenaeum in the Golden Age, 1632–1704.* Brill's Studies in Intellectual History 179. Leiden: Brill, 2009.

Mieth, Dietmar. "Die Moralenzyklika, die Fundamentalmoral und die Kommunikation in der Kirche." In *Moraltheologie im Abseits? Antwort auf die Enzyklika "Veritatis splendor"*, pp. 9–24. Quaestiones disputatae 153. Freiburg: Herder, 1994.

Mieth, Dietmar. "Wissenschaft, Religion und Kontingenz." In *Philosophie et théologie: Festschrift Emilio Brito*, edited by Éric Gaziaux, pp. 395–412. Bibliotheca Ephemeridum Theologicarum Lovaniensium 206. Leuven: Leuven University Press, 2007.

Mieth, Dietmar. "Science, Religion, and Contingency." In *The Contingent Nature of Life: Bioethics and the Limits of Human Existence*, edited by Marcus Düwell, Christoph Rehmann-Sutter and Dietmar Mieth, pp. 53–67. International Library of Ethics, Law, and the New Medicine 39. Berlin: Springer, 2008.

Migne, Jacques-Paul, ed. […] *Joannis Chrysostomi* […] *opera omnia* […]. *Tomus sextus*. Patrologiae cursus completus […]; Series Graeca […] 56. Paris, 1859.

Miniati, Stefano. *Nicholas Steno's Challenge for Truth: Reconciling Science and Faith*. Diss. phil., Pisa, 2009. Filosofia, storia, scienze sociali 18. Milan: FrancoAngeli, 2009.

Mirto, Alfonso. "Alessandro Segni e i suoi corrispondenti." *AMAT 67=N. S. 53* (2002), pp. 193–214.

Mirto, Alfonso. "Lettere di Stefano Gradi ai Fiorentini: Viviani, Dati, Redi, Leopoldo e Cosimo III de' Medici." *StSec. 49* (2008), pp. 371–404.

Moe, Harald. "When Steno Brought New Esteem to Glands." In *Nicolaus Steno 1638–1686: A Re-Consideration by Danish Scientists*, edited by Jacob E. Poulsen and Egill Snorrason, pp. 51–96. Gentofte: Nordisk Insulinlaboratorium, 1986.

Nathanson, Bernard N. *The Hand of God: A Journey from Death to Life by the Abortion Doctor Who Changed His Mind*, 2nd ed. Washington, DC: Regnery, 2001.

Nida-Rümelin, Julian. "Wissenschaftsethik." In *Angewandte Ethik: Die Bereichsethiken und ihre theoretische Fundierung. Ein Handbuch*, 2nd ed. Stuttgart: A. Kröner, 2005.

Nordström, Johan. "Antonio Magliabechi och Nicolaus Steno: Ur Magliabechis brev till Jacob Grovonius." *Lych. 1962* (published in 1963), pp. 1–42.

Numbers, Ronald L. "Science without God: Natural Laws and Christian Beliefs." In *When Science & Christianity Meet*, edited by David C. Lindberg and Ronald L. Numbers, pp. 265–285. Chicago: University of Chicago Press, 2003.

Olden-Jørgensen, Sebastian. "Nicholas Steno and René Descartes: A Cartesian Perspective on Steno's Scientific Development." In *The Revolution in Geology from the Renaissance to the Enlightenment*, edited by Gary D. Rosenberg, pp. 149–157. Memoir 203. Boulder, CO: Geological Society of America, 2009.

Olsen, Birger Munk, ed. *Nicolas Sténon: Œuvres choisies. Traduit du latin et de l'italien, présenté et annoté*. Classiques du Nord [Lumières] 17. Paris: Les Belles Lettres, 2010.

Overmann, Manfred. *Der Ursprung des französischen Materialismus: Die Kontinuität materialistischen Denkens von der Antike bis zur Aufklärung*. Diss.

phil., Cologne, 1991. Europäische Hochschulschriften 20,395. Frankfurt: P. Lang, 1993.

Papasoli, Benedetta. "Il soggiorno parigino di Niccolò Stenone (1664–65)." In *Niccolò Stenone 1638–1686: Due giornate di studio. Firenze 17–18 novembre 1986*, edited by Francesco Adorno and Carlo De Filippi, pp. 97–117. Biblioteca di storia della scienza 27. Florence: Olschki, 1988 = *FutUom. 14,1–2* (1987), pp. 97–117.

Parent, André. "Niels Stensen: A 17th Century Scientist with a Modern View of Brain Organization." *CJNS 40* (2013), pp. 482–492.

Paulli, Simon Jr. *Quadripartitum botanicum de simplicium medicamentorum facultatibus, in usus medicinae candidatorum, praxin medicam, Deo benedicente, auspicaturorum; nec non artis pharmaceutices studiosorum concinnatum ex veterum et recentiorum decretis, ac observationibus, cum medicis, tum anatomicis, itemque multis, chymica principia ac humaniora studia spectantibus, refertum. Additis dosibus purgantium magnopere desideratis, ex probatissimis practicis collectis. Una cum appendice & indicibus necessariis.* Strasbourg, 1667.

Paullini, Christian Franz. *De starcutero, famosissimo ggnante [corr.: gigante] boreali, dissertatio curiosa ad virum celeberrimum d[ominum] Nic[olaum] Stenonis, episcopum postea Titiopolitanum, & vicar[ium] apostolic[um], cujus cura & impensis prodiit Florentiae.* Florence, 1677 (lost).

Paullini, Christian Franz. *Cynographia curiosa seu canis descriptio, juxta methodum & leges illustris Academiae Naturae Curiosorum adornata, multisq[ue] curiosis, raris, jucundis & stupendis naturae artisque observationibus, secretis & quaestionibus referta, et mantissa curiosa ejusdem argumenti, complectente Joh[annis] Caji libell[um] de Canibus Britannicis, & Joh[annis] Henr[ici] Meibom[ii] epist[olam] de ΚΥΝΟΦΟΡΑ, aucta.* Nuremberg, 1685.

Paullini, Christian Franz. *Talpa, juxta methodum & leges Imperialis Academiae Leopoldinae descripta, variisq*[ue] *observationibus & curiositatibus conspersa.* Frankfurt, 1689.

Paullini, Christian Franz. *ΜΟΣΧΟΚΑΡΥΟΓΡΑΦΙΑ, seu nucis moschatae curiosa descriptio, historico-physico-medica, multis rarioribus naturae & artis observationibus, amoenis curiositatibus, & selectis memorabilibus illustrata & confirmata. Cum indicibus & autoribus vita.* Frankfurt, 1704 (*with appendix*: Vita, studia et gloria Pauliniana, fida crena descripta ab Esaia Dahlborn, Ph. & M. D. Anno 1703).

Pechlin, Johann Nikolaus. *Programma Anatomae cadaveris foeminae Aethiopicae praemissum.* Kiel, 1675 (lost).

[Pechlin, Johann Nikolaus]. *Jani Philadelphi consultatio desultoria de optima Christianorum secta, et vitiis Pontificiorum. Prodromus religionis medici.* Padua (corr. Amsterdam), 1688.

Pecquet, Jean. *Experimenta nova anatomica, quibus incognitum hactenus chyli receptaculum, & ab eo per thoracem in ramos usque subclavios vasa lactea deteguntur. Dissertatio anatomica de circulatione sanguinis et chyli motu. Huic secundae editioni, quae emendata est, illustrata, aucta, accessit de thoracicis lacteis dissertatio, in qua Io[annis] Riolani responsio ad eadem experimenta nova anatomica refutatur; & inventis recentibus canalis Virsungici demonstratur usus; & lacteum ad mammas a receptaculo iter indigitatur. Sequuntur gratulatoriae clarissimorum virorum cum prius editae, sed auctiores, tum recens additae ad authorem epistolae. Quibus & adjungitur brevis destructio, seu litura responsionis Riolani ad ejusdem Pecqueti experimenta*, 2nd ed. Paris, 1654.

Perrini, Paolo, Giuseppe Lanzino, and Giuliano Francesco Parenti. "Niels Stensen (1638–1686): Scientist, Neuroanatomist, and Saint." *NSurg. 67* (2010), pp. 3–9 (with two 'comments' on p. 9, of which the second: Goodrich, [*Comment*]).

Petschenig, Michael, and Michaela Zelzer, ed. *Sancti Ambrosi opera*, 2nd ed. Vol. 5, *Expositio psalmi CXVIII*. Corpus scriptorum ecclesiasticorum Latinorum 62. Vienna, Österreichische Akademie der Wissenschaften, 1999.

Pichery, Eugène, ed. *Jean Cassien: Conférences VIII–XVII*. Sources chrétiennes 54. Paris: Cerf, 1958.

Pieper, Josef. "Thomas von Aquin: Leben und Werk." In *Josef Pieper: Werke in acht Bänden*. Vol. 2, *Darstellungen und Interpretationen: Thomas von Aquin und die Scholastik*, edited by Berthold Wald, pp. 153–298. Hamburg: Felix Meiner, 2001.

Pius V, ed. *Catechismus, ex decreto Concilii Tridentini, ad parochos [...]*. Rome, 1566.

Pius X, ed. *Rituale Romanum: Pauli V Pontificis Maximi jussu editum a Benedicto XIV. et a Pio X. castigatum et auctum cui accedunt benedictionum et instructionum appendices duae*. Editio typica. Regensburg: Pustet, 1913.

Pomata, Gianna. "Praxis Historialis: The Uses of Historia in Early Modern Medicine." In *Historia: Empiricism and Erudition in Early Modern Europe*, edited by Gianna Pomata and Nancy G. Siraisi, pp. 105–146. Cambridge, MA: MIT Press, 2005.

Pompey, Heinrich. *Die Bedeutung der Medizin für die kirchliche Seelsorge im Selbstverständnis der sogenannten Pastoralmedizin: Eine bibliographisch-historische Untersuchung bis zur Mitte des 19. Jahrhunderts*. Diss. theol., Würzburg, 1966. Untersuchungen zur Theologie der Seelsorge 23. Freiburg: Herder, 1968.

Pontificia Academia Scientiarum, ed. *Blessed Niels Stensen and His Memorial Plaque in the Pontifical Academy of Sciences*, Vatican City: Ex Aedibus Academicis in Civitate Vaticana, 1989.

Pontifical Council for Pastoral Assistance to Health Care Workers, ed. *Charter for Health Care Workers*. Boston: Pauline Books & Media, 1995.

Porcarelli, Andrea. "Il rapporto tra filosofia e medicina nella storia del pensiero." In *Etica dell'atto medico*, edited by Abelardo Lobato, pp. 42–101. Philosophia 7. Bologna: Edizioni Studio Domenicano, 1991.

Portal, Antoine. *Histoire de l'anatomie et de la chirurgie, contenant L'origine & les progrès de ces Sciences; avec un Tableau Chronologique des principales Découvertes, & un Catalogue des ouvrages d'Anatomie & de Chirurgie, des Mémoires Académiques, des Dissertations insérées dans les Journaux, & de la plupart des Theses qui ont été soutenues dans les Facultés de Médecine de l'Europe. Tome troisième*. Paris, 1770.

Pott, Sandra. *Medizin, Medizinethik und schöne Literatur: Studien zu Säkularisierungsvorgängen vom frühen 17. bis zum frühen 19. Jahrhundert.* Säkularisierung in den Wissenschaften seit der Frühen Neuzeit 1. Berlin: De Gruyter, 2002.

Potter, Van Rensselaer. "Bioethics: the Science of Survival." *PBM 14* (1970), pp. 127–153.

Potter, Van Rensselaer. *Bioethics: Bridge to the Future.* Englewood Cliffs, NJ: Prentice-Hall, 1971.

Potter, Van Rensselaer. "Humility with Responsibility – A Bioethic for Oncologists: Presidential Address." *CanRes. 35* (1975), pp. 2297–2306.

Prosperi, Adriano. "Science and the Theological Imagination in the Seventeenth Century: Baptism and the Origins of the Individual." In *Christianity and Community in the West: Essays for John Bossy*, edited by Simon Ditchfield, pp. 206–231. Aldershot: Ashgate, 2001.

Putz, Reinhard V. "Der Leichnam in der Anatomie." *ZME 45* (1999), pp. 27–32.

Quattrin, Nevio. "Presenza di Stenone." *StenCath. 7* (1961), pp. 95–97.

Quattrin, Nevio. *Nicola Stenone scienziato e santo (1638–1686) nel III centenario di sua morte.* I quaderni dell'Accademia Olimpica 15. Vicenza: Accademia Olimpica, 1987.

Rager, Günter. "Gründung, Entwicklung und Gegenwart des Anatomischen Instituts." In *50 Jahre Anatomie an der Universität Freiburg 1938–1988 = 50 ans d'anatomie a l'Université de Fribourg 1938–1988*, pp. 12–32. Fribourg: Academic Press Fribourg, 1990.

Rager, Günter: "Institut für Anatomie und spezielle Embryologie." In *Histoire de l'Université de Fribourg Suisse 1889–1989: institutions, enseignement, recherches = Geschichte der Universität Freiburg Schweiz: Institutionen, Lehre und Forschungsbereiche.* Vol. 2, *Les Facultés = Die Fakultäten*, edited by Roland Ruffieux et al., pp. 858–862, 914–915. Fribourg: Academic Press Fribourg, 1991.

Rager, Günter: *Die Person: Wege zu ihrem Verständnis.* Studien zur theologischen Ethik 115. Fribourg: Academic Press Fribourg, 2006.

Redi, Francesco. *Osservazioni intorno agli animali viventi che si trovano negli animali viventi.* Florence, 1684.

[Redi, Francesco]. *Lettere di Francesco Redi Patrizio Aretino. Tomo terzo.* Florence, 1795.

[Redi, Francesco]. *Opere di Francesco Redi Gentiluomo Aretino e Accademico della Crusca.* Vol. 7, *Dalla Società Tipografica de' Classici Italiani contrada del Cappuccio.* Milano, 1811.

Redi, Francesco. *Esperienze intorno alla generazione degl'insetti.* Edited by Walter Bernardi. Biblioteca della scienza italiana 12. Florence: Giunti, 1996.

Reis, Hans. *Das Lebensrecht des ungeborenen Kindes als Verfassungsproblem.* Tübingen: Mohr, 1984.

Reusch, Franz Heinrich. *Der Index der verbotenen Bücher: Ein Beitrag zur Kirchen- und Literaturgeschichte*, vol. 2, bk 1. Bonn: M. Cohen, 1885. Reprint, Aalen: Scientia, 1967.

Richter, Paul. *Der Beginn des Menschenlebens bei Thomas von Aquin.* Diss. theol., Vienna, 2007. Studien der Moraltheologie 38. Münster: Lit, 2008.

Riva, Alessandro, and Francesca Testa Riva. "Niels Stensen (Niccolò Stenone) and His First Scientific Offspring: The Salivary Glands." In *Tenth European Anatomical Congress: Florence, September 17–21, 1995. Proceedings of the Symposium on Salivary Glands dedicated to Niels Stensen (Niccolò Stenone)*, edited by Alessandro Riva and Bernard Tandler. Lisse: Swets & Zeitlinger, 1996.

Riva, Alessandro, Bernhard Tandler, Masataka Murakami, and Martin C. Steward, ed. *Proceedings of the Second Symposium on Salivary Glands, Dedicated to Niels Steensen (Niccolò Stenone), Cagliari, May 23–25, 1997.* Lisse: Swets & Zeitlinger, 1998.

Roger, Jacques. *The Life Sciences in Eighteenth-Century French Thought*, edited by Keith R. Benson. Translated by Robert Ellrich. Stanford, CA: Stanford University Press, 1997.

Rome, Remacle. "Nicolas Sténon (1638–1686): Anatomiste, Géologue, Paléontologiste, Cristallographe, Vicaire Apostolique des régions Nordiques." *RQS, 5ième série, tome 17 = 127* (1956), pp. 517–572.

Rome, Remacle. "Nicolas Sténon et la 'Royal Society of London'." *Osiris 12* (1956), pp. 244–268.

Roob, Helmut. "Die Ärzte Herzog Ernsts des Frommen." *GothaM 1994*, pp. 27–35.

Rößler, Hole. "Der anatomische Blick und das Licht im theatrum: Über Empirie und Schaulust." In *Spuren der Avantgarde: Theatrum anatomicum. Frühe Neuzeit und Moderne im Kulturvergleich*, edited by Helmar Schramm, Ludger Schwarte and Jan Lazardzig, pp. 97–128. Theatrum Scientiarum 5. Berlin: De Gruyter, 2011.

Rothschuh, Karl E. *Konzepte der Medizin in Vergangenheit und Gegenwart.* Stuttgart: Hippokrates-Verlag, 1978.

Rovenius, Philippus. *Reipublicae Christianae libri duo, tractantes de variis hominum statibus, gradibus, officiis, & functionibus in Ecclesia Christi, & quae in singuils* [corr.: singulis] *amplectenda, quae fugienda sint [...]. Accessit ejusdem auctoris Tractatus de missionibus instituendis.* Antwerp, 1668 (corr. 1648).

Rowe, Katherine. "'God's Handy Worke': Divine Complicity and the Anatomist's Touch." In *The Body in Parts: Fantasies of Corporeality in Early Modern Europe*, edited by David Hillman and Carla Mazzio, pp. 285–309. New York: Routledge, 1997.

Ruestow, Edward G. *The Microscope in the Dutch Republic: The Shaping of Discovery*, 2nd ed. Cambridge: Cambridge University Press, 2004.

Ruffieux, Roland et al., ed. *Histoire de l'Université de Fribourg Suisse 1889–1989: institutions, enseignement, recherches. Geschichte der Universität Freiburg Schweiz: Institutionen, Lehre und Forschungsbereiche.* Vol. 3, *Personnes, dates et faits. Personen, Daten und Fakten.* Fribourg: Academic Press Fribourg, 1992.

Ruttkay, László. "Jessenius als Professor in Wittenberg: Zum 350. Todesjahr von Jessenius." *OrvosK 62/63* (1971), pp. 13–55.

Rütz, Lisbeth. "Let som videnskabsmand, svær som helgen: Seminar søgte nye synsvinkler på Stensen." *KO, 39th year, no. 17* (2013), p. 5.

Sacra Congregatio pro Causis Sanctorum, ed. *Osnabrugen: Beatificationis et canonizationis Servi Dei Nicolai Stenonis Episcopi Titiopolitani (†1686) positio super introductione causae et super virtutibus ex officio concinnata.* Sacra Congregatio pro Causis Sanctorum; Officium Historicum 38. Rome: Sacra Congregatio pro Causis Sanctorum, 1974.

Sacra Congregatio Rituum, ed. *Osnabrugen: Beatificationis et canonizationis Servi Dei Nicolai Stenonis Episcopi Titopolitani* [sic!] *(1638–1686) vota theologorum censorum super Servi Dei scriptis.* Rome: Tipografia Guerra e Belli, 1967.

Sajner, Josef, and Václav Pačes. "Niels Stensens tödliche Krankheit: Eine pathographische Studie." *MhJ* 6 (1971), pp. 53–63.

Sarteschi, Friderico Nicolao. *De scriptoribus congregationis clericorum regularium Matris Dei [...].* Rome, 1753.

Sauser, Gustav. *Die Geburt des ärztlichen Ethos aus dem Geiste der Anatomie: Antrittsrede von Dr. Dr. Mr. Gustav Sauser o. Prof. der Anatomie, Histologie und Embryologie gehalten anläßlich der Inauguration zum Rector magnificus des Studienjahres 1948/1949 am 20. November 1948 in der Aula der Leopold-Franzens-Universität zu Innsbruck.* Innsbruck: Tyrolia, c. 1948.

Savio, Pietro. "Ricerche sull'anatomico Guglielmo Riva." *BSBS* 66 (1968), pp. 229–267.

Schefer, Hubert W. "Das Berufsethos des Arztes Paracelsus." *Gesnerus; supplement 42* (1990).

Schepelern, Henrik Ditlev, ed. *Olai Borrichii Itinerarium 1660–1665: The Journal of the Danish Polyhistor Ole Borch.* Vol. 1, *Nov. 1660–Oct. 1661.* Copenhagen: Reitzels Forlag, 1983.

* Scherz, Gustav, ed. *Nicolai Stenonis epistolae et epistolae ad eum datae* [...] *cum prooemio ac notis Germanice scriptis.* Vol. 1. Copenhagen: Nyt Nordisk Forlag, 1952.

* Scherz, Gustav, ed. *Nicolai Stenonis epistolae et epistolae ad eum datae* [...] *cum prooemio ac notis Germanice scriptis.* Vol. 2. Copenhagen: Nyt Nordisk Forlag, 1952.

Scherz, Gustav, ed. *Im Rufe der Heiligkeit: Zeugnisse zur Fama Sanctitatis Niels Stensens.* Freiburg: Herder, 1953.

Scherz, Gustav. *Vom Wege Niels Stensens: Beiträge zu seiner naturwissenschaftlichen Entwicklung.* Acta historica scientiarum naturalium et medicinalium 14. Copenhagen: Munksgaard, 1956.

[Scherz, Gustav]. "Jesuit durch Niels Stensen." *StenCath.* 2 (1956), pp. 10–12.

[Scherz, Gustav]. "Rötenbeck und Stensen." *StenCath.* 3 (1957), pp. 84–86.

[Scherz, Gustav]. "Johann Valentin Wille und Stensen." *StenCath.* 4 (1958), pp. 20–21.

Scherz, Gustav, ed. *Nicolaus Steno and His Indice.* Acta historica scientiarum naturalium et medicinalium 15. Copenhagen: Munksgaard, 1958.

Scherz, Gustav. "Ein kostbares Stensenmanuskript." *StenCath.* 4 (1958), pp. 23–26.

Scherz, Gustav. "Danmarks Stensen-manuskript." *FoF* 5–6 (1958/1959), pp. 19–33.

Scherz, Gustav. "Stensenverbindungen aus Ex-libris." *StenCath. 6* (1960), pp. 60–62.

Scherz, Gustav, ed. "Niels Stensen's First Dissertation." *JHMAS 15* (1960), pp. 247–264 (with 12 unpaginated plates).

Scherz, Gustav, ed. *Pionier der Wissenschaft: Niels Stensen in seinen Schriften.* Acta historica scientiarum naturalium et medicinalium 17. Copenhagen: Munksgaard, 1963.

Scherz, Gustav, ed. *Nicolaus Steno's Lecture on the Anatomy of the Brain.* Copenhagen: Nyt Nordisk Forlag, 1965.

Scherz, Gustav. "Niels Stensens anatomische Forschung." In *Frühe Anatomie: Eine Anthologie*, edited by Robert Herrlinger and Fridolf Kudlien. Stuttgart: Wissenschaftliche Verlagsgesellschaft, 1967.

Scherz, Gustav, ed. *Steno and Brain Research in the Seventeenth Century: Proceedings of the International Historical Symposium on Nicolaus Steno and Brain Research in the Seventeenth Century held in Copenhagen 18–20 August 1965.* Analecta Medico-Historica Academiae Internationalis Historiae Medicinae 3. Oxford: Pergamon, 1968.

Scherz, Gustav, ed. *Steno: Geological Papers.* Acta historica scientiarum naturalium et medicinalium 20. Odense: Odense University Press, 1969.

Scherz, Gustav. *Niels Stensen: Eine Biographie.* Vol. 1, *1638–1677.* Edited by Franz Peter Sonntag. Leipzig: St. Benno-Verlag, 1987.

Scherz, Gustav. *Niels Stensen: Eine Biographie.* Vol. 2, *1677–1686.* Edited by Franz Peter Sonntag. Leipzig: St. Benno-Verlag, 1988.

Scherz, Gustav. "Nicolaus Steno the Humanist." In *Nicolaus Steno the Humanist.* Date and place of publication not specified, pp. 295–302.

Schiffers, Norbert. *Fragen der Physik an die Theologie: Die Säkularisierung der Wissenschaft und das Heilsverlangen nach Freiheit.* Habil. theol., Münster, **1966. Düsseldorf: Patmos, 1968.

Schirren, Carl. *Niels Stensen – Forscher, Gelehrter, Bischof: Dokumentation. Ärztetag am 16. November 1988 in der Stadthalle Osnabrück.* Kirche im Gespräch 3. Osnabrück, n.d. Mach. ms.

Schleiner, Winfried. *Medical Ethics in the Renaissance.* Washington, DC: Georgetown University Press, 1995.

Schleusener-Eichholz, Gudrun. *Das Auge im Mittelalter*, vol. 2. Diss. phil., Münster, c. 1984. Münstersche Mittelalter-Schriften 35,2. Munich: W. Fink, 1985.

Schmidt, Bernward. *Virtuelle Büchersäle: Lektüre und Zensur gelehrter Zeitschriften an der römischen Kurie 1665–1765.* Römische Inquisition und Indexkongregation 14. Paderborn: Schöningh, 2009.

Schmidt-Herrling, Eleonore. *Die Briefsammlung des Nürnberger Arztes Christoph Jacob Trew (1695–1769) in der Universitätsbibliothek Erlangen.* Katalog der Handschriften der Universitätsbibliothek Erlangen 5. Erlangen: Universitätsbibliothek, 1940.

Schmitz, Theodor. "Kardinal Bengsch und die 'Königsteiner Erklärung'." In *Adnotationes in iure canonico: Festgabe Franz X. Walter zur Vollendung des 65.*

Lebensjahres, edited by Elmar Güthoff and Karl-Heinz Selge, pp. 42–63. Fredersdorf: Rodak, 1994.

Schmitz, Werner. "Am schönsten ist, was wir nicht wissen können: Der Arzt und Bischof Niels Stensen wird seliggesprochen." *DtÄBl. 85* (1988), pp. A-2820–A-2821.

Schöllgen, Werner. *Die soziologischen Grundlagen der katholischen Sittenlehre.* Handbuch der katholischen Sittenlehre 5. Düsseldorf: Patmos, 1953.

Schrader, Justus, ed. *Observationes et historiae[,] omnes & singulae e Guiljelmi Harvei libello de generatione animalium excerptae, & in accuratissimum ordinem redactae. Item Wilhelmi Langly de generatione animalium observationes quaedam. Accedunt ovi faecundi singulis ab incubatione diebus factae inspectiones; ut et observationum anatomico-med[icinalium] decades quatuor; denique cadavera balsamo condiendi modus.* Amsterdam, 1674.

Schulz, Barbara. "Das Problem der Besessenheit aus medizinischer Sicht." Diss. med., Bonn, 1974. Mach. ms.

Schulze, Christian. *Medizin und Christentum in Spätantike und frühem Mittelalter: Christliche Ärzte und ihr Wirken.* Habil. med., Bochum, 2003. Studien und Texte zu Antike und Christentum = Studies and Texts in Antiquity and Christianity 27. Göttingen: Mohr Siebeck, 2005.

Schumacher, Gert-Horst, and Heinzgünther Wischhusen. *Anatomia Rostochiensis: Die Geschichte der Anatomie an der 550 Jahre alten Universität Rostock. Auf der Grundlage von Richard N. Wegner zur Geschichte der anatomischen Forschung an der Universität Rostock.* Berlin: Akademie-Verlag, 1970.

Schurig, Martin. *Muliebria historico-medica, hoc est partium genitalium muliebrium consideratio physico-medico-forensis, qua pudendi muliebris partes tam externae, quam internae, scilicet uterus cum ipsi annexis ovariis et tubis Fallopianis, nec non varia de clitoride et tribadismo, de hymene et nymphotomia seu feminarum circumcisione et castratione selectis et curiosis observationibus traduntur.* Dresden, 1729.

[Segneri, Paolo]. *Opere del padre Paolo Segneri della Compagnia di Gesù.* Vol. 6, bk. 2., *Il Cristiano instruito nella sua legge.* Edited by Vincenzo Morano. Naples, 1857.

Sekretariat der Deutschen Bischofskonferenz, ed. *Acta Pauli P*[a]*p*[ae] *VI. Litterae Encyclicae de propagatione humanae prolis recte ordinanda* [...]. = *Akten Papst Paul VI. Enzyklika* [**** *"Humanae vitae"*] *Papst Pauls VI. über die rechte Ordnung der Weitergabe menschlichen Lebens* [...] [**** *Lateinisch – deutsch*]. *Von den deutschen Bischöfen approbierte Übersetzung* [...]. Trier: Paulinus, 1968.
**** *Front cover.*

Sella, Domenico. *Italy in the Seventeenth Century.* London: Longman, 1997.

Shackelford, Jole. *A Philosophical Path for Paracelsian Medicine: The Ideas, Intellectual Context, and Influence of Petrus Severinus: 1540–1602.* Acta scientiarum naturalium et medicinalium 46. Copenhagen: Museum Tusculanum Press, 2004.

Shapin, Steven. "Descartes the Doctor: Rationalism and Its Therapies." *BJHS 33* (2000), pp. 131–154.

Short, Roger V. "The Magic and Mystery of the Oocyte: Ex Ovo Omnia." In *Biology and Pathology of the Oocyte: Its Role in Fertility and Reproductive Medicine*, edited by Alan O. Trounson and Roger G. Gosden, pp. 3–10. Cambridge: Cambridge University Press, 2003.

Sinibaldi, Giacomo. *Apollo bifrons medicas, & amenas dissertationes Latino, & Aetrusco sermone promiscuas exponens*. Rome, 1690.

Slottved, Ejvind and Ditlev Tamm. *The University of Copenhagen: A Danish Centre of Learning since 1479*. Copenhagen: University of Copenhagen, 2009.

Sobiech, Frank. *Herz, Gott, Kreuz: Die Spiritualität des Anatomen, Geologen und Bischofs Dr. med. Niels Stensen (1638–86)*. Diss. theol., Münster, 2003. Westfalia sacra 13. Münster: Aschendorff, 2004.

Sobiech, Frank. "Weihbischof Niels Stensen (1638–86): ein pastoraler Reformer. Seine Stellung zur Hexenverfolgung." *SpeeJ 12* (2005), pp. 109–126.

Sobiech, Frank. "Niels Stensen (1638–1686) und der Bergbau: Seine Reise durch Tirol, Niederungarn, Böhmen und Mitteldeutschland 1669–1670 im Spiegel seiner Theologie." In *Bergbau und Religion: Schwazer Silber. 6. Internationaler Montanhistorischer Kongress Schwaz 2007. Tagungsband*, edited by Wolfgang Ingenhaeff and Johann Bair, pp. 287–304. Innsbruck: Berenkamp, 2008.

Sobiech, Frank. "Nicholas Steno's Way from Experience to Faith: Geological Evolution and the Original Sin of Mankind." In *The Revolution in Geology from the Renaissance to the Enlightenment*, edited by Gary D. Rosenberg, pp. 179–186. Memoir 203. Boulder, CO: Geological Society of America, 2009.

Sobiech, Frank. *Radius in manu Dei: Ethos und Bioethik in Werk und Rezeption des Anatomen Niels Stensen (1638–1686)*. Westfalia sacra 17. 2nd ed. Münster: Aschendorff, 2014.

Sobiech, Frank. "The 'Capella Stenoniana' in Florence: The Tomb of Blessed Niels Stensen (1638–1686)." *ACM 85* (2015), pp. 73–76.

Sobiech, Frank. "Science, Ethos, and Transcendence in the Anatomy of Nicolaus Steno." *NCBQ 15* (2015), pp. 107–126.

Sobiech, Frank. "Simplicity of Faith, Intuition and Giordano Bruno. Nicolas Steno's Florentine diary and his philosophy lessons with Ferdinando III de' Medici: New insights from BNCF, Gal. 291." *KS* 2016, pp. 250-268.

[Société Internationale de Chirurgie, ed.]. *XVIᵉ Congrès de la Société Internationale de Chirurgie: Sous le haut patronage de Sa Majesté le Roi Frederik IX de Danemark*. [**** Guide du Congrès]. *Siège du Congrès: L'École des Hautes Études Techniques du Danemark / 10, Østervold*. [**** Copenhague,] *24.–30. juillet 1955*. [Copenhagen, 1955].

**** *Front cover.*

Sodi, Manlio, and Juan Javier Flores Arcas, ed. *Rituale Romanum: Editio Princeps (1614)*. Monumenta Liturgica Concilii Tridentini 5. Vatican City: Libreria Editrice Vaticana, 2004.

Sørensen, Peder. *Idea medicinae philosophicae, fundamenta continens totius doctrinae Paracelsicae, Hippocraticae, & Galenicae*. Basel, 1571.

Spedding, James, Robert Leslie Ellis, and Douglas Denon Heath. *The Works of Francis Bacon[,] Baron of Verulam, Viscount St. Alban, and Lord High Chancellor of England*. Vol. 6.1, *Literary and Professional Works*. London: Longman, 1861. Reprint, Stuttgart: F. Frommann, 1963.

Spinoza, Baruch de. *Opera posthuma. Quorum series post praefationem exhibetur*. [Amsterdam], 1677 [corr. 1678].

Spitzer, Beatrix, ed. *Paolo Zacchia: Die Beseelung des menschlichen Fötus. Buch IX, Kapitel 1 der Quaestiones medico-legales*. Cologne: Böhlau, 2002.

Spruit, Leen, and [Giusep]pina Totaro. *The Vatican Manuscript of Spinoza's Ethica*. Brill's Studies in Intellectual History 205 / Brill's Texts and Sources in Intellectual History 11. Leiden: Brill, 2011.

Staudacher, Peter, ed. *Platon: ΘΕΑΙΤΗΤΟΣ, ΣΟΦΙΣΤΗΣ, ΠΟΛΙΤΙΚΟΣ – Theaitetos, Der Sophist, Der Staatsmann*. Platon; Werke in acht Bänden 6. Darmstadt: Wissenschaftliche Buchgesellschaft, 1970 (6th ed. 2011).

Steiger, Johann Anselm. *Medizinische Theologie: Christus medicus und theologia medicinalis bei Martin Luther und im Luthertum der Barockzeit. Mit Edition dreier Quellentexte*. Studies in the History of Christian Traditions 121. Leiden: Brill, 2005.

Stelzenberger, Johannes. *Die Beziehungen der frühchristlichen Sittenlehre zur Ethik der Stoa: Eine moralgeschichtliche Studie*. Habil. theol., Würzburg, 1930. Munich: Max Hueber, 1933. Reprint, Hildesheim: Olms, 1989.

Stensen, Niels. *Disputatio anatomica de glandulis oris, & nuper observatis inde prodeuntibus vasis prima. Quam, divina favente gratia, sub praesidio viri clarissimi d[omini] Johannis van Horne, medicin[ae] doct[oris] anatomiae & chirurgiae professoris celeberrimi, placido eruditorum examini subjicit Nicolaus Stenonis, Hafnia-Danus. Ad-diem 6. Iulii, loco horisque solitis*. Leiden, 1661[a].

Stensen, Niels. *Disputatio anatomica de glandulis oris, & nuper observatis inde prodeuntibus vasis secunda. Quam, divina favente gratia, sub praesidio viri clarissimi d[omini] Johannis van Horne, medicin[ae] doct[oris] anatomiae & chirurgiae professoris celeberrimi, placido eruditorum examini subjicit Nicolaus Stenonis, Hafnia-Danus. Ad-diem 9. Iulii, loco horisque solitis pomerid[ianis]*. Leiden, 1661[b].

Stensen, Niels. *Apologiae prodromus, quo demonstratur, judicem Blasianum & rei anatomicae imperitum esse, & affectuum suorum servum*. Leiden, 1663.

Stensen, Niels. *Elementorum myologiae specimen, seu musculi descriptio geometrica. Cui accedunt canis carchariae dissectum caput, et dissectus piscis ex canum genere*. Florence, 1667.

Stephan, Joachim. "10 Observationes aus dem kasuistischen Sammelwerk des Johann Schenck von Grafenberg (1530–1598). " Diss. med., Erlangen-Nürnberg, 1967. Mach. ms.

Sterpellone, Luciano. "Niccolò Stenone, la scienza e la fede." *GPMO 10,7* (2008), pp. 44–45.

Stockhorst, Stefanie. "Unterweisung und Ostentation auf dem anatomischen Theater der Frühen Neuzeit: Die öffentliche Leichensektion als Modellfall des theatrum mundi." In *Zergliederungen – Anatomie und Wahrnehmung in der Frühen Neuzeit*, edited by Albert Schirrmeister = *ZSpr. 9,1/2 (2005)*, pp. 271–290. Frankfurt: Klostermann, 2005.

Storer, Norman W. Introduction to *The Sociology of Science: Theoretical and Empirical Investigations*, 2nd ed., by Robert K. Merton, pp. xi–xxxi. Chicago: University of Chicago Press, 1974.

Strkalj, Goran. "Niels Stensen and the Discovery of the Parotid Duct." *IJM 31* (2013), pp. 1491–1497.

Stroppiana, Luigi. "Giovanni Maria Lancisi." *ScMedIt-D 8* (1959), pp. 5–14.

Swammerdam, Jan. *Miraculum naturae sive uteri muliebris fabrica, notis in d[omini] Joh[annis] van Horne prodromum illustrata, & tabulis, a clariss[imis] expertissimisque viris cum ipso archetypo collatis, adumbrata. Adjecta est nova methodus, cavitates corporis ita praeparandi, ut suam semper genuinam faciem servent. Ad illustriss[imam] Regiam Societatem Londinensem.* Leiden, 1672.

Sylvius, Frans de le Boë. *Epistola apologetica, improbas aeque ac ineptas Antonii Deusingii, aliorumque ejusdem farinae hominum cavillationes atque calumnias summatim perstringens.* Leiden, 1664.

Sylvius, Frans de le Boë. *Disputationum medicarum decas, primarias corporis humani functiones naturales, nec non febrium naturam, ex anatomicis, practicis & chimicis experimentis deductas, complectens. Annexis 1. Epistola apologetica contra Antonium Deusingium. 2. de affectus epidemii, anno 1669 Leidae grassantis, causis naturalibus, 3. de hominis cognitione, binis orationibus. Omnibus ad Leidense exemplar fideliter conformatis. Editio tertia, copioso rerum ac verborum catalogo locupletata.* Jena, 1674.

[Sylvius, Frans de le Boë]. *Opera medica, tam hactenus inedita, quam variis locis & formis edita; nunc vero certo ordine disposita, & in unum volumen redacta. Editio nova, cui accedunt casus medicinales annor[um] 1659[,] 60 & 61 quos ex ore cl[arissimi] Sylvii calamo excepit Joachimus Merian.* Utrecht, 1695.

Tafi, Angelo. *I vescovi di Arezzo dalle origini della diocesi (sec. III) ad oggi.* With a preface by Prof. Alberto Fatucchi. Cortona: Calosci, 1986.

Ten Have, Henk A. M. J. "Potter's Notion of Bioethics." *KIEJ 22* (2012), pp. 59–82.

Thielicke, Helmut. *Being Human … Becoming Human: An Essay in Christian Anthropology.* Translated by Geoffrey W. Bromiley. Garden City, NY: Doubleday, 1984.

Thomas-Morus-Bildungswerk Schwerin, ed. *Diener der Wahrheit: Ausgewählte Vorträge, Buchbeiträge und Predigten zum Leben und Wirken von Niels Stensen. Dokumentation.* Thomas-Morus-Bildungswerk Schwerin; Schriftenreihe 19. Schwerin: Thomas-Morus-Bildungswerk, 2011.

Thomas-Morus-Bildungswerk Schwerin, ed. *Auf Pilgerreise mit Niels Stensen: Dokumentation.* Thomas-Morus-Bildungswerk Schwerin; Schriftenreihe 21. Schwerin: Thomas-Morus-Bildungswerk, 2013.

Thümmel, Hans Georg, ed. *Geschichte der Medizinischen Fakultät Greifswald: Geschichte der Medizinischen Fakultät von 1456 bis 1713 von Christoph Helwig d. J. und das Dekanatsbuch der Medizinischen Fakultät von 1714 bis 1823.* Beiträge zur Geschichte der Universität Greifswald 3. Stuttgart: Steiner, 2002.

Tjomsland, Anne. "Niels Stensen: His Tercentenary." *AMH 10* (1938), pp. 491–507.

Toellner, Richard. "Zum Begriff der Autorität in der Medizin der Renaissance." In *Humanismus und Medizin*, edited by Rudolf Schmitz and Gundolf Keil, pp. 159–179. Deutsche Forschungsgemeinschaft; Mitteilung der Kommission für Humanismusforschung 11. Weinheim: Acta Humaniora, 1984.

Toellner, Richard. "Der Körper des Menschen in der philosophischen und theologischen Anthropologie des späten Mittelalters und der beginnenden Neuzeit." In *Gepeinigt, begehrt, vergessen: Symbolik und Sozialbezug des Körpers im späten Mittelalter und in der frühen Neuzeit*, edited by Klaus Schreiner and Norbert Schnitzler, pp. 131–146. Munich: W. Fink, 1992.

Tognocchi a Terrinca, Antonio. *Genealogicum et honorificum theatrum etrusco-minoriticum [...] anno Domini MDCLXXX elaboratum.* Florence, 1682.

Tolmer, Léon. "Une page d'histoire des sciences 1661–1669: Vingt-deux lettres inédites d'André de Graindorge à P.-D. Huet publiées et annotées." *MANC, n.s. 10* (1942), pp. 243–337.

Tolmer, Léon. *Pierre-Daniel Huet (1630–1721): Humaniste – physicien.* Bayeux: Colas [1949].

Trillitzsch, Winfried, ed. *Seneca im literarischen Urteil der Antike: Darstellung und Sammlung der Zeugnisse.* Vol. 2, *Quellensammlung (Testimonien).* Amsterdam: Hakkert, 1971.

Tubbs, R. Shane, Nicholas Gianaris, Mohammadali M. Shoja, Marios Loukas, and Aaron A. Cohen[–]Gadol. "'The Heart Is Simply a Muscle' and First Description of the Tetralogy of 'Fallot': Early Contributions to Cardiac Anatomy and Pathology by Bishop and Anatomist Niels Stensen (1638–1686)." *IJC 154* (2012), pp. 312–315.

Tubbs, R. Shane, Martin M. Mortazavi, Mohammadali M. Shoja, Marios Loukas, and Aaron A. Cohen-Gadol. "The Bishop and Anatomist Niels Stensen (1638–1686) and His Contributions to Our Early Understanding of the Brain." *ChNS 27* (2011), pp. 1–6.

[University of Oklahoma Libraries, ed.]. *An Exhibition Concerning Nicolaus Steno (1638–1686): In Celebration of the Addition of the Fifty-Thousandth Volume to the History of Science Collections of the University of Oklahoma Libraries. Norman, Oklahoma / February, 1982.* [Norman, OK, 1982] (unpaginated, 16 pages).

Usener, Hermann, ed. *Epicurea: Editio stereotypa editionis primae (MDCCC-LXXXVII).* Stuttgart: B. G. Teubner, 1966.

Van Almeloveen, Theodor Jansson. *Inventa nov-antiqua. Id est brevis enarratio ortus & progressus artis medicae; ac praecipue de inventis vulgo novis, aut nuperrime in ea repertis. Subjicitur ejusdem rerum inventarum onomasticon.* Amsterdam, 1684.

Van Besien, Liliane and Yves Van Besien. "L'étonnant destin de Nicolas Stenon (1638–1686): The Astonishing Destiny of Niels Steensen." *Actes SFHAD (XIXᵉ congrès Paris, 2009) 14* (2009), pp. 78–81. Full convention publication under: http://www.bium.univ-paris5.fr/sfhad/vol14/2009.pdf.

Van Dülmen, Richard. *Kultur und Alltag in der Frühen Neuzeit*, 4th ed. Vol. 1, *Das Haus und seine Menschen: 16.–18. Jahrhundert.* Munich: C.H. Beck, 2005.

Venard, Marc. "Fragen der Ethik." In *Die Zeit der Konfessionen (1530–1620/1630)*, edited by Heribert Smolinsky, pp. 1173–1198. Die Geschichte des Christentums 8. Freiburg: Herder, 1992.

Vilar, Johannes. "Was weiß die Postmoderne von der Sinnfrage? Selbstverwirklichung und Menschenbild." *FKTh. 22* (2006), pp. 241–274.

Visser, Jan. *Rovenius und seine Werke: Beitrag zur Geschichte der nordniederländischen katholischen Frömmigkeit in der ersten Hälfte des 17. Jahrhunderts.* Diss. theol., Bern, 1965. Van Gorcum's Historische Bibliotheek 79. Assen: Van Gorkum & Comp., 1966.

Visser, Jan. "La relation entre Jansénius et Rovenius." In *L'image de C. Jansénius jusqu'à la fin du XVIIIe siècle: Actes du colloque. Louvain, 7–9 novembre 1985*, edited by Edmond J. M. van Eijl, pp. 43–51. Leuven: Leuven University Press, 1987.

Vugs, Josephus Gerardus. "Leven en Werk van Niels Stensen (1638–1686): Onderzoeker van het Zenuwstelsel." Diss. med., Leiden, 1968.

Walsh, James J. *Catholic Churchmen in Science: Sketches of the Lives of Catholic Ecclesiastics Who Were among the Great Founders in Science.* Philadelphia: American Ecclesiastical Review, 1906.

Weber, Helmut. *Allgemeine Moraltheologie: Ruf und Antwort.* Graz: Styria, 1991.

Weber, Helmut. *Spezielle Moraltheologie: Grundfragen des christlichen Lebens.* Graz: Styria, 1999.

Wegner, Richard N. *Das Anatomenbildnis: Seine Entwicklung im Zusammenhang mit der anatomischen Abbildung.* Basel: Schwabe, [1939].

Werl, Dionysius von. *Philanthon vindicatus sive Hermannus Conringius ob andabaticam suam anno 1677 Helmstadii editam discussionem, praetensamq[ue] pag. 329 demonstrationem juste, sed tamen misericorditer castigatus.* Hanover, 1678.

White, Yvonne A. R., Dori C. Woods, Yasushi Takai, Osamu Ishihara, Hiroyuki Seki, and Jonathan L. Tilly. "Oocyte Formation by Mitotically Active Germ Cells Purified from Ovaries of Reproductive-Age Women." *NatMed. 18* (2012), pp. 413–421.

Wicklein, Eva-Maria. *Nicolaus Steno nach seiner Konversion im Jahre 1667.* Diss. med., Hamburg, 1991. Publikationen der Katholischen Akademie Hamburg; Reihe Wissenschaft 2. Hamburg: Katholische Akademie, 1992.

[Willius, Johann Valentin]. *De philiatrorum Germanorum itineribus: dissertationes tres, quas in amorem popularium, & exanthlatarum molestiarum dulcem memoriam, festivo calamo, mente sincera, doctus usu, periculo cautus, horis otiosis scripsit Joachimus Vitus Wigandus, Windecensis philiater. Opus posthumum.* Freiburg, 1678.

Willoh, Karl. *Geschichte der katholischen Pfarreien im Herzogtum Oldenburg.* A. *Dekanat Vechta-Neuenkirchen. Die Pfarren Vechta und Wildeshausen.* Vol. 3. Cologne: n.d. [1898]. Reprint, Osnabrück: Wenner, 1975.

Willoh, Karl. *Geschichte der katholischen Pfarreien im Herzogtum Oldenburg.* B. *Dekanat Cloppenburg. Die Pfarren Altenoythe, Barssel, Bösel, Cappeln, Cloppenburg-Crapendorf, Emsteck, Essen, Friesoythe.* Vol. 4. Cologne: n.d. [1898]. Reprint, Osnabrück: Wenner, 1975.

Windle, Bertram C. A. "Nicolaus Stensen (1638–1687 [sic!])." In *Twelve Catholic Men of Science*, pp. 45–68. London: Catholic Truth Society, 1914 (with an unpaginated portrait of Stensen between pages 44 and 45).

Winstrup, Peder and Janus Laurentius Ulmivallius. *Disputatio philologica et philosophica de usu linguarum et disciplinarum philosophicarum in theologia [...].* Copenhagen, 1633.

Wischhusen, Heinz-Günther, Gerhard Schlegel, and Emil Ehler. "Biographie und Inaugural-Dissertation (1661–1662) des Niels Stensen (1638–1686)." *DG 37* (1982), pp. 2064–2069 = in *Diener der Wahrheit: Ausgewählte Vorträge, Buchbeiträge und Predigten zum Leben und Wirken von Niels Stensen. Dokumentation*, edited by Thomas-Morus-Bildungswerk Schwerin, pp. 14–27. Thomas-Morus-Bildungswerk Schwerin; Schriftenreihe 19. Schwerin: Thomas-Morus-Bildungswerk, 2011.****

**** *Pages from the book publication are cited in parentheses.*

Wolf, Hubert, and Bernward Schmidt. *Benedikt XIV. und die Reform des Buchzensurverfahrens: Zur Geschichte und Rezeption von "Sollicita ac provida".* Römische Inquisition und Indexkongregation 13. Paderborn: Schöningh, 2011.

Wolf-Heidegger, Gerhard, and Anna Maria Cetto. *Die anatomische Sektion in bildlicher Darstellung.* Basel: Karger, 1967.

Wollgast, Siegfried. *Zur Geschichte des Promotionswesens in Deutschland.* Bergisch Gladbach: Grätz, 2001.

Wright-St Clair, Rex E. "Go to the Bedside." *JRCGP 21* (1971), pp. 443–452.

Ziggelaar, August, ed. *Chaos: Niels Stensen's Chaos-Manuscript Copenhagen, 1659.* Acta historica scientiarum naturalium et medicinalium 44. Copenhagen: Danish National Library of Science and Medicine, 1997.

Ziggelaar, August: "Fra Niels Stensens udvikling: Personlig krise og glæde over Guds skaberværk. Om Niels Stensens vej over naturvidenskaben til Gud." *Cath. (Kbh.).* Under: http://catholica.dk/fileadmin/Catholica_files/1329/1329_Fra_Niels_Stensens_udvikling.pdf (Article no. #1329-24.05.2013; 9 pages).****

**** *Abbreviated version:* Ziggelaar, August. "Personlig krise og glæde over Guds skaberværk. Uddrag af artikel om Niels Stensens vej fra naturvidenskaben til Gud." *KO, 39th year, no. 8* (2013), p. 8.

Zwierlein, Eduard. *Begegnung und Verantwortung: Ärztliches Ethos und Medizinische Ethik.* Würzburg: Königshausen & Neumann, 2007.

Zwinger, Theodor. Παρατηρήσεων *sive observationum medicarum, rararum, novarum, admirabilium, & monstrosarum volumen in tomis septem.* 7 vols., Basel and Freiburg i. Br., 1584–1597.

Abbreviations

Abbreviations for Citations of Stensen Editions

The thematic introductions to the works in *BIBLIOGRAPHY, C* marked with a preceding asterisk (*) are cited in the usual fashion. The source materials within these works are cited in abbreviated form according to the following scheme:

from Larsen/Scherz, *Opera theologica*, vol. 1: respective abbreviation***, p. …, line …

from Larsen/Scherz, *Opera theologica*, vol. 2: respective abbreviation***, p. …, line …

from Maar, *Opera philosophica*,** vol. 1: respective abbreviation***, p. …, line …

from Maar, *Opera philosophica*,** vol. 2: respective abbreviation***, p. …, line …

from Scherz, *Epistolae*, vol. 1: E [= Epistola]*** … (respective number), p. …, line …

from Scherz, *Epistolae*, vol. 2: E [= Epistola]*** … (respective number) or Add. [= Additamentum]*** … (respective number), p. …, line …

** For English translations (without the appendices XXXIV–XXXVI) see Kardel/Maquet, *Biography and Original Papers* with the page numbers of Maar's edition on the margins respectively.

*** The following abbreviations are used for the source materials cited in this fashion:

Add. 1–49	Additamentum 1–49: Scherz, *Epistolae*, vol. 1, pp. 898–997
AnRaj.	De anatome rajae epistola: Maar, *Opera philosophica*, vol. 1, no. XVI
AProd.	Apologiae prodromus, quo demonstratur, judicem Blasianum & rei anatomicae imperitum esse, & affectuum suorum servum: Maar, *Opera philosophica*, vol. 1, no. XIII
AvCun.	Observationes anatomicae in avibus & cuniculis: Maar, *Opera philosophica*, vol. 1, no. IX = Scherz, *Epistolae*, vol. 1, E 9

(continued)

© Springer International Publishing Switzerland 2016
F. Sobiech, *Ethos, Bioethics, and Sexual Ethics in Work and Reception of the Anatomist Niels Stensen (1638-1686)*, Philosophy and Medicine 117, DOI 10.1007/978-3-319-32912-3

CanCap.	Canis carchariae dissectum caput: Maar, *Opera philosophica*, vol. 2, no. XXIII
DeConv.	De propria conversione epistola: Larsen/Scherz, *Opera theologica*, vol. 1, pp. 126–129 = Scherz, *Epistolae*, vol. 1, E 73
DefConv.	Defensio et plenior elucidatio epistolae de propria conversione: Larsen/Scherz, *Opera theologica*, vol. 1, pp. 380–437
DefScrut.	Defensio et plenior elucidatio scrutinii reformatorum: Larsen/Scherz, *Opera theologica*, vol. 1, pp. 260–289
DePhil.	Ad novae philosophiae reformatorem de vera philosophia epistola: Larsen/Scherz, *Opera theologica*, vol. 1, pp. 95–103
Discours	Discours sur l'anatomie du cerveau: Maar, *Opera philosophica*, vol. 2, no. XVIII
DucSal.	De prima ductus salivalis exterioris inventione & Bilsianis experimentis: Maar, *Opera philosophica*, vol. 1, no. I = Scherz, *Epistolae*, vol. 1, E 1
E 1–224	Epistola 1–224: Scherz, *Epistolae*, vol. 1, pp. 133–480
E 225–478	Epistola 225–478: Scherz, *Epistolae*, vol. 2, pp. 481–897
Embryo	Embryo monstro affinis Parisiis dissectus: Maar, *Opera philosophica*, vol. 2, no. XX
ExAc.	Extracts from Holger Jacobæus' Exercitia academica: Maar, *Opera philosophica*, vol. 2, no. XXXVI
Examen I	Examen responsionum brevium et extemporanearum ad quaestiones baronis de Reck, non quidem plane extemporaneum, sed nec plena otii libertate conscriptum: Larsen/Scherz, *Opera theologica*, vol. 1, pp. 302–331
GlandOc.	De glandulis oculorum novisque earundem vasis observationes anatomicae: Maar, *Opera philosophica*, vol. 1, no. V
GlandOr.	De glandulis oris & novis inde prodeuntibus salivae vasis: Maar, *Opera philosophica*, vol. 1, no. II
HepRed.	Responsio ad vindicias hepatis redivivi: Maar, *Opera philosophica*, vol. 1, no. IV
LyVar.	Lymphaticorum varietas: Maar, *Opera philosophica*, vol. 1, no. XII
MotCor.	Ex variorum animalium sectionibus hinc inde factis excerptae observationes circa motum cordis auricularumque & venae cavae: Maar, *Opera philosophica*, vol. 1, no. X
MuscAqu.	Historia musculorum aquilae: Maar, *Opera philosophica*, vol. 2, no. XXXII
MuscGland.	De musculis & glandulis observationum specimen: Maar, *Opera philosophica*, vol. 1, no. XV
Myol.	Elementorum myologiae specimen, seu musculi descriptio geometrica: Maar, *Opera philosophica*, vol. 2, no. XXII
NarVas.	De narium vasis: Maar, *Opera philosophica*, vol. 1, no. VI
NicPulv.	Cur nicotianae pulvis oculos clariores reddat. De lactea gelatina observatio: Maar, *Opera philosophica*, vol. 1, no. VIII = Scherz, *Epistolae*, vol. 1, E 7
NovMusc.	Nova musculorum & cordis fabrica: Maar, *Opera philosophica*, vol. 1, no. XIV = Scherz, *Epistolae*, vol. 1, E 13
Op.	[Opusculum 1–15]: Larsen/Scherz, *Opera theologica*, vol. 2, pp. 399–543
OvaViv. I	Observationes anatomicae spectantes ova viviparorum: Maar, *Opera philosophica*, vol. 2, no. XXV
OvaViv. II	Ova viviparorum spectantes observationes: Maar, *Opera philosophica*, vol. 2, no. XXVI

(continued)

OvPul.	In ovo & pullo observationes: Maar, *Opera philosophica*, vol. 2, no. XIX
ParHAg.	Parochorum hoc age […]: Larsen/Scherz, *Opera theologica*, vol. 2, pp. 12–52
PiscCan.	Historia dissecti piscis ex canum genere: Maar, *Opera philosophica*, vol. 2, no. XXIV
Prodromus	De solido intra solidum naturaliter contento dissertationis prodromus: Maar, *Opera philosophica*, vol. 2, no. XXVII
Prooemium	Prooemium demonstrationum anatomicarum in Theatro Hafniensi anni 1673: Maar, *Opera philosophica*, vol. 2, no. XXXI
Scrut.	Scrutinium reformatorum ad demonstrandum reformatores morum in ecclesia fuisse a Deo, reformatores fidei non fuisse a Deo: Larsen/Scherz, *Opera theologica*, vol. 1, pp. 112–120
Sermo	[Sermo 1–45]: Larsen/Scherz, *Opera theologica*, vol. 2, pp. 176–375
Spir.	[Opera spirituale 1–12]: Larsen/Scherz, *Opera theologica*, vol. 2, pp. 70–147
StenBrun. II	[Epistola ad Johannem Brunsmannum II]: Larsen/Scherz, *Opera theologica*, vol. 1, pp. 505–508
SudOr.	Sudorum origo ex glandulis. De insertione & valvula lactei thoracici & lymphaticorum: Maar, *Opera philosophica*, vol. 1, no. VII = Scherz, *Epistolae*, vol. 1, E 5
UtLep.	Uterus leporis proprium foetum resolventis: Maar, *Opera philosophica*, vol. 2, no. XXI
VarOb.	Variae in oculis & naso observationes novae &c.: Maar, *Opera philosophica*, vol. 1, no. III = Scherz, *Epistolae*, vol. 1, E 3
VesPul.	De vesiculis in pulmone. Anatome cuniculi praegnantis. In pulmonibus experimenta. De lacteis mammarum. In cygno observationes, &c.: Maar, Opera philosophica, vol. 1, no. XI = Scherz, *Epistolae*, vol. 1, E 11
VithHyd.	De vitulo hydrocephalo: Maar, *Opera philosophica*, vol. 2, no. XXVIII
VitTrans.	De vitelli in intestina pulli transitu epistola: Maar, *Opera philosophica*, vol. 1, no. XVII

Abbreviations for Periodicals and Encyclopedias

AAS	*Acta Apostolicae Sedis*. Rome, 1909–.
ACM	*Archivos de Cardiología de México*. México, 1930–.
Acta anat.	*Acta anatomica*. Basel, 1945/1946– (since 1999: Cells Tissues Organs)
Actes SFHAD	*Actes. Société française d'histoire de l'art dentaire*. Strasbourg, 1996–. Under: http://www.bium.univ-paris5.fr/sfhad/actes_deb.htm *and* http://www.biusante.parisdescartes.fr/sfhad/actes_deb.htm.
ADB	*Allgemeine Deutsche Biographie*. 55 vols. and 1 index vol. Edited by the Historische Commission bei der Königlichen Akademie der Wissenschaften (Munich). Leipzig, Duncker & Humblot, 1875–1912.
ÄLex.[3]	*Ärztelexikon: Von der Antike bis zur Gegenwart*, 3rd ed. Edited by Wolfgang Uwe Eckart and Christoph Gradmann. Heidelberg: Springer, 2006.

(continued)

AES	*Archives Européennes de sociologie*. Cambridge, 1960–.
AJP	*American Journal of Psychiatry*. Arlington, VA, 1844–.
AKGWG	*Abhandlungen der Königlichen Gesellschaft der Wissenschaften zu Göttingen*. Berlin, 1838/1841 (1843)–1892.
AL	*Arca Lovaniensis*. Leuven, 1972/1973–.
AMAT	*Atti e memorie dell'Accademia toscana di scienze e lettere la Colombaria*. Florence [1943–1950], n.s. 1951/1952–.
AMH	*Annals of Medical History*. New York, 1917–1942, n.s. 1929–1938.
AMPH	*Acta medica & philosophica Hafniensia*. Copenhagen: Vol. 1 (1671/1672), published 1673), Vol. 2 (1673), published 1675, Vol. 3 & 4 (1674/1675/1676), published 1677, Vol. 5 (1677/1678/1679), published 1680.
ANB	*American National Biography*. Edited by John A. Garraty and Mark C. Carnes. 24 vols. Oxford: Oxford University Press, 1999. 2 supplements: Oxford: Oxford University Press, 2002–2005.
AOS	*Actualités Odonto-Stomatologiques*. Paris, 1947–.
APS	*Aesthetic Plastic Surgery*. New York, 1976/1977–.
ASS	*Acta Sanctae Sedis*. Rome, 1865–1908.
AUn.	*Annali universali di Medicina*. Milano, 1817–1874, 4th ser. 1851–1871.
BBKL	*Biographisch-bibliographisches Kirchenlexikon*. Edited by Friedrich Wilhelm Bautz and Traugott Bautz. 14 vols. & 20 supplements. Hamm, 1975–.
BBL	*Braunschweigisches Biographisches Lexikon: 8. bis 18. Jahrhundert*. Edited by Horst-Rüdiger Jarck. Braunschweig: Appelhans, 2006.
BBU	*Bibliographie biographique universelle [...]*, 2nd ed. By Edouard-Marie Oettinger. 2 vols. Brussels: Stienon, 1854 (reprint, Hildesheim: Gerstenberg, 1971).
BDL II	*Die Bischöfe der deutschsprachigen Länder 1945–2001: Ein biographisches Lexikon*. Edited by Erwin Gatz. Berlin: Duncker & Humblot, 2002.
BEdtM	*Biographische Enzyklopädie deutschsprachiger Mediziner*. Edited by Dietrich von Engelhardt. 2 vols. Munich: Saur, 2002.
BHRR II	*Die Bischöfe des Heiligen Römischen Reiches 1448 bis 1648: Ein biographisches Lexikon*. Edited by Erwin Gatz. Berlin: Duncker & Humblot, 1996.
Bioethikos	*Revista Bioethikos*. São Camilo, 2007–. Under: http://www.saocamilo-sp.br/novo/publicacoes/publicacoes-bioethikos.php.
BiogrHM	*A Biographical History of Medicine: Excerpts and Essays on the Men and Their Work*. By John H. Talbott. New York: Grune & Stratton, 1970.
BJHS	*British Journal for the History of Science*. Cambridge, 1962/1963–.
BLÄ	*Biographisches Lexikon der hervorragenden Ärzte aller Zeiten und Völker*, 2nd ed. Edited by August Hirsch. 5 vols. and 1 supplement. Munich: Urban & Schwarzenberg, 1929–1935 (3rd ed. Munich: Urban & Schwarzenberg, 1962).
BLLi.	*Bratislavské lekárske Listy*. Bratislava, 1975–.
BMPh.	*Bibliographia medica & physica novissima: perpetuo continuanda sive conspectus primus catalogi librorum medicorum chymicorum, anatomicorum, chyrurgicorum, botanicorum ut & physicorum, &c. Quotquot currente hoc semisaeculo, id est ab anno reparatae salutis 1651 (inclusive) per universam Europam, in quavis lingua, orientali tum Graeca, Latina, Gallica, Hispanica, Italica, Anglica, Germanica & Belgica, aut novi aut emendatiores & auctiores typis prodierunt [...]*. By Cornelis van Beughem. Amsterdam, 1681.

(continued)

BNSten.	*Bibliographia Nicolai Stenonis.* By Michael Jensen. Mørke: Impetus, 1986.
BollFil.	*Bollettino filosofico.* Rome, 1935–1940.
BRM	*Bibliotheca realis medica, omnium materiarum, rerum, et titulorum, in universa medicina occurentium. Ordine alphabetico sic disposita, ut primo statim intuitu tituli, et sub titulis autores medici, justa velut acie collocati, in oculos statim et animos incurrant. Accedit index autorum copiosissimus.* By Martin Lipenius. Frankfurt, 1679.
BSBS	*Bollettino storico-bibliografico subalpino.* Turin, 1896–1934, n.s. 1935–.
BSFSN	*Bulletin de la Société Fribourgeoise des Sciences Naturelles = Bulletin der Naturforschenden Gesellschaft Freiburg.* Fribourg, 1879/1880–.
CanRes.	*Cancer Research.* Philadelphia, 1916–.
Cath. (Kbh.)	*Catholica. Et katolsk tidsskrift på nettet.* Copenhagen, 2004–. Under: http://www.catholica.dk.
CBMH	*Canadian bulletin of medical history = Bulletin canadien d'histoire de la médecine.* Waterloo, ON, 1984–.
CCHA-R	*Canadian Catholic Historical Association: Report/Rapport.* Ottawa, 1933/1934–1965 (changing titles since 1966; since 1990: CCHA Historical Studies).
CG-BNCF	*La Collezione galileiana della Biblioteca Nazionale di Firenze.* Compiled by Angiolo Procissi. 3 vols. Indici e cataloghi; Nuova serie 5. Rome: 1959–1994.
ChNS	*Child's Nervous System.* Heidelberg 1985–.
CHR	*The Catholic Historical Review.* Washington, DC, 1915/1916–.
Cimbria lit.	*Cimbria literata [...].* By Johannes Moller. 3 vols. Copenhagen, 1744.
CJNS	*Canadian Journal of Neurological Sciences.* Calgary, 1974–.
ClinCard.	*Clinical Cardiology.* Hoboken, NJ, 1978–.
Clio med.	*Clio medica.* Amsterdam, 1965/1966–.
DbÄ	*Die berühmten Ärzte,* 2nd ed. Edited by René Dumesnil and Hans Schadewaldt. Cologne: Aulis, 1966.
DBF	*Dictionnaire de biographie française.* Edited by Jules Balteau, Michel Prévost and Roman d'Amat. Hitherto 20 vols. with 122 fascicles. Paris: Letouzey at Ané, 1933–. Supplement, fascicle 1. Edited by Béatrice Wattel. Paris: Letouzey at Ané, 2009.
DBI	*Dizionario biografico degli Italiani.* Edited by Alberto Maria Ghisalberti et al. Hitherto 77 vols. Rome: Istituto della Enciclopedia Italiana, 1960–.
DBioVolt.	*Dizionario di Volterra: Storia e descrizione della città, personaggi e bibliografia.* Vol 3, *I personaggi e gli scritti: Dizionario biografico e bibliografico di Volterra.* By Angelo Marrucci. Curated by Lelio Lagorio. Ospedaletto: Pacini, 1997.
DBL³	*Dansk biografisk leksikon,* 3rd ed. Edited by Svend Cedergreen Bech et al. 16 vols. Copenhagen: Gyldendal, 1979–1984. Data supplements. Copenhagen: Gyldendal, 1984–1995.
DBPh.	*The Dictionary of Seventeenth-Century British Philosophers.* Edited by Andrew Pyle. 2 vols. Bristol: Thoemmes Press, 2000.
DBSM	*Dizionario biografico della storia della medicina e delle scienze naturali (liber amicorum).* Edited by Roy Porter. 4 vols. Milan: F.M. Ricci, 1985–1989.
DDBK	*Dokumente der Deutschen Bischofskonferenz.* Edited by Sekretariat der Deutschen Bischofskonferenz. Hitherto 2 vols. Cologne: Bachem, 1998–2010.

(continued)

DDPh.	*The Dictionary of Seventeenth and Eighteenth-Century Dutch Philosophers.* Edited by Wiep van Bunge et al. 2 vols. Bristol: Thoemmes Press, 2003.
DDPhS	*Dutch Medical Biography: A Biographical Dictionary of Dutch Physicians and Surgeons 1475–1975.* By Gerrit A. Lindeboom. Amsterdam: Rodopi, 1984.
DermK	*Dermatologie und Kosmetik.* Reinbek, 1960–at least 1991.
DFPh.	*The Dictionary of Seventeenth-Century French Philosophers.* Edited by Luc Foisneau. 2 vols. London: Thoemmes Continuum, 2008.
DG	*Das deutsche Gesundheitswesen: Zeitschrift für klinische Medizin. Organ der Gesellschaft für Klinische Medizin der DDR.* Berlin, 1946–1992 (from 1985: Zeitschrift für klinische Medizin: Organ der Gesellschaft für Klinische Medizin der DDR).
DH	*Compendium of Creeds, Definitions, and Declarations on Matters of Faith and Morals. Enchiridion symbolorum definitionum et declarationum de rebus fidei et morum,* 43rd ed. By Heinrich Denzinger. Edited by Robert Fastiggi and Anne Englund Nash for the English edition. San Francisco, CA: Ignatius Press, 2012.
DHCJ	*Diccionario histórico de la Compañía de Jesús: Biográfico-temático.* Edited by Charles E. O'Neill and Joaquín M.a Domínguez. 4 vols. Rome: Institutum Historicum Societatis Iesu, 2001.
DHCME	*Diccionario histórico de la ciencia moderna en España.* Edited by José M. López Piñero et al. 2 vols. Historia, ciencia, sociedad 180 [vol. 1] and 181 [vol. 2]. Barcelona: Península, 1983.
DHM	*Dictionnaire historique de la médecine ancienne et moderne […].* By Nicolas François Joseph Eloy. 4 vols. Mons, 1778 (reprint Brussels: Culture et civilisation, 1973).
DHVG	*Die Diözese Hildesheim in Vergangenheit und Gegenwart.* Hildesheim, 1927–.
DISF	*Dizionario interdisciplinare di scienza e fede: Cultura scientifica, filosofia e teologia.* Edited by Giuseppe Tanzella-Nitti and Alberto Strumia. 2 vols. Vatican City: Urbaniana University Press, 2002.
DMB	*Dictionary of Medical Biography.* Edited by William F. Bynum and Helen Bynum. 5 vols. Westport, CT: Greenwood Press, 2007.
DMW	*Deutsche medizinische Wochenschrift.* Stuttgart, 1875–.
DOAP	*Dizionario di opere anonime e pseudonime di scrittori italiani o come che sia aventi relazione all'Italia.* By Gaetano Melzi. 3 vols. Milano, 1848–1859. 2 supplements by Giambattista Passano (Ancona, 1887) and Emmanuele Rocco (Naples, 1888).
DPR	*Dictionnaire de Port-Royal.* Edited by Jean Lesaulnier and Antony McKenna. Dictionnaires & références 11. Paris: Champion, 2004.
DresdÄ	*Dresdner Ärzte: Historisch-biographisches Lexikon.* By Volker Klimpel. Dresdner Miniaturen 5. Dresden: Hellerau, 1998.
DRR I	*Drammaturgia romana: Repertorio bibliografico cronologico dei testi drammatici pubblicati a Roma e nel Lazio.* By Saverio Franchi. With contributions by Orietta Sartori. 2 vols. Sussidi eruditi 42 & 45. Rome: Edizioni di storia e letteratura, 1988.
DSB	*Dictionary of Scientific Biography.* Edited by Charles Coulston Gillespie. 16 vols. New York: Scribner, 1970–1980.
DSI	*Dizionario storico dell'Inquisizione.* Edited by Adriano Prosperi. 3 vols., 1 index vol. and 1 "Inserto iconografico". Pisa: Edizioni della Normale, 2010.

(continued)

DSp.	*Dictionnaire de spiritualité, ascétique et mystique: Doctrine et histoire.* Edited by Marcel Viller, Charles Baumgartner and André Rayez. 16 vols. and 1 index vol. Paris: Beauchesne, 1937–1995.
DSP	*Dizionario storico portatile [...].* By Jean-Baptiste Ladvocat. 4 vols. Naples, 1754–1755.
DtÄBl.	*Deutsches Ärzteblatt.* Cologne, 1872–.
DVÅ	*Dansk veterinærhistorisk Årbog.* Copenhagen, 1934–.
EBioeth.	*Encyclopedia of Bioethics.* Edited by Warren T. Reich. 4 vols. New York: Free Press, 1978.
EILS	*Encyclopedia of Italian Literary Studies.* Edited by Gaetana Marrone. 2 vols. New York: Routledge, 2007.
EJM	*European Journal of Morphology.* Lisse, 1956/1958–2005.
EMG	*Enzyklopädie Medizingeschichte.* Edited by Werner E. Gerabek et al. 3 vols. Berlin: De Gruyter, 2007.
EndokrG	*Zur Geschichte der Endokrinologie und Reproduktionsmedizin: 256 Biographien und Berichte.* Edited by Gerhard Bettendorf. Heidelberg: Springer, 1995.
EndokrI	*Endokrinologie[–]Informationen.* Stuttgart, 1977–.
ENZ	*Enzyklopädie der Neuzeit.* Edited by Friedrich Jaeger. 15 vols. and 1 index vol. Stuttgart: Metzler, 2005–2012.
EPapi	*Enciclopedia dei Papi,* 2nd ed. Edited by Massimo Bray. 3 vols. Rome: Istituto della Enciclopedia Italiana, 2008.
ESR	*Encyclopedia of the Scientific Revolution: From Copernicus to Newton.* Edited by Wilbur Applebaum. New York: Garland, 2000 (paperback edition 2008).
FachwA	*Die Fachwörter der Anatomie, Histologie und Embryologie: Ableitung und Aussprache,* 29th ed. Edited by Hermann Triepel, Hermann Stieve, Robert Herrlinger and Adolf Faller. Munich: J.F. Bergmann, 1978.
FdA	*Fortschritte der Andrologie. Advances in Andrology. Progrès en andrologie.* Berlin: Grosse, 1970–1990.
Fejér s. s.	*Defuncti secundi saeculi Societatis Jesu 1641–1740.* By Josephus Fejér. 5 vols. Rome: Curia Generalitia S.J., 1985–1990, mach. ms.
FKTh.	*Forum Katholische Theologie.* Rothenburg o. d. Tbr., 1985–.
FL	*Forfatterlexikon omfattende Danmark, Norge og Island indtil 1814.* By Holger Ehrencron-Müller. 12 vols. Copenhagen: Aschehoug, 1924–1935. 2nd supplement. Copenhagen: Aschehoug, 1939.
FoF	*Fund og forskning i Det Kongelige Biblioteks samlinger.* Copenhagen, 1954–.
FutUom.	*Il futuro dell'uomo.* Edited by Istituto Stensen. Florence, 1974–.
Gesnerus	*Gesnerus.* Basel, 1943/1944–.
GlobBioeth.	*Global Bioethics.* Florence, 1988–.
GothaM	*Gothaer Museumsheft.* Gotha, 1978–1999.
GPMO	*Il Giornale della Previdenza dei Medici e degli Odontoiatri.* Rome, 1999–.
Greg.	*Gregorianum.* Rome, 1920–.
GreifswaldK	*Greifswalder Köpfe: Gelehrtenporträts und Lebensbilder des 16.–18. Jahrhunderts aus der pommerschen Landesuniversität.* By Dirk Alvermann and Birgit Dahlenburg. Rostock: Hinstorff, 2006.
GVicenza	*Il Giornale di Vicenza.* Vicenza, 1915–.

(continued)

HBL	*Hannoversches Biographisches Lexikon: Von den Anfängen bis in die Gegenwart.* By Dirk Böttcher, Klaus Mlynek, Waldemar R. Röhrbein and Hugo Thielen. Hanover: Schlütersche, 2002.
HC	*Hierarchia Catholica medii* [et recentioris] *aevi [...].* Edited by Konrad Eubel et al. Hitherto 9 vols. Padua, 1898–.
HGK	*Handbuch Gelehrtenkultur der Frühen Neuzeit.* Vol. 1, *Bio-bibliographisches Repertorium.* Edited by Herbert Jaumann, Berlin: De Gruyter, 2004.
HS	*History of Science.* Cambridge, 1962–.
HWPh.	*Historisches Wörterbuch der Philosophie.* Edited by Joachim Ritter et al. 12 vols. and 1 index vol. Basel: Schwabe, 1971–2007.
IHR	*Intellectual History Review.* Abingdon, 2007–.
IJC	*International Journal of Cardiology.* Amsterdam, 1981/1982–.
IJM	*International Journal of Morphology.* Temuco, CL, 1983–.
IKZ	*Internationale Katholische Zeitschrift Communio.* Ostfildern, 1972–.
ILP	*Index librorum prohibitorum 1600–1966.* By Jesús Martínez de Bujanda. Index des livres interdits 11. Montréal: Médiaspaul, 2002.
ISRN-OG	*International Scholarly Research Network Obstetrics and Gynecology.* New York, 2011–. Under: http://www.hindawi.com/isrn/obgyn.
JHMAS	*Journal of the History of Medicine and Allied Sciences.* Cary, NC, 1946–.
JHNS	*Journal of the History of Neurosciences.* Philadelphia, 1992–.
JRCGP	*The Journal of the Royal College of General Practitioners.* London, 1958–.
KidW	*Die Kirche in der Welt: Wegweisung für die katholische Arbeit am Menschen der Gegenwart. Ein Loseblatt-Lexikon.* Münster, 1947–1961.
KIEJ	*Kennedy Institute of Ethics Journal.* Baltimore, 1991–.
Killy[2]	*Killy-Literaturlexikon: Autoren und Werke des deutschsprachigen Kulturraumes,* 2nd ed. Edited by Wilhelm Kühlmann. 12 vols. Berlin: De Gruyter, 2011.
KO	*Katolsk orientering.* Copenhagen, 1975–.
KS	*Kirkehistoriske samlinger.* Copenhagen, 1849–.
Lancet	*The Lancet.* London, 1823–.
LBioeth.	*Lexikon der Bioethik.* Edited by Wilhelm Korff. Gütersloh: Gütersloher Verlagshaus, 1998.
LFam.	*Lexikon Familie: Mehrdeutige und umstrittene Begriffe zu Familie, Leben und ethischen Fragen.* Edited by Päpstlicher Rat für die Familie. Paderborn: Schöningh, 2007.
LindR	*Lindenius renovatus, sive, Johannis Antonidae van der Linden de scriptis medicis libri duo: [...] a postremae editionis anno 1662 usque ad praesentem continuati, dimidio pene amplificati, per plurimum interpolati, & ab extantioribus mendis purgati.* By Georg Abraham Mercklin. Nuremberg, 1686.
LouvMéd.	*Louvain médical.* Leuven, 1970–.
LThK[3]	*Lexikon für Theologie und Kirche,* 3rd ed. Edited by Walter Kasper. 10 vols. and 1 index vol. Freiburg: Herder, 1993–2001.
Lych.	*Lychnos.* Uppsala, 1936–.
MANC	*Mémoires de l'Académie Nationale des Sciences, Arts, et Belles-Lettres de Caen.* Caen, 1754–, n.s. 1925–.
MhJ	*Medizinhistorisches Journal.* Stuttgart, 1966–.

(continued)

MNIR	*Mededelingen van het Nederlands* [Historisch] *Instituut te Rome.* 's-Gravenhage, 1921–2001/2002, n.s. 1974–1987.
NassA	*Nassauische Annalen.* Wiesbaden, 1827–.
NassB	*Nassauische Biographie. Kurzbiographien aus 13 Jahrhunderten*, 2nd ed. By Otto Renkhoff. Veröffentlichungen der Historischen Kommission für Nassau 39. Wiesbaden, 1992.
NatMed.	*Nature Medicine.* New York, 1995–.
NCBQ	*National Catholic Bioethics Quarterly.* Philadelphia, 2001–.
NCE²	*New Catholic Encyclopedia*, 2nd ed. Edited by Thomas Carson and Joann Cerrito. 14 vols. and 1 index vol. Detroit: Thomson/Gale, 2003. Hitherto 10 supplements 2009–2012, edited by Robert L. Fastiggi. Detroit: Thomson/Gale, 2009–2013.
NDB	*Neue Deutsche Biographie.* Edited by Historische Kommission bei der Bayerischen Akademie der Wissenschaften. Hitherto 25 vols. Berlin, 1953–.
NGL	*Nürnbergisches Gelehrten-Lexicon [...].* By Georg Andreas Will and Christian Conrad Nopitsch. 4 vols. Nuremberg, 1755–1758. 4 supplements = vols. 5–8. Altdorf, 1802–1808.
NJN	*National Jesuit News.* Philadelphia, 1971–.
NMAe.	*Neues Magazin für Aerzte.* Leipzig, 1779–1798.
NSurg.	*Neurosurgery.* Hagerstown, MD, 1977–.
Nuncius	*Nuncius.* Florence, 1986–.
NYT-BS	*The New York Times Biographical Service: A Compilation of Current Biographical Information of General Interest.* Ann Arbor, MI, 1970–2001.
ODNB²	*Oxford Dictionary of National Biography [...]*, 2nd ed. Edited by Henry C. G. Matthew and Brian Harrison. 60 vols. and 1 index vol. Oxford: Oxford University Press, 2004. Supplement 2001–2004, edited by Lawrence Goldman. Oxford: Oxford University Press, 2009.
OrvosK	*Orvostörténeti közlemények: Communicationes de historia artis medicinae.* Budapest, 1955–.
Osiris	*Osiris.* Chicago, 1936–.
PAPS	*Proceedings of the American Philosophical Society.* Philadelphia, 1838/1840–.
PBM	*Perspectives in Biology and Medicine.* Baltimore, 1957/1958–.
PhT	*Physics Today.* Melville, NY, 1948–.
PhTRS	*Philosophical Transactions* [of the Royal Society]. London, 1665/1666–1677/1679, 1683–1886, from 1887 A and B.
PSexF	*Personenlexikon der Sexualforschung.* Edited by Volkmar Sigusch and Günter Grau. Frankfurt: Campus, 2009.
PSLMA	*Proverbia sententiaeque latinitatis medii aevi: Lateinische Sprichwörter und Sentenzen des Mittelalters in alphabetischer Anordnung.* Edited by Hans Walther. Carmina medii aevi posterioris Latina 2. 5 vols. and 1 index vol. Göttingen: Vandenhoeck & Ruprecht, 1963–1969.
PsRep.	*Psychological Reports.* Missoula, MT, 1955–.
RDV	*Repertorium der diplomatischen Vertreter aller Länder seit dem Westfälischen Frieden (1648) [...].* Edited by Ludwig Bittner et al. 3 vols. Oldenburg: Stalling, 1936–1965.
RGG⁴	*Religion in Geschichte und Gegenwart: Handwörterbuch für Theologie und Religionswissenschaft*, 4th ed. Edited by Hans Dieter Betz et al. 8 vols. and 1 index vol. Tübingen: Mohr, 1998–2008.

(continued)

RIIK-P	*Prosopographie von Römischer Inquisition und Indexkongregation 1701–1813*. Vol. 2, *M–Z*. By Herman H. Schwedt. Edited by Hubert Wolf. Römische Inquisition und Indexkongregation 3,2. Paderborn: Schöningh, 2010.
RQS	*Revue des Questions Scientifiques*. Namur, 1877–.
ScMedIt-D	*Scientia medica Italica/Deutsche Ausgabe*. Rome, 1950–1960.
ScrittIt.	*Gli scrittori d'Italia [...]*. By Giammaria Mazzuchelli. 2 vols. in 6 bks. Brescia, 1753–1763.
SHBL	*Biographisches Lexikon für Schleswig-Holstein und Lübeck*. Edited by Schleswig-Holsteinische Landesbibliothek. 13 vols. Neumünster: Wachholtz, 1982–2011.
SOP	*Scriptores Ordinis Praedicatorum [...]*. By Jacobus Quétif and Jacobus Échard. 2 vols. Paris, 1719–1721. Several supplement fascicles by Thomas Bonnet, Rémy Coulon and Antoine Papillon. Lyon, 1885–1934.
SpeeJ	*Spee-Jahrbuch*. Trier, 1994–2009, 2010/2011–.
StenCath.	*Stenoniana Catholica*. Copenhagen, 1955–1961.
StSec.	*Studi secenteschi*. Florence, 1960–.
ThLex.	*Thomas-Lexikon*. By Ludwig Schütz. Paderborn: Schöningh, 2nd ed. 1895. Reprint, Stuttgart: F. Frommann, 1983.
ThPQ	*Theologisch-praktische Quartalschrift*. Regensburg, 1848–.
TRE	*Theologische Realenzyklopädie*. Edited by Gerhard Krause and Gerhard Müller. 36 vols. and 2 index vols. Berlin: De Gruyter, 1977–2007.
ZME	*Zeitschrift für medizinische Ethik*. Ostfildern, 1993–.
ZSpr.	*Zeitsprünge: Forschungen zur Frühen Neuzeit*. Frankfurt, 1997–.

General Abbreviations and Symbols

*	born
†	died
A.D.	anno Domini
art.	article; articulum
B.C.	before Christ
bk., bks.	book, books
BVerfG	Bundesverfassungsgericht (German Constitutional Court)
b/w	black/white
Cap.	caput; capitolo
col., cols.	column, columns
coll.	collection
corr.	correctum
CRS	Ordo Clericorum Regularium a Somascha
CRV	Congregatio Canonicorum Regularium Vindesemensis-Victorina
CSsR	Congregatio Sanctissimi Redemptoris
Diss.	dissertation
ed.	edited/editor(s)
eg	exempli gratia
et al.	et alii

(continued)

etc.	et cetera
e. V.	eingetragener Verein (registered association)
Fig., fig.	Figure(s)
fol.	folio
Fr.	Father
GDR	German Democratic Republic
Habil.	habilitation thesis
i. Br.	im Breisgau
ie	id est
i. Ü.	im Üchtland
i. W.	in Westfalen
Jr.	junior
Lib.	Liber
LXX	Septuaginta
mach. ms(s).	machine manuscript(s)
med.	in medicine
med. dent.	in dentistry
MIC	Congregatio Clericorum Regularium Marianorum sub titulo Immaculatae Conceptionis Beatae Virginis Mariae
ms.	manuscript
Msgr.	Monsignore
n., N.	note
n.d.	no date
no., nos.	number, numbers
n.p.	no place
n.s.	new series
o. d. Tbr.	ob der Tauber
OESA	Ordo Fratrum Eremitarum Sancti Augustini
OFM	Ordo Fratrum Minorum
OFMCap.	Ordo Fratrum Minorum Capuccinorum
OMin.	Ordo Fratrum Minorum
OP	Ordo Fratrum Praedicatorum
OSB	Ordo Sancti Benedicti
OSCl.	Ordo Sanctae Clarae
p., pp.	page, pages
pag.	pagina
par.	synoptic parallel
phil.	in philosophy
r	recto
sect.	sectio
ser.	series
SJ	Societas Jesu
Sr.	Senior, Sister
St.	Saint
st.n.	stili novi = Gregorian calendar

(continued)

st.v.	stili veteris = Julian calendar
s.v.	sub verbo
theol.	in theology
trans.	translated, translation, translator
v	verso
vol., vols.	volume, volumes
Vul.	Vulgata Sixto-Clementina

Index of Bible References[1]

---

[1] On 4 EZR 5:8 of the pseudepigraphical Apocalypse of Ezra, see footnote 789 above. Superscript numbers refer to footnotes.

© Springer International Publishing Switzerland 2016　　　　　　　　251
F. Sobiech, *Ethos, Bioethics, and Sexual Ethics in Work and Reception of the Anatomist Niels Stensen (1638-1686)*, Philosophy and Medicine 117, DOI 10.1007/978-3-319-32912-3

Index of Named Persons[1]

---

[1] This index contains all historical persons mentioned in the main text body, with the exception of biblical names and Niels Stensen. Superscript numbers refer to footnotes.

© Springer International Publishing Switzerland 2016   253
F. Sobiech, *Ethos, Bioethics, and Sexual Ethics in Work and Reception
of the Anatomist Niels Stensen (1638-1686)*, Philosophy and Medicine 117,
DOI 10.1007/978-3-319-32912-3